Analog Design and Simulation using OrCAD® Capture and PSpice®

Analog Design and Simulation using OrCAD® Capture and PSpice®

Second Edition

DENNIS FITZPATRICK

Newnes is an imprint of Elsevier
The Boulevard, Langford Lane, Kidlington, Oxford OX5 1GB, United Kingdom
50 Hampshire Street, 5th Floor, Cambridge, MA 02139, United States

Notices
Knowledge and best practice in this field are constantly changing. As new research and experience broaden our understanding, changes in research methods, professional practices, or medical treatment may become necessary.

Practitioners and researchers must always rely on their own experience and knowledge in evaluating and using any information, methods, compounds, or experiments described herein. In using such information or methods they should be mindful of their own safety and the safety of others, including parties for whom they have a professional responsibility.

To the fullest extent of the law, neither the Publisher nor the authors, contributors, or editors, assume any liability for any injury and/or damage to persons or property as a matter of products liability, negligence or otherwise, or from any use or operation of any methods, products, instructions, or ideas contained in the material herein.

Library of Congress Cataloging-in-Publication Data
A catalog record for this book is available from the Library of Congress

British Library Cataloguing-in-Publication Data
A catalogue record for this book is available from the British Library

ISBN: 978-0-08-102505-5

For information on all Newnes publications
visit our website at https://www.elsevier.com/books-and-journals

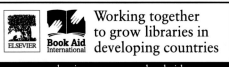

Working together
to grow libraries in
developing countries

www.elsevier.com • www.bookaid.org

Publisher: Mara Conner
Acquisition Editor: Tim Pitts
Editorial Project Manager: Katie Chan
Production Project Manager: Anitha Sivaraj
Designer: Mark Rogers

Typeset by SPi Global, India

CONTENTS

Preface   *xi*
Instructions   *xiii*

1. Getting Started   **1**

    1.1 Starting Capture   1
    1.2 Creating a PSpice Project   2
    1.3 Symbols and Parts   8
    1.4 PSpice Modeling Applications   18
    1.5 Design Templates   19
    1.6 Demo Designs   20
    1.7 Exporting Capture Designs   21
    1.8 Saving a Project   23
    1.9 Summary   26
    1.10 Exercises   27
    1.11 Extra Library Work   32

2. DC Bias Point Analysis   **35**

    2.1 Netlist Generation   38
    2.2 Displaying Bias Points   43
    2.3 Save Bias Point   44
    2.4 Load Bias Point   45
    2.5 Exercises   45

3. DC Analysis   **53**

    3.1 DC Voltage Sweep   54
    3.2 Markers   54
    3.3 Exercises   59

4. AC Analysis   **69**

    4.1 Simulation Parameters   70
    4.2 AC Markers   72
    4.3 Exercises   72

5. Parametric Sweep   **79**

    5.1 Property Editor   79
    5.2 Exercises   83

6. Stimulus Editor   **99**

    6.1 Stimulus Editor Transient Sources   100
    6.2 User-Generated Time-Voltage Waveforms   106
    6.3 Simulation Profiles   107
    6.4 Exercise   107

7. Transient Analysis   **117**

    7.1 Simulation Settings   118
    7.2 Scheduling   118
    7.3 Check Points   119
    7.4 Defining a Time-Voltage Stimulus Using Text Files   120
    7.5 Exercises   122

8. Convergence Problems and Error Messages   **131**

    8.1 Common Error Messages   131
    8.2 Establishing a Bias Point   132
    8.3 Convergence Issues   133
    8.4 Simulation Settings Options   134
    8.5 Exercises   136

9. Transformers   **141**

    9.1 Linear Transformer   141
    9.2 Nonlinear Transformer   142
    9.3 Predefined Transformers   144
    9.4 Exercises   144

10. Monte Carlo Analysis   **151**

    10.1 Simulation Settings   152
    10.2 Adding Tolerance Values   155
    10.3 Exercises   156

11. Worst Case Analysis   **165**

    11.1 Sensitivity Analysis   166
    11.2 Worst Case Analysis   167
    11.3 Adding Tolerances   168
    11.4 Collating Functions   168
    11.5 Exercise   169

12. Performance Analysis   **177**

    12.1 Measurement Functions   177

12.2 Measurement Definitions   177
12.3 Exercises   180

13. Analog Behaviorial Models   **185**

13.1 ABM Devices   185
13.2 Exercises   192

14. Noise Analysis   **197**

14.1 Noise Types   197
14.2 Total Noise Contributions   198
14.3 Running a Noise Analysis   199
14.4 Noise Definitions   200
14.5 Exercise   203

15. Temperature Analysis   **209**

15.1 Temperature Coefficients   209
15.2 Running a Temperature Analysis   210
15.3 Exercises   212

16. Adding and Creating PSpice Models   **217**

16.1 Capture Properties for a PSpice Part   217
16.2 PSpice Model Definition   219
16.3 Subcircuits   221
16.4 Model Editor   223
16.5 Exercises   231

17. Transmission Lines   **243**

17.1 Ideal Transmission Lines   243
17.2 Lossy Transmission Lines   244
17.3 Exercises   247

18. Digital Simulation   **259**

18.1 Digital Device Models   259
18.2 Digital Circuits   260
18.3 Digital Simulation Profile   262
18.4 Displaying Digital Signals   263
18.5 Exercises   264

19. Mixed Simulation   **275**

19.1 Exercises   276

20. Creating Hierarchical Designs   **283**

20.1 Hierarchical Ports and Off-Page Connectors   285
20.2 Hierarchical Blocks and Symbols   287
20.3 Passing Parameters   289
20.4 Hierarchical Netlist   290
20.5 Exercises   291

21. Magnetic Parts Editor   **311**

21.1 Design Cycle   311
21.2 Exercises   312

22. Test Benches   **331**

22.1 Selection of Test Bench Parts   332
22.2 Unconnected Floating Nets   333
22.3 Comparing and Updating Differences Between
     the Master Design and Test Bench Designs   335
22.4 Exercises   336

23. Advanced Analysis   **347**

23.1 Introduction   347

24. Sensitivity Analysis   **351**

Introduction   351
24.1 Absolute and Relative Analysis   352
24.2 Example   353
24.3 Assigning Component and Parameter Tolerances   354
24.4 Exercises   357

25. Optimizer   **367**

Introduction   367
25.1 Optimization Engines   367
25.2 Measurement Expressions   368
25.3 Specifications   368
25.4 Exercises   369

26. Monte Carlo   **379**

26.1 Introduction   379
26.2 Exercise   382

27. Smoke Analysis   **385**

    27.1 Passive Smoke Parameters   387
    27.2 Active Smoke Parameters   392
    27.3 Derating Files   395
    27.4 Example 1   398
    27.5 Exercises   401
    27.6 Example 2   404
    27.7 Example 3   414
    27.8 Example 4   419

Appendix   *425*
Index   *427*

PREFACE

The Cadence/OrCAD family of Electronic Design Automation (EDA) software provides a complete design flow from schematic entry to circuit simulation through to PCB layout.

The circuit is drawn using the Capture or Capture CIS schematic editor and circuit simulations are performed using PSpice. The schematic diagram is translated into a printed circuit board design using Cadence Allegro or PCB Editor which has replaced OrCAD Layout. The book has been written incorporating the features in the latest Cadence/OrCAD 17.2 software release and can be used with previous software releases. The majority of the circuits in the book can be simulated with the latest demo Lite software version that is free and not time limited. OrCAD Lite incorporates all of the key features of the full version and is only limited by the component count and complexity of the circuits. More information about the products can be found from the OrCAD and Cadence websites. The free OrCAD Lite can be downloaded from the OrCAD websites or you can request a DVD.

http://www.orcad.com/products/orcad-lite-overview

This book will benefit anybody with an interest in using the Cadence/OrCAD professional simulation software for the design and analysis of electronic circuits. The book provides a practical hands-on approach to using the software and at the end of each chapter there are exercises with step-by-step instructions to complete.

Thanks are due to the technical staff at the University of West London, Keith Pamment and Seth Thomas for reviewing the simulation exercises, Taranjit Kukal and Alok Tripathi from Cadence for reviewing the technical aspects of the book, and Parallel-Systems UK for their support.

This second edition of the book incorporates many of the new features that have been included with the latest software releases up to 17.2. Again big thanks to Alok Tripathi from Cadence Design Systems for his help in ensuring the technical correctness of the text and reviewing the new chapters and features. Many thanks also to Keith Pamment from City University, London and to Bob Doe from Parallel-Systems for reviewing and providing feedback for the new chapters and features. Many thanks.

INSTRUCTIONS

Throughout the book, bold type will indicate tool-specific keywords and also which menus to select, for example, the menu selection to create a new project is shown below.

The instruction sequence will be, **File > New > Project**. The words in bold indicate which successive menus to select from the top tool bar as shown above.

A right mouse button click will be written as **rmb**. For example, select the part in the schematic and **rmb > rotate**.

Bold type is also used to name any dialog box and windows that may appear. For example the **Create PSpice Project** window is shown below.

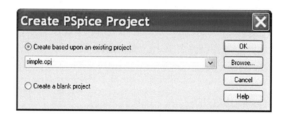

LIMITS OF ORCAD DEMO LITE DVD

The latest OrCAD demo version is OrCAD Lite 17.2 that is free to download or order from the OrCAD and Cadence websites:

http://www.orcad.com/products/orcad-lite-overview

OrCAD 17.2 Lite includes all the key features of the full version and is limited only by the size and design complexity. You can use the Lite version for most of the exercises in this book but there are some limits that have changed from previous versions.

OrCAD 17.2 Lite

For PSpice simulation, circuits are limited to 75 nodes, 20 transistors, no subcircuit limits, 65 digital primitives, 10 transmission lines (ideal or nonideal) with no more than four pairwise coupled lines.

The Model editor is limited to diode models for device characterization and parameterized part creation. Model Import Wizard only supports two pin parts and models.

All PSpice libraries, including parameterized libraries from the full version, are included.

There is no limit to stimulus generation using the Stimulus editor.

The Magnetic Parts Editor can only be used to design power transformers. The Magnetic Parts Editor supplied database cannot be edited and only contains a single magnetic core.

You cannot use Level 3 of Core model (Tabrizi), MOSFET BSIM 3.2 or MOSFET BSIM models.

The PSpice DMI models are not supported in the Lite Version of the simulator.

IBIS import is not supported.

Device model interface (DMI) is not supported.

ADVANCED ANALYSIS

Smoke analysis—Can only use diodes, resistors, transistors, and capacitors.

Optimizer—Can only use Random and MLSQ engines. Random engine is limited to 5 runs only. Only a maximum of two component parameters can be optimized. Limited to only one measurement specification and one curve-fit specification. Only one error calculation method is supported for optimizing curve-fit specification.

Parametric plotter—Can sweep the values of only two design and/or model parameters. Only Linear sweep is supported. A maximum of 10 sweeps are allowed. Can evaluate the influence of changing parameter values only on one measurement expression or a trace.

Display Plot is not available.

Monte Carlo/Worst Case analysis—Only one measurement specification is allowed. A maximum of three devices with tolerance are supported. A maximum of 20 Monte Carlo runs are supported.

Sensitivity analysis—Only one measurement specification is allowed. A maximum of three devices with tolerance are supported. A maximum of 20 runs are supported.

Encrypted parameterized models cannot be simulated.

CHAPTER 1

Getting Started

Chapter Outline

1.1.	Starting Capture	1
1.2.	Creating a PSpice Project	2
1.3.	Symbols and Parts	8
	1.3.1 Symbols	8
	1.3.2 Parts	10
	1.3.3 Search for Parts	12
	1.3.4 Quick Place of PSpice Components	16
1.4.	PSpice Modeling Applications	18
1.5.	Design Templates	19
1.6.	Demo Designs	20
1.7.	Exporting Capture Designs	21
1.8.	Saving a Project	23
	1.8.1 Saved Designs	24
	1.8.2 Find and Replace Text Utility	24
	1.8.3 Password Protection	25
1.9.	Summary	26
1.10.	Exercises	27
	Exercise 1	27
	Exercise 2	29
1.11.	Extra Library Work	32

For those of you who are familiar in setting up projects and drawing schematics in Capture, you may want to skip this chapter as this chapter is for those of you who have little or no experience of using Capture. The chapter will describe how to start Capture and how to set up the project type and libraries for PSpice simulation. The chapter will also introduce some of the features found in the latest software releases. At the end of each chapter there are some exercises to do and as you go through the book, each chapter will build upon the exercises from previous chapters.

1.1. STARTING CAPTURE

Circuit diagrams for PSpice simulation are drawn in either Capture or Capture component information system (CIS) schematic editor. The CIS option allows you to select and place components from a component database instead of selecting and placing components from a library. For this book, it does not matter if the circuits are drawn in Capture or Capture CIS.

If you have the software installed under the OrCAD name, launch Capture or Capture CIS, by clicking on:

Start > Program Files > OrCAD xx.x > Capture

or

Start > Program Files > OrCAD xx.x > Capture CIS

where xx.x is the version number, i.e., 10.5, 11.0, 15.5, 15.7, 16.0, 16.2, 16.3, 16.5, or 16.6.

For example:

Start > All Programs > Cadence > OrCAD 16.6 Lite > OrCAD Capture CIS Lite

Start > All Programs > Cadence > Release 16.5 > Capture

If you have the Cadence software installed, the products are installed under the Allegro platform name. In this case, only Capture CIS is available and is branded as Design Entry CIS.

Start > Program Files > Allegro SPB 16.6 > Design Entry CIS

For the 17.2 release:

Start > All Programs > Cadence Release 17.2-2016 > OrCAD Lite Products > Capture CIS Lite

or

Start > All Programs > Cadence Release 17.2-2016 > OrCAD Products > Capture

1.2. CREATING A PSPICE PROJECT

New designs started in Capture will automatically create a project file (.opj) that references the design file (.dsn) and associated project files such as the libraries and output report files. The design file encapsulates the schematic folders and associated schematic pages. For example, you may want to have separate schematics for each stage of your design rather than have one complete circuit. An advantage of this hierarchical design approach provides selective simulation of individual circuits. Alternatively, you can have a flat design where more than one page is associated with just one schematic folder. Either way, one schematic folder known as the root folder, signified by a "/" and one associated page will be created initially for the project. Other schematic folders and pages can be added at a later stage.

Before the circuit diagram is drawn, the project type and libraries attached to the project need to be set up. This can be done by clicking on the **New Project** icon on the **Start Page** under the **Getting Started** heading. For previous software releases, select from the top tool bar, **File > New > Project.**

Note

Previous projects can be selected from the File menu or the **Recent Files** list on the **Start Page**.

If the Start Page does not appear, then from the top tool bar, select: **Help > Start Page**. To close the Start Page, right mouse click on the Start Page tab and select **Close**.

In the **New Project** window (Fig. 1.1) you enter the name of the project and then you have a choice of one of four project types.

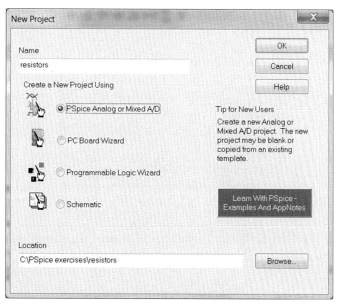

Fig. 1.1 Creating a new project.

- **Analog or Mixed A/D** is used for PSpice simulations.
- **PC Board Wizard** is used for schematic to PCB projects.
- **Programmable Logic Wizard** is used for CPLD and FPGA designs.
- **Schematic** is used for schematic and wiring diagrams.

When you select a Project type, the **Tip for New Users** gives a brief explanation of the project type. For PSpice projects, select **Analog or Mixed A/D**. This will activate the PSpice menu on the top toolbar in Capture. In release 17.2, the New Project Dialog Box incorporates the Learn with PSpice link that provides access to Examples and AppNotes.

It is recommended that a new directory location (folder) should be created for each new project. This can be done by clicking on the **Browse** button in Fig. 1.1 that opens up Windows Explorer allowing you to create and name the project folder. If you have a previous software release to 17.2, then when you click on the **Browse** button in Fig. 1.1, the **Select Directory** window appears as shown in Fig. 1.2.

Fig. 1.2 Creating a project folder location.

By selecting the **Create Dir…** button, the **Create Directory** window (Fig. 1.3) appears that allows you to name the directory (folder).

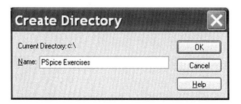

Fig. 1.3 Creating the project folder.

Tip

If you intend to migrate your design to a PCB layout or implement an FPGA, then it is recommended that you do not include white spaces or other reserved characters such as * . / and \ in the path and project name. Use underscores instead of white spaces.

The newly created folder, PSpice Exercises, in this example, will appear in the **Select Directory** window. However, you must highlight the folder and click OK in order to ensure that the folder is active. The folder will then appear as an "open" yellow folder as shown in Fig. 1.4. This allows you the option to add further subfolders by clicking on **Create Dir** and following the same procedure as above. If no folders are to be added, then just click on OK.

Fig. 1.4 The project folder has been selected.

The project folder location will then appear in the Location box of the Project Manager (see Fig. 1.1). See exercises at the end of the chapter.

An alternative method of creating the project folder is to type in the folder location path directly into the Location box in the Project manager in Fig. 1.1 and Capture will automatically create the folder.

Tip
Version 17.2 by default will save project path names and design names in uppercase. To turn this off select:
Options > Preferences > More Preferences
In the **Extended Preferences Setup** select **Design and Libraries** and un-tick:
Save design name as UPPERCASE

Note

It is a common mistake to create a project folder in the **Select Directory** window and not select the folder. Make sure you double click on the new folder name in the **Select Directory** window (Fig. 1.4).

The next dialog box to appear is the **Create PSpice Project** window which sets up the project for PSpice simulation (Fig. 1.5).

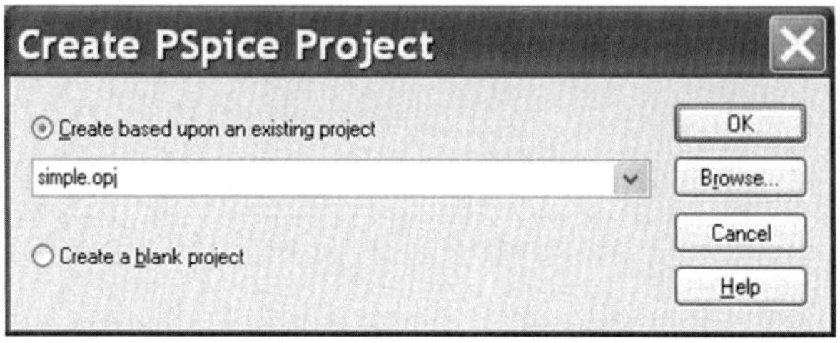

Fig. 1.5 Create PSpice Project.

The pull down menu option allows you to select preconfigured Capture-PSpice.

The pull down menu option allows you to select preconfigured Capture-PSpice projects and a selection of configured libraries. In previous Lite versions, a default eval.olb library was available for projects. Now all PSpice libraries are included with the Lite version.

There is also an option to create updated versions of an existing project, i.e., to create a newer version 2 based upon the original version 1 project. You select the function, **Create based upon an existing project** and then **Browse** to select an existing project which will copy the project and all its associated files into the new project. This is similar to using the **File > Save As** function from the **File** menu.

If the **Create a blank project** option is selected, then no Capture-PSpice libraries are added to the project. The libraries can be added later. This will be demonstrated in one of the exercises at the end of this chapter.

When a new project is created, a **Project Manager** window is created as shown in Fig. 1.6. In this example, the **simple** library option consisting of 5 default libraries has been selected. In this example, the resistors.dsn file has been expanded to show the SCHEMATIC1 folder and the underlying PAGE1 in which you draw the circuit diagram.

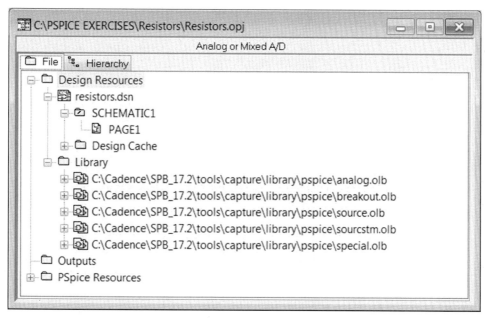

Fig. 1.6 Project Manager showing the Capture parts libraries and their location.

The Project Manager also shows the absolute path to the libraries. Remember that these are Capture parts libraries which define the graphics for the parts. They are not the PSpice model libraries. The preinstalled Capture libraries can be found, depending on the OrCAD or Cadence software version you are using, for example:

<software install path> OrCad > OrCAD_16.6_Lite > tools > capture > library > pspice

or

<software install path> Cadence > SPB_16.6 > tools > capture > library > pspice

Normally the **<software install path>** is on the **C:** drive, for example:

C:\Cadence\SPB_17.2\tools\capture\library\pspice

Tip

If the **Project Manager** window is not displayed, select from the top tool bar, **Window > <project name>.opj** file as shown in Fig. 1.7. Here the project name is **resistors**. Note the project name file extension .opj.

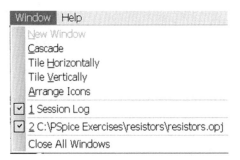

Fig. 1.7 Displaying the Project Manager window.

Alternatively, click on either of the Project manager icons.

1.3. SYMBOLS AND PARTS

Before drawing a schematic diagram, it is useful to know the difference between a part and a symbol.

Parts are graphical objects where the part name represents for example a 2N3906 transistor, an LF411 opamp, etc. Connecting wires drawn to Capture parts are known as nets and they have net names known as a net alias. Capture will assign a default net alias to a wire but the default name can be changed. It is always good practice to rename default nets to more meaningful names such as out, clock, +5 V, etc.

Symbols are also graphical parts but the wires connected to symbols inherit the name of the symbol. For example, when placing a "0" symbol to represent a ground connection, the connecting wire will take on the net name of "0." To define a +5V connection you can use a generic VCC_CIRCLE symbol and rename it to +5 V. All wires connected to the +5 V symbol will take on a net name of +5 V. There are many different symbols you can use to define power, ground, and digital logic level connections and you can rename them accordingly.

1.3.1 Symbols

Symbols also differ from parts in that they are not placed from the **Place Part** menu in Capture. Instead you have to select the symbol from the **Place** menu. Fig. 1.8 shows the top portion of the **Place** menu.

Fig. 1.8 Place menu.

The **Place** menu also shows the corresponding short cut keys. For example to place a Power symbol, press **F** and the **Place Power** menu appears as shown in Fig. 1.9.

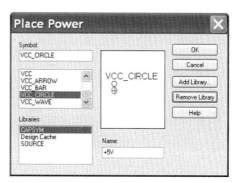

Fig. 1.9 Place Power menu.

In the **Place Power** menu in Fig. 1.9, a **VCC_CIRCLE** symbol has been selected and its name has been changed to +5 V. Any wires (nets) connected to +5 V will take on the net name +5 V.

Other symbols include hierarchical ports and off-page connectors which allow signals to be connected together throughout the design. These will be discussed in a later Chapter.

There are two symbol libraries, **source** and **capsym**. **Capsym** contains all the analog ground and power symbols. The **source** library also contains the analog 0V symbol and also the digital **$D_HI** and **$D_LO** symbols that are used to set a digital level of "HI" or "LO" on a wire or pin of an IC.

1.3.2 Parts

To place a part, select, **Place > Part**. Fig. 1.10A shows the **Place Part** window for earlier Capture versions and Fig. 1.10B shows the **Place Part** window in later versions.

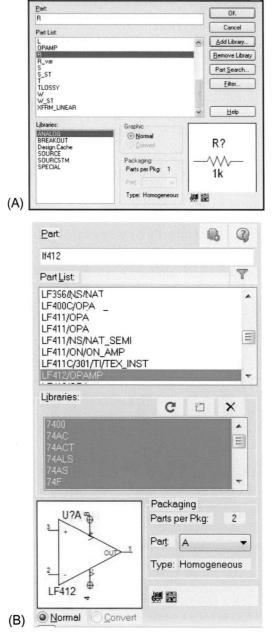

(A)

(B)

Fig. 1.10 Place part menu: (A) version 16.0, (B) version post-16.3, and

(Continued)

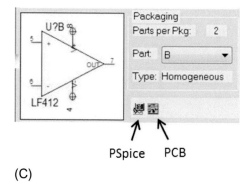

(C)

Fig. 1.10, cont'd (C) parts per package.

Although the two **Place Part** windows look different, they have the same functionality in that they display the list of libraries available and the parts available in those libraries. They also provide a part search function. In Fig. 1.10A, only the analog library has been highlighted and so only those parts for that library are shown in the **Parts List**.

In Fig. 1.10B, all the libraries have been highlighted and so you see the name of the part (LF412) and the library it comes from (OpAmp). If you place the cursor over any part in the **Parts List**, a tool tip rectangular bar appears giving the absolute path name to the library part. The graphical description for the LF412 shows that there are 2 parts per package and part A has been selected. Selecting B will show a similar part with different pin numbers (see Fig. 1.10C). The part is homogenous in that all parts in the package are the same and have the same number of pins. A solid-state relay with a coil and a switch is an example of a heterogeneous type in which the parts in the package are not the same.

The green PSpice icon indicates that a PSpice model is attached to the LF412 and can therefore be simulated. The red icon indicates a PCB footprint is attached to the part.

Note

Batteries, voltage sources, and current sources are found in the **source** library from the **Place Part** menu (**Place > Part**) and are not to be confused with the power symbols (VCC_circle, 0V, etc.) from the **capsym** library (**Place > Power** or **Place > Ground**) that are effectively used to "invisibly" connect wires with the same net name together.

In the **Place Power** or **Place Ground** window (Fig. 1.11), there is a **source** library that contains only the digital HI and LO and ground 0V symbols.

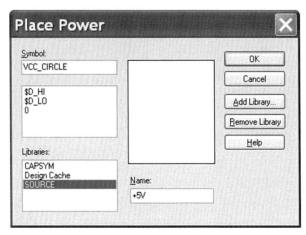

Fig. 1.11 The source library for Place Power contains the digital HI and LO and ground 0V symbols.

To recap, symbols are placed from the **Place** menu and parts are placed from the **Place > Part** menu. Also note that both **Part** libraries and **Symbol** libraries have an .olb extension and are the Capture graphical parts.

1.3.3 Search for Parts

In the **Place Part** window dialog box (Fig. 1.10A) there is a **Search for Part** function that enables you to search for parts in any of the preinstalled and other custom-created libraries. If searching for standard vendor parts, it is recommended that wild-card characters be used as different semiconductor vendors tend to append their own numbers and letters to industry standard part numbers; for example, LF412CN/NOPB, LF412CP, LF412ACN/NOPB, LF412CPE4, LF412CDR, etc. So to search for an LF412 you would use LF412* where "*" is a wildcard so any characters or numbers after LF412 are ignored during the search. The "?" character is used to ignore individual characters, for example to look for a logic gate function based on the 7408, irrespective of family type, you could use 74??08. Fig. 1.12 shows the search results for an AD648 of which there are three types, the AD648A, AD648B, and AD648C.

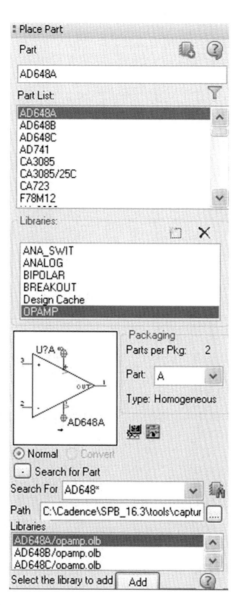

Fig. 1.12 Search for Part.

In 17.2 and later versions of 16.6, a database of available PSpice parts has been added that enables you to search for parts not only by name but also by categories. Categories are very useful if you are looking for a specific type of part, i.e., a FET input operational amplifier or a single supply operational amplifier. The **PSpice Part Search** window shown in Fig. 1.13 is opened by selecting:

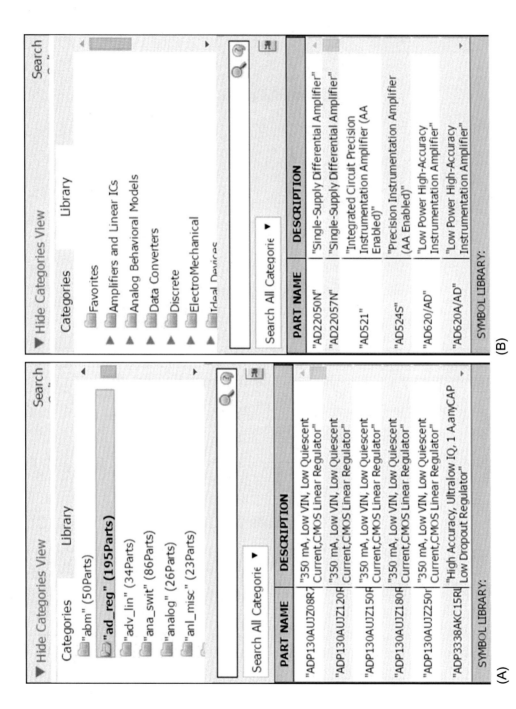

Fig. 1.13 PSpice Part Search by (A) libraries and (B) categories.

Place > PSpice Component > Search

Fig. 1.13A shows the installed libraries that are available and the parts within the libraries. In Fig. 1.13B, the library parts are shown as categories, i.e., Pressure sensors, Operational amplifiers, data converters, etc.

You can search for a part in all categories or just one selected category (see Fig. 1.14). Selecting the Symbol Viewer icon, symbols can be viewed before being placed in the schematic page. The Symbol Viewer updates every time you select a new part. You can only select and place one part at a time.

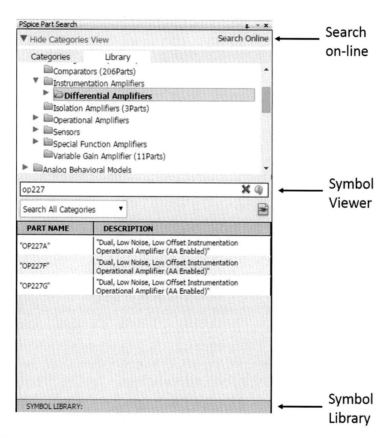

Fig. 1.14 PSpice Search Part by categories.

When you select the Search Online icon, the online search facility opens up the OrCAD Capture Marketplace website that acts as a portal to search vendor websites for Capture parts, IBIS, Verilog, VHDL, and PSpice models.

1.3.4 Quick Place of PSpice Components

From release 16.6, there is a new feature which allows you to quickly place generic PSpice parts rather than search for vendor-specific semiconductor device numbers. Parts are placed from the top tool bar, **Place > PSpice Component** menu as shown in Fig. 1.15.

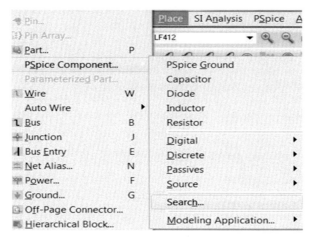

Fig. 1.15 Placing a PSpice Component.

The first five components represent the most common components used in a schematic. The PSpice Ground (Hot key—G) is from the source library and the discrete components are from the analog library. Table 1.1 summarizes the list of available Quick Place PSpice components and their library of origin.

Components from the **breakout** library are generic semiconductor parts with default names and parameters that can be customized using the PSpice Model Editor. In the demo and Lite software versions, only diodes can be edited in the Model Editor. The Model Editor is introduced in a later chapter.

If you know the actual semiconductor device number, then it may be best to find and use that known semiconductor device from the PSpice libraries. The generic parts are useful if you are new to electronics and you just want to get a circuit up and running quickly. Customizing semiconductor parts from the **breakout** library is mainly used for more advanced circuits where specific semiconductor characteristics are required.

Table 1.1 List of available Quick Place PSpice components

Components		Capture library	Simulation ready
Digital			
Gates	AND,OR,NAND,NOR,XOR,INV	dig_prim	√
Flip–Flop	D,JK,RS,T	dig_prim	√
ADC	8 bit, 10 bit, 12 bit	breakout	√
DAC	8 bit, 10 bit, 12 bit	breakout	√
Discrete			
	Diode, NPN, PNP, NPN Darlington, PNP Darlington, NMOS, PMOS, Power NMOS, Power Diode, N-JFET, P_JFET, IGBT, GAsFET	breakout.olb	√
	OpAmp	analog.olb	√
Passives			
	R, C, L, TLine ideal, TLine lossy	analog.olb	√
	Potentiometer, coupling	breakout.olb	√
Source			
Controlled sources	VCVS, VCCS, CCVS, CCCS	analog.olb	√
Current sources	AC, DC, pulse, sine, exponential, FM sine	source.olb	√
Voltage sources	AC, DC, pulse, sine, exponential, FM sine	source.olb	√

Note

If you are using the demo 16.6 OrCAD Lite version, be careful not to use the PSpice Component digital gates and flip-flops as they are from the gate and latch libraries. These devices do not have PSpice models and therefore cannot be stimulated. In the full version of 16.6 and subsequent versions afterward (17.2), the digital gates and flip-flops are from the dig_prim library and can be simulated.

Tip

The **eval** library which is available in both full and demo Lite software releases contains general purpose industry standard analog and digital semiconductors.

1.4. PSPICE MODELING APPLICATIONS

New PSpice models are normally created in the Model Editor (see Chapter 16). However, from release 16.6, the PSpice Modeling Application provides a quick method to create PSpice models in Capture. The application can be found under the **Place** menu; **Place > PSpice Components > Modeling Applications.**

For example, to create a PSpice inductor model, select **Passives** then **Inductors** from the pull down menu (see Fig. 1.16).

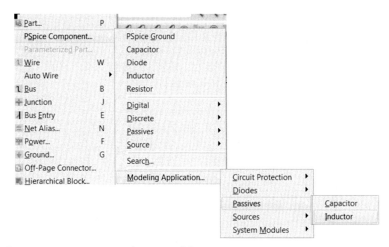

Fig. 1.16 Creating a new passive inductor model.

You can then add the manufacturer's data for the inductor parameters such as tolerance, temperature coefficients, and parasitics as shown in Fig. 1.17.

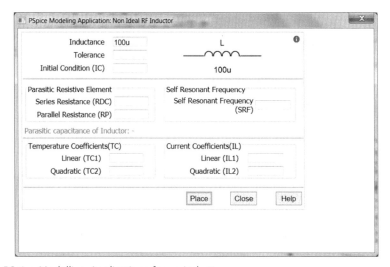

Fig. 1.17 PSpice Modelling Applications for an inductor.

Table 1.2 lists the PSpice components that can be created in Capture using PSpice Modelling Applications.

Table 1.2 Modelling Applications PSpice models

Device	Model type	Model
Circuit protection	Transient voltage suppressors	Transient voltage suppressors with lead inductance
Diodes	Zener	
	LED	
Passives	Capacitor	
	Inductor	
Sources	Independent sources	Pulse, sine, DC, exponential, FM, impulse, three phase, noise
	Piecewise linear sources (PWL)	Voltage, current
System modules	Switch	Time controlled, voltage controlled, current controlled
	Transformer	Two winding, custom tap, centre tap, flyback, forward, forward with reset winding
	VCO	Sinusoidal, triangular, square

Tip
The shortcut to select PSpice Model Application, Independent Sources is Shift-R.

1.5. DESIGN TEMPLATES

Capture release versions of 16.3 to 16.6, included design templates of complete electronic circuits and topologies including simulation profiles for analog, digital, mixed, and switched mode power supplies. You select any of these design templates from the pull down menu in the **Create PSpice Project** dialog box when you create a new project see Fig. 1.18.

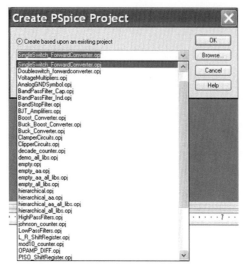

Fig. 1.18 Available design templates.

Fig. 1.19 shows the design template for a single switch forward converter which includes the schematic and explanatory text.

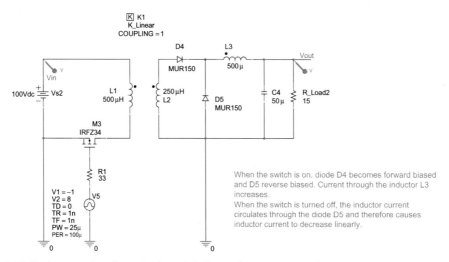

Fig. 1.19 Design template for a single switch forward converter topology.

1.6. DEMO DESIGNS

From release 17.2 onward, the **Open Demo Designs** window provides access to all types of design examples included with the installed software. Over 150 demo designs can be accessed by; **File > Open > Demo Designs** (Fig. 1.20).

Fig. 1.20 Demo designs.

The designs are complete working circuits and include circuit equations and theory.

1.7. EXPORTING CAPTURE DESIGNS

From the 16.3 version of Capture, the schematic design can be exported to a PDF file with searchable text. This is especially useful when sending designs to those who do not have the OrCAD software installed but have Adobe Reader installed. To export a PDF file, select **File > Export > PDF.**

In 17.2 the PDF export function generates intelligent PDF files that allow you to navigate hierarchical designs, provides a list components and nets, and displays component properties. A list of schematics, components, and net names appear in the PDF bookmark window. When you select a component or a net name, the corresponding component or net name is highlighted in the design schematic. Selecting a component or hierarchical block displays the attached properties. For hierarchical blocks, selecting **descend** displays the underlying schematic design.

Two files are needed to export an intelligent PDF, the Postscript (PS) driver and the Postscript to PDF converter (PDD). In 17.2, the PS driver from Microsoft is already setup at installation and is seen in the Devices and Printer list as ORCADPS_17.2. If not, the Microsoft PS driver can be installed using the Microsoft "Add Printer Wizard."

Devices and Printers > Add a printer
Add local printer > Create a new port > Local port
Enter a port name > (anything you like, i.e., OrCAD)
Install the printer driver > Generic > MS Publisher Color Printer
Replace the current driver (if asked)
Printer name > (call it anything you like, i.e., OrCADPS_17.2)
Do not share this printer
Do not set as the default printer (un-tick box)

If all goes well, the OrCADPS_17.2 will be added to the printer list.

Alternatively, the Adobe Distiller PS driver can be downloaded. (http://www.adobe.com/support/downloads/product.jsp?platform=windows& product=pdrv).

A recommended PDD converter file is available for free download from Ghostscript. https://ghostscript.com/download/. An alternative PDD converter is Acrobat Distiller that is also available for free download.

Note

The 17.2 installation automatically installs a Microsoft PS driver that is seen as OrCADPS_17.2 in the Devices and Printer list. You only need to install a PDD converter such as Ghostscript or Acrobat Distiller.

When you first start **File > Export > PDF**, the dialog box provides a link to the online Ghostscript converter. You just download the executable file and browse to the file path.

Other PDD converters can be used as long as they support pdfmark (bookmarks).

In 17.2, Capture designs can be exported to a single HTML file (**File > Export Design > HTML**) that can be viewed in a web browser such as Google Chrome. You have the option to select all or individual schematics and pages as shown in Fig. 1.21.

In 17.2, Capture designs and libraries can also be exported to XML and 17.2 also supports the import of XML design files.

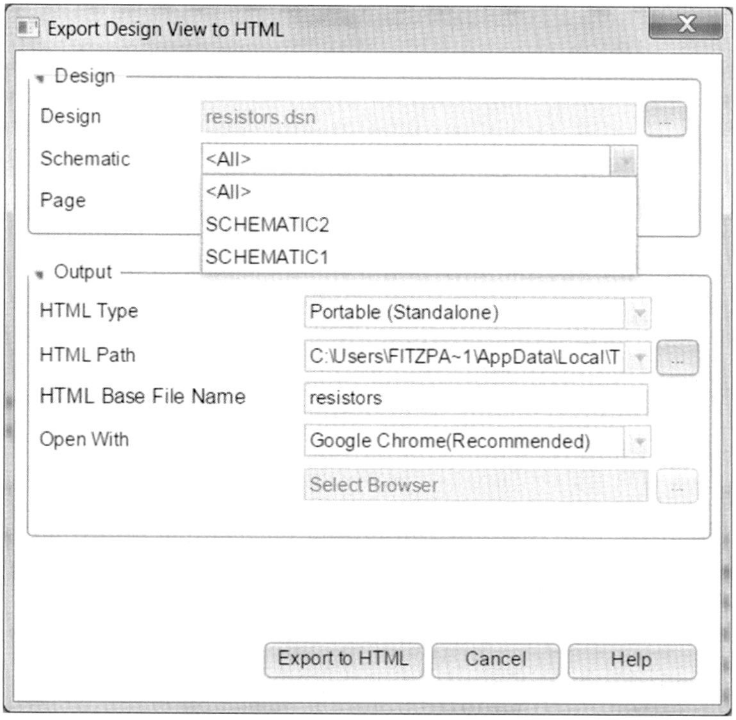

Fig. 1.21 Exporting Capture design to HTML.

1.8. SAVING A PROJECT

The quickest way to save a project and all its associated files and referenced files to its original location is to use the Save command either by **File > Save o**r using Control-S or clicking on the Save icon.

The Project Manager does not have to be active. You can save the project while working in a schematic page. If you want to rename the project or save the project to another location, then you can use the **Save Project As** command, **File > Save Project As.** The Save Project As windows dialog box allows you to rename the project and enter or browse to a project location (see Fig. 1.22). However, you need to activate the Project Manager (click on the dsn file) for the **Project Save As** command to become available in the File menu.

The Design file and individual libraries in the Project Manager can be saved using the Save As command from the File menu or by clicking the right mouse button (rmb) to activate the context menu and selecting **Save As**.

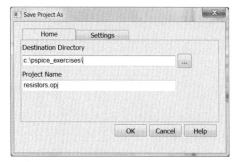

Fig. 1.22 Saving the project with another name or another location.

1.8.1 Saved Designs

From 16.6 onward, pages that are not saved in the design are marked with an asterisk. Fig. 1.23 shows that Page 2 in the RC schematic has not been saved and hence the schematic **rc** folder and design file, **rc.dsn** are also shown as not having been saved.

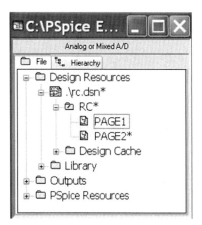

Fig. 1.23 Unsaved pages are marked with an asterisk.

Note

From 16.6, you can open designs created in earlier versions of OrCAD without having to update the design. The designs will only be updated to 16.6 if you save the design.

1.8.2 Find and Replace Text Utility

Capture already has a Find and Replace function (Edit > Replace) of text but in 17.2, the Find and Replace Text Utility can find and replace text from global names, nets, ports, comments, and of page connectors (see Fig. 1.24). The report section provides information on the text type and location, whereas a log file gives a summary of the action taken. You can search the entire design or the current page you are working on and then select from the top tool bar, **Tools > Utilities > Find ReplaceText**.

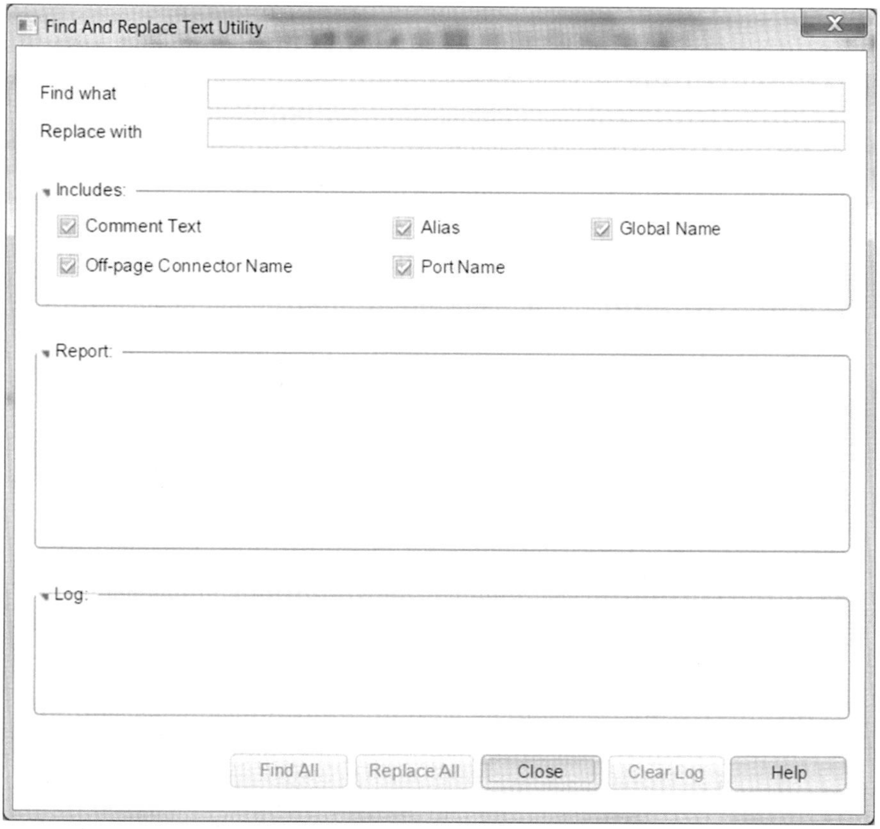

Fig. 1.24 Find and replace text.

1.8.3 Password Protection

In 17.2, Capture designs can be password protected but a word of warning, lost passwords cannot be retrieved even by Cadence. This function is not available in the 17.2 Lite version only in the full version. Passwords can be set by either **Design > Set Password** or highlighting the dsn file in the Project Manager and **rmb > Set Password**. Other options include removing or changing the password.

1.9. SUMMARY

Fig. 1.25 shows the Project Manager created for a PSpice project and the visual checks that can be made to ensure that a PSpice project has been setup correctly. One common mistake is not selecting and highlighting the project folder that is created (see Fig. 1.4).

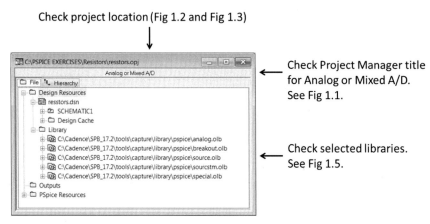

Fig. 1.25 Project Manager set up for a PSpice project.

Another common mistake when creating a project is that the wrong project type has been created, for example you see PCB instead of Analog or Mixed A/D in the Project Manager title. One way around this is to create a new project (of the correct type) and copy and paste the .dsn file from the previous Project Manager into the new Project Manager. From version 16.3 onward, you can now change the project type by highlighting **Design Resources** in the Project Manager and right mouse button, **rmb > Change Project Type**, see Fig. 1.26.

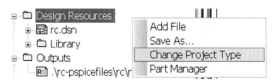

Fig. 1.26 Changing project type.

1.10. EXERCISES

Exercise 1

You will create a new PSpice project as discussed in section 1.2 and name it **resistors**. The project will be created in a folder called, for example, C:\PSpice Exercises\resistors and will be configured with the **simple** five default libraries.

1. You can select the New Project icon 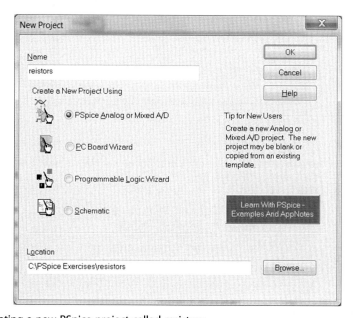 New Project, or select **File > New > Project**. Enter **resistors** for the **Name** and select **Analog or Mixed A/D**. In **Location**, enter **C:\PSpice exercises\resistors** or you can browse to your own folder location. Check your entries with Fig. 1.27 and then click on OK.

Fig. 1.27 Creating a new PSpice project called resistors.

Note

You can also use the **Browse** button to create and name the project folder.

---

2. In the **Create PSpice Project**, select **simple.opj** as shown in Fig. 1.28 and click on OK.

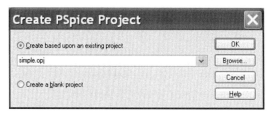

Fig. 1.28 Selecting the simple.opj project template.

3. The Project Manager window will appear as shown in Fig. 1.29.

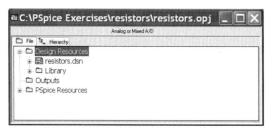

Fig. 1.29 Project Manager window.

4. Expand **resistors.dsn** by double clicking on it to open SCHEMATIC1 folder (Fig. 1.30).

Fig. 1.30 Schematic1 folder.

5. Double click on SCHEMATIC1 to open PAGE1 and double click on PAGE1 to open up the schematic page (Fig. 1.31).

Fig. 1.31 Page 1 folder.

6. When you first open the schematic page, you will see some preplaced text and two voltage sources preplaced. Delete the sources and text by drawing a box around the sources and text and press the delete the key.

Exercise 2

You will draw the resistor network shown in Fig. 1.32.

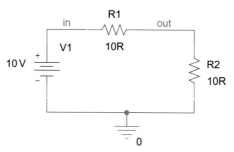

Fig. 1.32 Simple resistor circuit.

1. To place a resistor, select **Place > Part** and select an R from the analog.olb library. Double click on the R and the resistor will attach to the cursor. In previous versions you click on **OK** and the **Place Part** menu disappears. From release 16 onward, when you double click on a part or click on the **Place Part** , the menu will remain open.
   When you place the first resistor in the schematic, another resistor will be attached to the cursor, click on the right mouse button, **rmb > Rotate** or press R on the keyboard and place the second resistor. To exit place part mode, **rmb > End Mode** or press escape.
   Whenever a part is selected, there is a **rmb** context menu for place part options.
   P is the hotkey to place a part or you can select the Place Part icons depending on which software release you have   .
2. For R1 and R2, double click on the default resistor value of 1k and change its value to 10R.
3. Place the voltage source which can be found in the **source** library. Change its voltage to 10 V.
4. To place a ground symbol, **Place > Ground** (or press G) or click on the icon   and select the 0 V symbol from the capsym.olb library (Fig. 1.33).

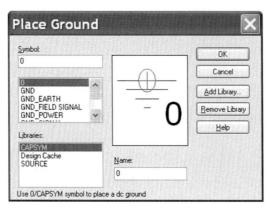

Fig. 1.33 Placing a 0 V ground symbol.

5. To draw a wire, **Place > Wire** (or select the wire icon ⌐ ⌐ or press W). You can always zoom in by pressing the "I" key on the keyboard or "O" for zoom out.

Note

To exit out of wire mode, you can press escape (**Esc**) on the keyboard or press W on the keyboard which toggles wire mode on and off. If you make a mistake you can always select the undo icon ↩ .

Note

There are new features available from version 16.3 onward to automatically connect wires to two or more points and a feature to automatically connect wires to a bus which is described in the Chapter on Digital Simulation. To automatically draw a wire, select **Place > Auto Wire > Two Points** (Fig. 1.34) or click on the icon 𝑓 . Click on the first wire point and then click on the second point.

Auto Wire		Two Points	
⌐ Bus	B	𝑓 Multiple Points	
✛ Junction	J	𝑓 Connect to Bus	

Fig. 1.34 Auto Wire allows for the automatic connection of wires and busses.

---

6. Capture automatically labels each wire connection also known as a node with a node number which by default is not displayed on the schematic. However, you can assign your own labels to wire nodes which will give meaning to a node, i.e., **input** or **output** and is useful when you want to analyze different nodes in a circuit. These labels are known as "Net Alias" and are placed on a wire by highlighting a wire and then selecting, **Place > Net Alias** (or select the net alias icon **N1** abc or press N).

7. Save the project by selecting **File > Save**.

Note

Parts can be pushed together such that when you move the parts away, wires are automatically connected. If not already set, select, **Options > Preferences > Miscellaneous > Wire Drag** and check, **Allow component move with connectivity changes** on as shown in Fig. 1.35. This also allows parts to be moved onto a wire and be connected.

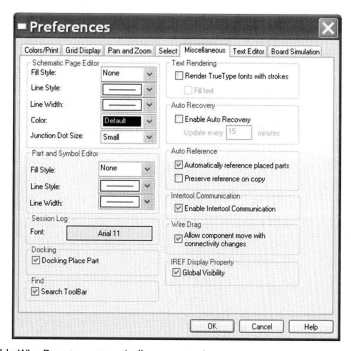

Fig. 1.35 Enable Wire Drag to automatically connect wires to parts.

1.11. EXTRA LIBRARY WORK

1. Select **Place Part** to open the **Place Part** window and **Add Library** by clicking on the ⬜ icon, see Fig. 1.36. In previous versions, you just click on **Add Library** (see Fig. 1.10A).

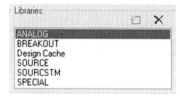

Fig. 1.36 Adding a library.

2. The **Browse File** window (Fig. 1.37) will open. Make sure you are in the **<install path> tools > capture > library > pspice** and select the **ana_swit.olb** library and click on **Open**.

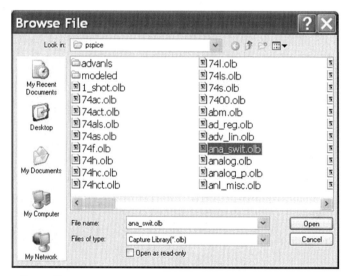

Fig. 1.37 Browse library files.

3. Close **Place Part.** Has the library been added to the library list in the Project Manager? Expand the Library folder.
4. In the Project Manager select the **Library** folder and **right mouse button > Add File** (Fig. 1.38).

Fig. 1.38 Adding a library to the Project Manager.

The **Browse File** window will open as in step 2. Make sure you are in the **<install path> tools > capture > library > pspice** and select the **1_shot.olb** library and click on **Open**.

5. Select **Place Part**. Has the **1_shot.olb** library been added to the list of **Libraries**?
6. In **Place Part** select all the libraries and click on **Remove Library**. Note which libraries are now available.

As you create new projects, any libraries added to previous projects in **Place Part** will be added to the list of available libraries. However, these libraries are not added to the Project Manager. Only libraries added via the **Library** folder in the Project Manager will be added to the configured list of Project libraries and can only be deleted via the Project Manager.

From 16.2 onward, when you select **Place Part**, the menu appears on the right hand side of the schematic and reduces the available size of the schematic page. However, at the top right of the **Place Part** menu, there is a thumbtack icon (Fig. 1.39) which gives you the option to effectively hide the Place Part menu. When you select the thumbtack icon, the Place Part menu disappears and the words Place Part appears as shown in Fig. 1.40. The menu re-appears when you move the mouse inside the Place Part window box and retracts when you move the mouse back to the left.

Thumbtack

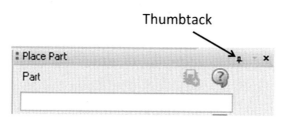

Fig. 1.39 Hiding the Place Part menu.

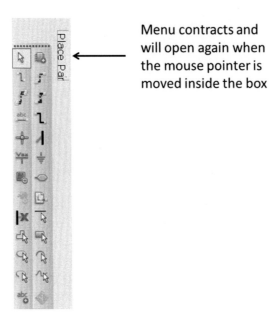

Menu contracts and
will open again when
the mouse pointer is
moved inside the box

Fig. 1.40 Place Part menu contracted.

CHAPTER 2

DC Bias Point Analysis

Chapter Outline

2.1. Netlist Generation   38
2.2. Displaying Bias Points   43
2.3. Save Bias Point   44
2.4. Load Bias Point   45
2.5. Exercises   45
     Exercise 1   45

When you connect a battery or a power supply to a circuit, the circuit voltages and currents effectively settle down to what is known as a DC steady-state condition. This is also known as the operating point or Bias Point of a circuit under steady-state conditions. In PSpice, the Bias Point analysis calculates the node voltages and currents through the devices in the circuit. For example, for a simple common emitter transistor amplifier, the Bias Point analysis will calculate the base, emitter, and collector bias voltages and the base, collector, and emitter quiescent currents.

Bias Point analysis will also take into account any voltage sources applied to the circuit and any initial conditions set on devices or nodes in the circuit. For example, you may want to preset a capacitor to a known voltage or set an initial digital state, a logic "1" or "0" on a digital device.

The calculated Bias Point voltages and currents are also used as a "starting point" for the other circuit analysis calculations. For example, when you run a transient (time) or an AC (frequency) analysis, PSpice automatically runs a Bias Point analysis first. However, the Bias Point analysis can be turned off for special cases in which a DC steady-state solution cannot be found. This is especially useful in the case of an oscillator which relies on the fact that it has no steady-state condition.

With the Bias point enabled, the output file will provide a list of all the analog and digital node voltages, the currents and total power of all voltage sources in the circuit, and a list of small signal parameters for all devices in the circuit. There is an option to suppress the bias information in the PSpice simulation profile.

Analog Design and Simulation Using OrCAD Capture and PSpice
https://doi.org/10.1016/B978-0-08-102505-5.00002-1

Note

With a Bias Point simulation, all capacitors are implemented as open circuit and all inductors are implemented as short circuits in order to calculate the DC Bias Point.

The RC circuit in Fig. 2.1 is based upon the resistor circuit in Chapter 1. A capacitor, C1 has been added in parallel with R2. With the circuit drawn, a PSpice simulation profile needs to be set up. The settings are accessed from the top tool bar, **PSpice > New Simulation Profile.** This is where the different analysis types for DC, AC, transient, and Bias Point are selected. By default, **Bias Point** is selected.

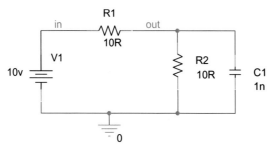

Fig. 2.1 RC circuit ready for DC Bias Point Analysis.

Fig. 2.2 shows the default PSpice Simulation Profile for a DC Bias Point Analysis. For this example, the default settings are used.

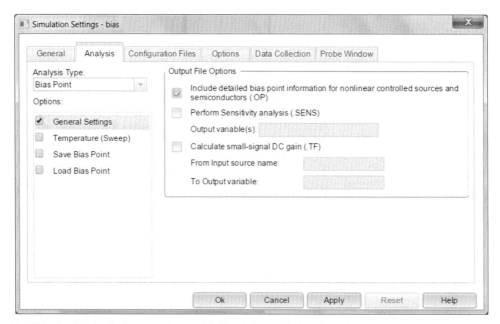

Fig. 2.2 Simulation Settings windows with Bias Point analysis selected.

Note

The latest 17.2 version of the Simulation Settings window shown in Fig. 2.2 for Bias settings now has an HTML look to it but has the same functionality.

To run the simulation, you select **PSpice > Run** or you can select the play button.

Note

When you run a simulation, the **PSpice Netlist Generation** dialog box shown in Fig. 2.3 appears.

Fig. 2.3 PSpice Netlist Generation.

In previous versions when you first run a simulation, the **Undo Warning** dialog box shown in Fig. 2.4 will appear. This just states that you will not be able to undo or redo any previous actions. Just check the **Do not show this box again** as shown in Fig. 2.4 and click on **Yes**.

Fig. 2.4 Undo Warning.

In 17.2 the appearance of the PSpice Simulation Profile Settings has changed to an HTML look.

2.1. NETLIST GENERATION

The circuit diagram drawn in Capture is represented as a netlist of all the components and their respective connections to other components. This netlist is automatically generated when you run the simulation and can be seen in the **Outputs** folder in the Project Manager, see Fig. 2.5.

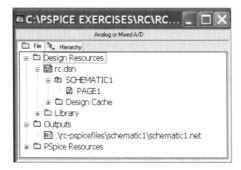

Fig. 2.5 The Schematic1.net netlist file has been generated in the **Outputs** folder.

Note

By default the circuit diagram is called SCHEMATIC1 and so the generated netlist is called schematic1.net as seen in Fig. 2.5. To rename the schematic, select the **SCHEMATIC1** name, right mouse click, and select **Rename** as shown in Fig. 2.6, entering **RC** as the new schematic name.

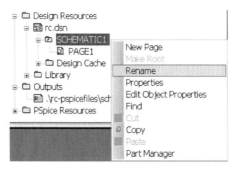

Fig. 2.6 Renaming SCHEMATIC1 to RC.

When you run the simulation an **rc.net** netlist will appear in the outputs folder. Fig. 2.7 shows the original **schematic1.net** and the newly generated **rc.net**. Note that Capture is not case sensitive. Any schematics named in upper case will generate lower case named netlist files (Fig. 2.7).

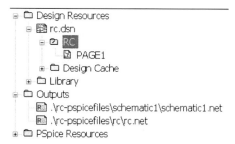

Fig. 2.7 Renamed RC schematic and generated rc.net netlist.

To open the netlist file, double click on the name and the netlist will appear as shown in Fig. 2.8.

```
1:  * source RC
2:  V_V1          IN 0 10V
3:  R_R1          IN OUT  10R TC=0,0
4:  R_R2          0 OUT  10R TC=0,0
5:  C_C1          OUT 0  1n  TC=0,0
6:
```

Fig. 2.8 resistors.net netlist.

The netlist in Fig. 2.8 shows that resistor R1 is connected between nodes, IN and OUT, has a value of 10 ohms, and a zero temperature coefficient (TC). The voltage source, V1, is connected between IN and 0 V and has a value of 10 V. The **R_** indicates that this is a flat netlist compared to a hierarchical netlist. Resistor properties and hierarchical circuits will be discussed in more detail in a later chapter. For now, the netlist defines the net connections between R1, R2, and V1 and any defined net names.

When you run a PSpice simulation, PSpice is launched and the PSpice environment window appears. However, there is no graphical output for a DC Bias Point analysis as no waveform data is calculated and hence no traces are available to be plotted. Instead you can look in the **Output File** to see the calculation results or view on the schematic, the bias voltages, currents, and instantaneous power for the circuit components. From the top toolbar in Capture or PSpice, select **View > Output File** and scroll to the bottom of the file (see Fig. 2.9). Because the nets have been named, you can easily see the calculated output voltages and currents. Remember, during a DC Bias Point analysis, capacitors are open circuit, hence the RC circuit is behaving as a simple potential divider with R1 and R2 dividing the 10 V DC voltage down to 5 V on the "out" net.

```
**** INCLUDING RC.net ****
* source RC
V_V1          IN 0 10V
R_R1          IN OUT  10R TC=0,0
R_R2          0 OUT   10R TC=0,0
C_C1          OUT 0   1n  TC=0,0

**** RESUMING bias.cir ****
.END

**** 05/15/11 11:03:59 ******* PSpice 16.3.0 (June 2009) ****** ID# 0 ********

** Profile: "RC-bias"  [ C:\PSPICE EXERCISES\RC\RC-PSpiceFiles\RC\bias.sim ]


****     SMALL SIGNAL BIAS SOLUTION      TEMPERATURE =   27.000 DEG C


****************************************************************************

  NODE   VOLTAGE      NODE   VOLTAGE      NODE   VOLTAGE      NODE   VOLTAGE


(   IN)   10.0000  (  OUT)    5.0000



      VOLTAGE SOURCE CURRENTS
      NAME           CURRENT

      V_V1           -5.000E-01

      TOTAL POWER DISSIPATION   5.00E+00   WATTS
```

Fig. 2.9 Bias Point analysis Output file.

Note
Capture is not case sensitive when naming nets.

A new feature (in release 16.3) as seen in Fig. 2.9 is the use of colors in the output file to highlight the different syntax such as text, component values, comments, expressions, and keywords. The default colors can be changed by selecting, **Options > Preferences > Text Editor** as shown in Fig. 2.10.

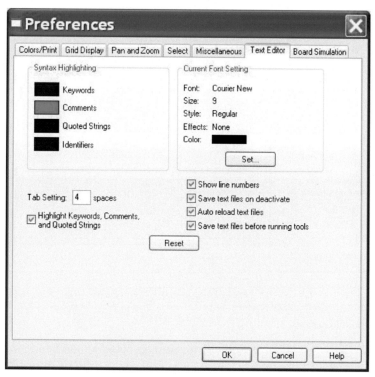

Fig. 2.10 Text Editor default colors and fonts.

The **Output File** also highlights any errors and warnings that may have occurred and is useful to determine where the errors occur. See the exercises at the end of the chapter for an example.

As mentioned previously, the output file provides a report on the analog and digital node voltages and device currents together with a list of small signal parameters for the devices in the circuit. There is the option to suppress the bias information reported in the output file. Under the **Options** tab in the PSpice Simulation profile (Simulation Settings), select **Output** file in the **Category**: box as shown in Fig. 2.11 and uncheck the **(NOBIAS)** option. There is also the option to list all the devices (LIST) in the circuit, summarizing their connecting nodes, values, models, and other parameters. Fig. 2.11 shows the other available options available for the output file. The **Reset** button resets all the options back to their default settings.

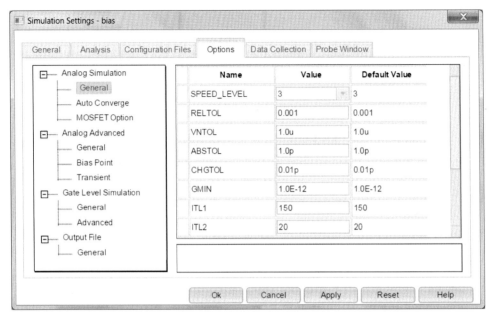

Fig. 2.11 Output file options.

Fig. 2.12 shows the PSpice Simulation Settings window with previous software versions. The new 17.2 layout makes it easier to navigate and access all available options.

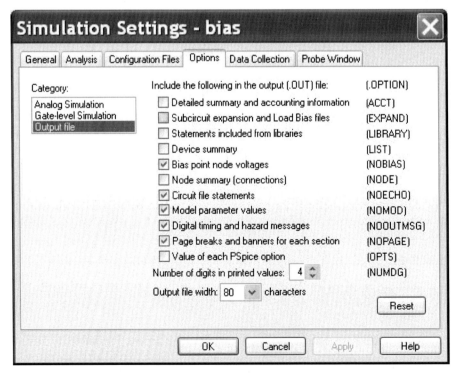

Fig. 2.12 Output file options with previous software releases.

2.2. DISPLAYING BIAS POINTS

After a simulation is run, the bias voltage, current, and power values can be displayed on the schematic. In Capture, select **PSpice > Bias Points > Enable** or select the bias display icons which have changed in appearance in release 16.3. See Figs. 2.13 and 2.14.

Fig. 2.13 Bias display icons pre-16.3.

Fig. 2.14 Bias Display icons in 16.3.

Fig. 2.15 shows the displayed bias voltages, currents, and instantaneous power for the resistor RC circuit.

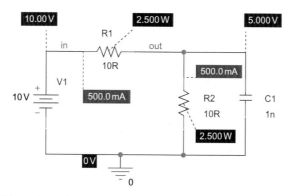

Fig. 2.15 Displayed bias point voltages, currents, and power.

The number of significant digits displayed for bias points can be changed by selecting, **PSpice > Bias Points > Preferences** as shown in Fig. 2.16. Up to 10 Precision digits can now be displayed.

Bias Point Preferences

Displayed Precision 4

Print	Color	Font	
✓		Arial 7	Current
✓		Arial 7	Voltage
✓		Arial 7	Power

OK

Cancel

Fig. 2.16 Changing the bias point display preferences.

Individual bias values can be turned on and off for voltage, current, or power. For example, if you select a wire net, the voltage icon will be activated **▯v** which will enable you to toggle the value for bias voltage on or off for a selected wire.

If you select a component pin, the current icon will be activated **▯I** which will enable you to toggle the value for bias current on or off for a selected part.

If you select a component, the instantaneous icon will be activated **▯⊦** which will enable you to toggle the value for bias power on or off for a selected part.

> **Tip**
> You may need to press F5 in order to refresh the display after you turn bias display on and off.

2.3. SAVE BIAS POINT

You can save and reuse the Bias Point data from a simulation which is useful if you have to run a number of simulations on a large circuit which has a long simulation run time. This is assuming that the circuit netlist, i.e., the connectivity of the components has not changed. Remember that other analyses use the calculated results from a Bias Point analysis so when you resimulate the circuit, assuming that the netlist has not changed, the initial Bias Point calculations can be saved and reused thus reducing the simulation run time. Saving the Bias Point is also useful when a simulation fails to converge to a solution.

In the **Simulation Profile Settings**, select Bias Point analysis and then select **Save Bias Point**. As you can see in Fig. 2.17, the data from the Bias Point analysis is saved in a file called saved_bias_point.txt. You select the **Browse…** button to select or create the folder in which to save the file.

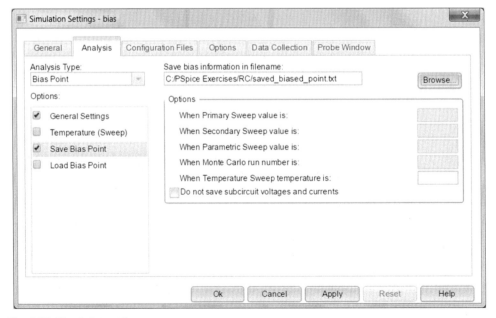

Fig. 2.17 Bias Point settings.

The Bias Point saved data contains node voltages and digital states for all the devices in the circuit, the total power and current supplied by any voltage sources, and a list of model parameters for the devices in the circuit.

Note

When saving Bias Point data, it is a good idea to add a .txt extension to the Bias Point data filename so you can readily open the file in a text editor such as Wordpad or Notepad.

2.4. LOAD BIAS POINT

Saved Bias Point analysis data is loaded by selecting the **Load Bias Point** option in the simulation profile. Fig. 2.18 shows a previously saved Bias Point data file being selected. Bias Point information can also be saved and used for a DC Sweep and a Transient analysis.

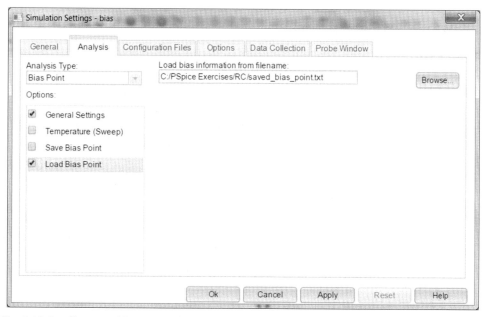

Fig. 2.18 Loading saved Bias Point data.

2.5. EXERCISES

Exercise 1

1. The circuit in Fig. 2.19 is based upon the resistor circuit in Chapter 1. Add a 1n capacitor, from the **analog** library, in parallel with R2.

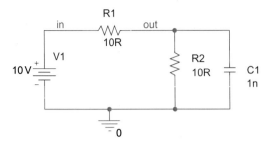

Fig. 2.19 RC circuit.

2. Delete the 0 V volt symbol and resimulate. A warning message will appear (Fig. 2.20) asking you to check the Session Log which is normally open at the bottom of the screen in Capture.

Fig. 2.20 Warning message.

You may have to expand the window upwards to see the complete message. If the session log is not visible, it can be found from the top tool bar, **Window > Session** and should contain the following message.

WARNING [NET0129] Your design does not contain a Ground (0) net.

Your reported net number shown will be different to the earlier one. The PSpice window will then open and display the output file which will report that a number of numbered nodes are floating (see Fig. 2.21).

```
**** INCLUDING RC.net ****
* source RC
V_V1          IN N00593 10V
R_R1          IN OUT   10R TC=0,0
R_R2          N00593 OUT  10R TC=0,0
C_C1          OUT N00593  1n  TC=0,0

**** RESUMING bias.cir ****
.END

ERROR -- Node IN is floating
ERROR -- Node N00593 is floating
ERROR -- Node OUT is floating
```

Fig. 2.21 Output file reporting floating node errors.

3. In the RC circuit, reconnect the ground symbol and resimulate. There should be no errors.

Note

PSpice automatically numbers nodes to wires in your circuit unless you assign a net name to them. In the previous output file, every node is floating because node 0 has not been assigned. In PSpice as with other Spice simulation software tools, there must be a 0 V node in the circuit otherwise nodes will be reported as floating in the output file.

In the **capsym** library there are other ground symbols as shown in Fig. 2.22. For PSpice simulations, make sure that you select the ground symbol with the **0** showing. The other symbols can be placed in the circuit to show the difference between ground connections as long as there is a 0 V node in the circuit.

Fig. 2.22 Ground symbols.

4. Create a bias point simulation profile based upon the initial bias point. **PSpice > New Simulation Profile** or click on the icon 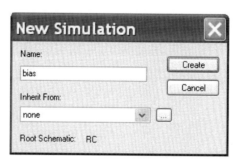 and enter bias for the simulation **Name**, see Fig. 2.23 and click on **Create**.

Fig. 2.23 Creating a bias point simulation profile.

5. The dialog box in Fig. 2.24 will appear telling you that a simulation profile of the same name already exists. New projects include a default Bias Point simulation profile called Bias.

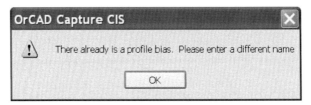

Fig. 2.24 A simulation profile with the same name already exists.

Click on OK and Capture will automatically present you with a new name of bias1 (Fig. 2.25). Click on **Create**.

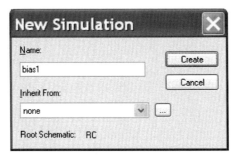

Fig. 2.25 The simulation profile name is automatically incremented to bias1.

6. The simulation settings window will appear. Fig. 2.26 shows the 17.2 version of the simulation settings that has the same functionality. If not already set by default, select the **Analysis type** to **Bias Point**. Click on **Apply** but do not exit.

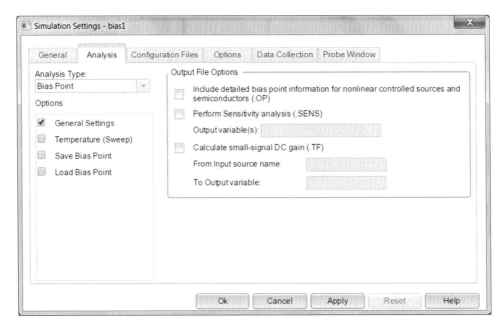

Fig. 2.26 Setting the Bias Point analysis.

7. If you are using 17.2 or later, select the Options tab and then select **Output File >
General**. Uncheck NOBIAS and check LIST as shown in Fig. 2.27.

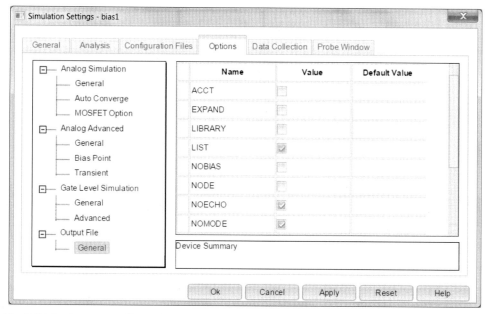

Fig. 2.27 Setting output file options in 17.2.

If you are using previous versions to 17.2; under the **Options** tab, select **Output File**
in the **Category:** box and uncheck (NOBIAS) and check (LIST), see Fig. 2.28. Click
on OK.

Fig. 2.28 Setting output file options previous to 17.2.

8. Run the simulation. When PSpice launches, select **View > Output** file. The output file shows a summary of the resistors, capacitor, and voltage source used in the circuit as shown in Fig. 2.29. The output voltages and currents are not reported in the output file.

```
****  RESISTORS

NAME                    NODES        MODEL      VALUE        TC1        TC2        TCE

R_R1            IN      OUT                    1.00E+01
R_R2             0      OUT                    1.00E+01


****  CAPACITORS

NAME                    NODES        MODEL      VALUE    In. Cond.      TC1        TC2

C_C1            OUT        0                   1.00E-09


****  INDEPENDENT SOURCES

NAME                    NODES     DC VALUE   AC VALUE   AC PHASE

V_V1            IN         0   1.00E+01   0.00E+00   0.00E+00   degrees
```

Fig. 2.29 Summary list of circuit devices.

9. Create another new PSpice simulation profile but this time the bias point simulation settings will be inherited from the previous bias1 point simulation profile. In the **New Simulation** window, enter **bias2** for the **Name**. Click on the pull down **Inherit From** menu as shown in Fig. 2.30, select bias1 and click on **Create**. In the simulation settings, you will see the Bias Point analysis is selected. Select **Options > Output file** and you will see LIST is checked and NOBIAS is unchecked. Close on OK to close the new simulation profile.

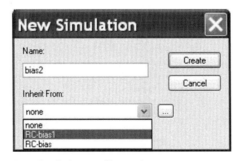

Fig. 2.30 Inheriting an existing simulation profile settings.

You now have created three bias point simulation profiles, bias, bias1, and bias2.

10. If not already displayed, open the **Project Manager** window by either selecting, **Window > <project path>\RC.opj** file (Fig. 2.31) or click on the icon ![icon] ![icon] .

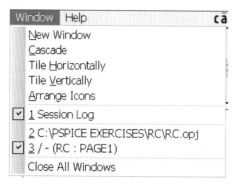

Fig. 2.31 Selecting the Project Manager.

The Project Manager will be displayed as shown in Fig. 2.32.

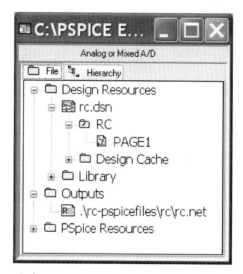

Fig. 2.32 Project Manager window.

11. In the **Project Manager** window, expand the **PSpice Resources > Simulation Profiles**. The three bias point analysis simulation profiles that have been created are listed in the Project Manager (Fig. 2.33) and are also displayed on the left-hand side of the top tool bar (Fig. 2.34) in which a pull down menu lists the simulation profiles, bias, bias1, and bias2.

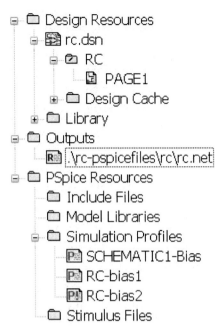

Fig. 2.33 Bias profiles listed in Project Manager.

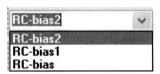

Fig. 2.34 List of bias profiles.

Note that bias2 is selected which appears in the Project Manager with a red icon to its left. This indicates that this is the current or active profile. You can select another profile in the pull down list or alternatively select the profile in the Project Manager and **right mouse button (rmb) > Make Active**.

12. Select bias1 in the pull down profile list in Fig. 2.34 and note the change in the Project Manager Simulation Profiles section. Bias1 is now the active profile.

Note

This is a useful way to switch between different profiles of different simulation settings or different analysis types.

CHAPTER 3

DC Analysis

Chapter Outline

3.1. DC Voltage Sweep   54
3.2. Markers   54
3.3. Exercises   59
    Exercise 1   59
    Exercise 2   63

The DC analysis calculates the circuit's bias point over a range of values when sweeping a voltage or current source, temperature, a global parameter, or a model parameter. The swept value can increase in a linear or a logarithmic range or can be a list of increasing values.

This is useful, for example, if you want to see the circuit response for a change in the supply voltage or to see how a change in a resistor value affects the circuit response. The DC Sweep also allows for nested sweeps such that one of two variables is kept constant while sweeping the other variable. For example, the characteristic transistor $I_C - V_{CE}$ curve plots the collector current against the collector-emitter voltage for fixed values of base current. The DC Sweep will then contain two variables, the collector-emitter voltage V_{CE} and the base current I_B. The base current is the secondary sweep, while the V_{CE} is the primary sweep. The collector current I_C is recorded by successive voltage sweeps of the collector-emitter voltage V_{CE} for stepped values of base current producing a series of curves as shown in Fig. 3.1.

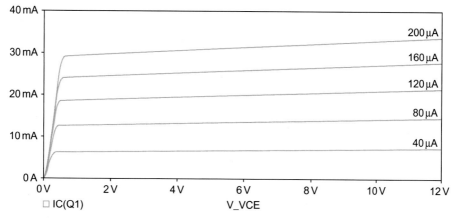

Fig. 3.1 Nested sweep for a transistor characteristic curve.

Analog Design and Simulation Using OrCAD Capture and PSpice
https://doi.org/10.1016/B978-0-08-102505-5.00003-3

3.1. DC VOLTAGE SWEEP

As with other analysis types, a PSpice simulation profile needs to be created. For a DC Sweep Analysis, select **PSpice > New Simulation Profile** and select **DC Sweep** for the Analysis type. Make sure the **Sweep Variable** is set to **Voltage source**. The **Name** is the reference designator for the voltage source which in this case is **V1**. The sweep type is set to Linear, starting from 0 to 10 V in steps of 1 V. You can also enter a list of voltages in the **Value list**, for example; 1 2 4 5 99 100 as long as the voltages are increasing in value. Fig. 3.2 shows the simulation profile for a DC linear sweep of V1 from 0 to 10 V in steps of 1 V.

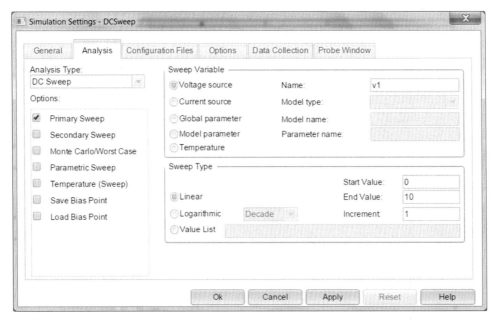

Fig. 3.2 DC Sweep simulation settings.

Note
You can rename the reference designator for the voltage source to anything you like as long as the first character is a V. For example, the voltage source could be named, Vsupply. This also applies to other components such as resistors, especially when you want to define for example a load resistor, RL.

3.2. MARKERS

Markers are used to record the voltages on nodes or currents through components and are accessed from the PSpice menu. They enable data to be automatically displayed as a waveform in the PSpice waveform viewer which is known as the **Probe** window. To add markers select, **PSpice > Markers** and then you have a choice of placing current, voltage, or differential voltage markers as shown in Fig. 3.3. The Advanced markers are primarily used with an AC analysis and will be covered in the chapter on AC analysis.

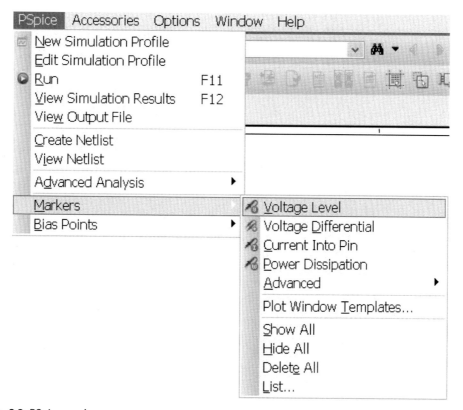

Fig. 3.3 PSpice markers.

Fig. 3.4A shows the icons for the markers in version 16.2 and Fig. 3.4B the markers from 16.3 versions onwards.

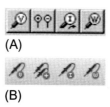

Fig. 3.4 PSpice marker icons: (A) version 16.2 and (B) version 16.3.

Voltage markers are placed on wires whereas current markers must be placed on a component pin. A message (Fig. 3.5) will appear if you try to place a current marker on a wire instead of a component pin. The component pin is a different color to a wire and in version 16.3, component pins are noticeably thinner than the connecting wires. For power markers, the markers are placed on the body of the device.

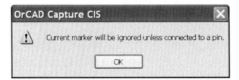

Fig. 3.5 Warning message if you try to place a current marker on a wire.

Note

You can only add markers to a circuit after you have set up a PSpice Simulation Profile and not before. It is a common mistake to add markers to a circuit and then set up a new simulation profile and have the markers disappear when the profile is set up.

Fig. 3.6 shows the addition of two voltage markers on the resistor circuit which will record the voltage at nodes "**in**" and "**out**" respectively and automatically display their voltage traces in the PSpice Probe waveform window.

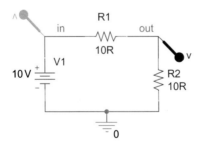

Fig. 3.6 Addition of two voltage markers.

When the simulation is run, PSpice will launch and **Probe**, will plot the two voltage traces at nodes **in** and **out**. The x-axis is the swept voltage V(in) and the y-axis is the resultant voltage V(out). Note that the respective traces have the same color as the markers placed in the circuit. See Figs. 3.6 and 3.7.

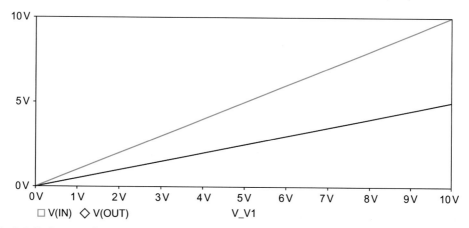

Fig. 3.7 Probe waveform viewer displaying the voltages at nodes, **in** and **out**.

Note

When you first place markers, their color is initially gray. After a simulation is run, the markers will change color and the respective waveform in **Probe,** will reflect the marker's color. If you delete a trace name in Probe, the marker on the schematic will turn gray. Double clicking on the marker, the marker color will be restored and the trace name will re-appear in Probe.

In release 16.3, you can now change the Probe background color which by default is black. From the top tool bar in PSpice, select, **Tools > Options > Color Settings**.

The waveform trace color in Probe can be changed by highlighting the trace and **right mouse click > Properties**, for release 16.2 (Fig. 3.8) or **right mouse click > Trace Property** in release 16.3. Both actions will similarly display the **Trace Properties** box shown in Fig. 3.9.

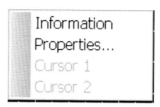

Fig. 3.8 Trace options for 16.2.

In the **Trace Properties** box (Fig. 3.9), the trace color, pattern, width, and symbol can be changed.

Fig. 3.9 Changing trace properties.

Note

From version 16.3, you can now have the functionality to **right mouse button** on a trace to open a context-sensitive menu which displays all the related trace options as shown in Fig. 3.10.

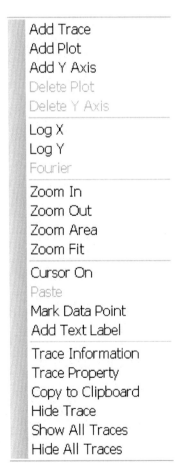

Fig. 3.10 Trace menu options.

From 16.3 you can now change the foreground colors such as the axis and grid lines and also the background color for Probe. **Tools > Options > Color Settings** as shown in Fig. 3.11.

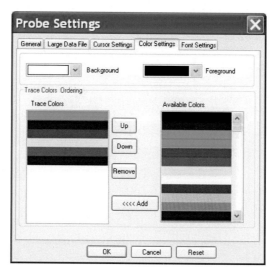

Fig. 3.11 Changing colors in Probe.

3.3. EXERCISES
Exercise 1

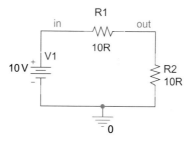

Fig. 3.12 Resistor network.

The resistor network circuit in Fig. 3.12 is a potential divider where the ratio of the output voltage to the input voltage is given by

$$\frac{V_{out}}{V_{in}} = \frac{R1}{R1+R2}$$

(3.1)

which can be written as:

$$V_{out} = V_{in} \frac{R1}{R1 + R2} \qquad (3.2)$$

Therefore the output voltage is determined by the ratio of the fixed values of the resistors R1 and R2. If R1 = R2, the output voltage is equal to half of the input voltage.

1. Draw the resistor circuit in Fig. 3.12 and name the nodes as shown by selecting **Place > Net Alias** (Fig. 3.13).

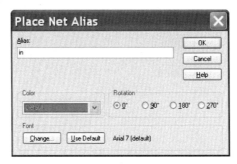

Fig. 3.13 Placing Net Alias.

2. Create a PSpice simulation profile, **PSpice > New Simulation Profile** and for the **Analysis type**, select **DC Sweep**. Select the **Sweep variable** as the voltage source V1 and set up a **linear** sweep from 0 to 10 V in steps of 1 V, see Fig. 3.14. Click on OK.

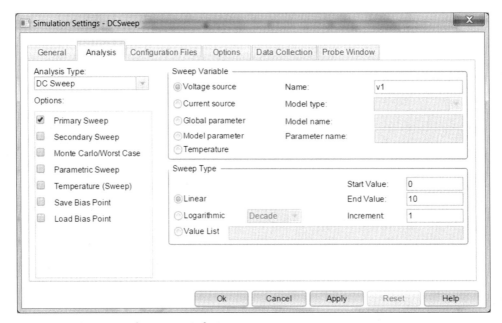

Fig. 3.14 DC Sweep simulation setting for V1.

3. Place voltage markers on nodes, **in** and **out** (Fig. 3.15).

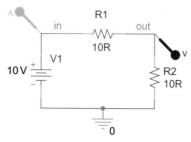

Fig. 3.15 Placing voltage markers on the nodes.

4. Run the simulation ▶.
5. PSpice will launch and Probe will display the two traces for V(in) and V(out). Fig. 3.16 shows the trace colors are the same as the marker colors in the circuit diagram in Fig. 3.15. Note that the circuit is acting as a potential divider in that the output voltage is half of the input voltage.

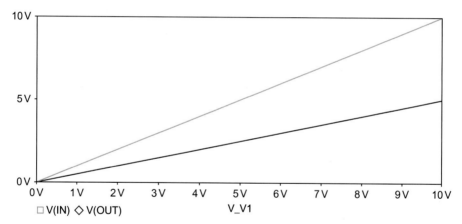

Fig. 3.16 Output voltage response of resistor circuit for a DC Sweep of V_{in}.

6. Delete the ground symbol and re-simulate. You should see the warning message dialog box (Fig. 3.17) and a message will be displayed asking you to check the Session Log.

Orcad Capture ✕

Warnings were reported, check Session Log   OK

☐ Do not show this dialog again

Fig. 3.17 Warning message.

7. The Session Log is normally open at the bottom of the screen, if not, the session log can be found from the top tool bar, **Window > Session Log**. The warning message will read:

WARNING [NET0129] Your design does not contain a Ground (0) net.

8. Click on OK in the Warning message (Fig. 3.15) and PSpice will launch. If the Output file is not displayed, then select, **View > Output File**. The output file is shown in Fig. 3.18. Your circuit will have different node numbers. The removal of a 0 V symbol was covered in the exercises at the end of Chapter 2 and is mentioned here again as a reminder that in order for the analog voltages to be calculated, a 0 V node must exist in the circuit otherwise nodes will be reported as floating.

```
* source RC SWEEP
V_V1          IN N00555 10V
R_R1          IN OUT   10R TC=0,0
R_R2          N00555 OUT   10R TC=0,0

**** RESUMING "DC Sweep.cir" ****
.END

ERROR -- Node IN is floating
ERROR -- Node N00555 is floating
ERROR -- Node OUT is floating
```

Fig. 3.18 Output file showing floating node error messages.

9. Re-connect a 0 V symbol. **Place > Ground** or press G on the keyboard and select a 0 V symbol.
10. Remove resistor R2 from the circuit and simulate.
11. PSpice will start and display the output file with an error message:

```
ERROR -- Less than 2 connections at node out
```

12. The Error message relates to no DC path to ground at node **out**, i.e., the end of the resistor R2 is floating. One other requirement in PSpice is that every node in a circuit must have a DC path to ground. If you need to simulate a node as open circuit, then you can always connect a large value resistor, for example, 100 G ohm or 1T ohm from a node to ground which will provide a DC path to ground without adversely affecting the bias conditions of the circuit.

Similarly, a short circuit can be implemented by a very small resistance of say 1μ ohm or less.

Exercise 2

A nested sweep will be performed to display the transistor characteristic family of curves. You will set up the VCE for the transistor as the primary sweep and the base current as the secondary sweep.

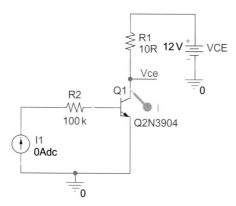

Fig. 3.19 Transistor circuit.

In 17.2 you can search for the Q2N3904 transistor using, **Place > PSpice Compo- nent > Search** and entering Q2N3904 in the search field. See Chapter 1, Section 1.3.3, Fig. 1.13. Alternatively follow the following steps.
1. Draw the circuit in Fig. 3.19. The transistor can be found in the **bipolar** library. In the **Place Part** menu, select the Add Library icon  as shown in Fig. 3.20 or in previous versions, click on **Add Library**.

Fig. 3.20 Add Library.

This will open up the **Browse File** window shown in Fig. 3.21. Scroll along or alternatively, type in bipolar.olb in the **File name** field. Select bipolar.olb and click on **Open**.

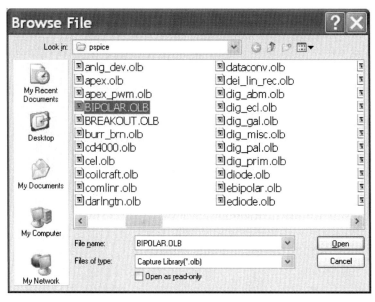

Fig. 3.21 PSpice-Capture libraries.

2. The bipolar library will now be added to the list of libraries in the Place Part menu. Select the bipolar library and type in Q2N3904 (not case sensitive) in the Part box, see Fig. 3.22. Double click on the q2n3904 transistor and place in the schematic page.

Fig. 3.22 Q2N3904 selected from the bipolar library.

Note

The Q2N3904 is available in the OrCAD demo version.

3. Place the rest of the components as shown in Fig. 3.19. The dc current source, idc is from the **source** library.
4. You will set up a nested DC Sweep where the VCE will be the primary sweep and the base current the secondary sweep.
5. Create a simulation profile, **PSpice > New Simulation Profile** and select the **Analysis type** to **DC Sweep**. For the **Primary Sweep**, which is shown by default, select the **Sweep variable** to be voltage the source **VCE**. The **Sweep type** is **Linear** with a **Start value** of 0 V, an **End value** of 12 V and an **Increment** of 0.1 V, see Fig. 3.23. Click on **Apply** but do not exit the Simulation Profile.

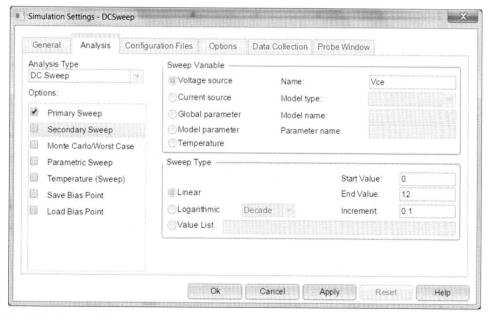

Fig. 3.23 Primary DC Sweep settings.

6. In the **Options** box, select the **Secondary Sweep**. The **Sweep variable** is the current source I1. The **Sweep type** is **Linear** with a **Start value** of 40 μA and an **End value** of 200 μA, the **Increment** is 40 μA. Make sure the **Secondary Sweep** box is checked and click on OK, see Fig. 3.24.

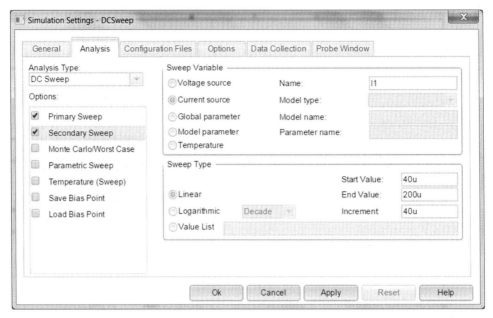

Fig. 3.24 Secondary DC Sweep settings.

7. Place a current marker on the collector pin of the transistor and simulate.
8. You should see the characteristics curves shown in Fig. 3.25.

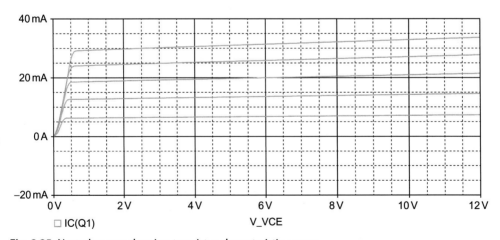

Fig. 3.25 Nested sweep showing transistor characteristic curves.

9. Select **Plot > Axis Settings > YAxis** and change the **Data Range** to **User Defined** and enter a range from 0 mA to 40 mA. Click on OK and see the change.

10. Select **Plot > Axis Settings > YGrid** and uncheck **Automatic** and set the **Major Spacing** to 10 m. Click on OK and see the change.
11. Select **Plot > Axis Settings > XGrid** and uncheck both **Major** and **Minor Grids** to **None**. Click on OK and see the change.
12. Select **Plot > Axis Settings > YGrid** and uncheck both **Major** and **Minor Grids** to **None**. Click on OK and see the change.
13. Select **Plot > Label > Text,** change the font color to silver and add the base currents as shown in Fig. 3.26.

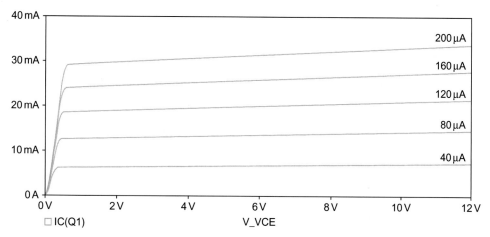

Fig. 3.26 Nested sweep showing transistor characteristic curves for Ic against Vce for fixed values of Ib.

CHAPTER 4

AC Analysis

Chapter Outline

4.1. Simulation Parameters 70
4.2. AC Markers 72
4.3. Exercises 72
    Exercise 1 72
    4.3.1 Twin T notch filter 76

The AC analysis is used to calculate the frequency and phase response of a circuit by frequency sweeping an AC source connected to the circuit. The AC sweep analysis is a linear analysis and calculates what is known as the small signal response of a circuit over a range of frequencies by replacing any nonlinear circuit device models with linear models. The DC bias point analysis is run prior to the AC analysis and is used to effectively linearize the circuit around the DC bias point. It must be noted that the AC analysis does not take into account effects such as clipping. You will have to run a transient analysis to determine these effects.

To perform an AC analysis, the independent voltage source V_{AC} or current source I_{AC} (Fig. 4.1A) from the **source** library is used. However, any independent voltage source which has an AC property attached to the part can be used as an input to the circuit. Fig. 4.1B shows the properties attached to the V_{AC} part as displayed in the Property Editor.

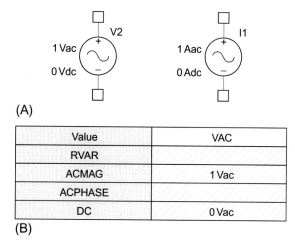

(A)

Value	VAC
RVAR	
ACMAG	1 Vac
ACPHASE	
DC	0 Vac

(B)

Fig. 4.1 Independent V_{AC} and I_{AC} sources: (A) capture parts and (B) V_{AC} properties.

Analog Design and Simulation Using OrCAD Capture and PSpice
https://doi.org/10.1016/B978-0-08-102505-5.00004-5

By default, the magnitude of the V_{AC} source is 1 V. In calculating the frequency response of a circuit, you are normally looking to calculate the gain and phase response of the circuit. Since the circuit gain is given by the ratio of V_{out} to V_{in}, setting V_{in} to 1 V, the gain or transfer function of the circuit will be equal to the output voltage, V_{out}.

4.1. SIMULATION PARAMETERS

One example in which an AC analysis is used to determine the frequency response of a circuit is the notch filter, which attenuates a narrow band of unwanted frequencies—for instance, the removal of the mains frequency which can lead to unwanted "hum" in an audio amplifier. One common implementation of a twin T notch filter is shown in Fig. 4.2, where the notch frequency is given by:

$$f_o = \frac{1}{2\pi RC} \tag{4.1}$$

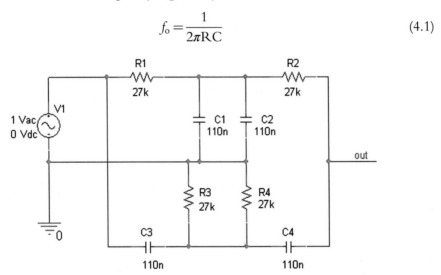

Fig. 4.2 Implementation of a twin T notch filter.

Fig. 4.3 shows the notch filter response of the circuit where the output is attenuated by −60 dB at the notch frequency of 53 Hz.

To set up an AC analysis, a PSpice simulation profile needs to be created: **PSpice > New Simulation** Profile. In Fig. 4.4, the **Analysis type** is set to **AC Sweep/Noise** and has been set up for a logarithmic frequency sweep starting from 1 Hz to 100 kHz. You have the choice to sweep the frequency linearly over the whole range or logarithmically either in decades or in octaves. If you want to use a linear range, note that the **Total Points** is applied over the whole frequency range compared to, say, a decade range where the number of points applies to each decade.

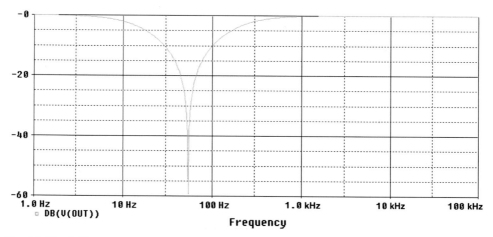

Fig. 4.3 Notch filter response.

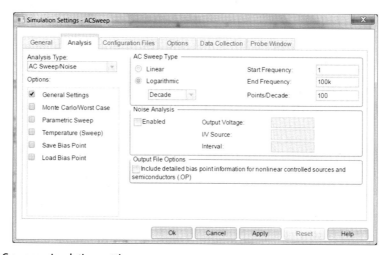

Fig. 4.4 AC sweep simulation settings.

Note

It is a common mistake to mix up the total number of points for linear and decade (or octave). So if your AC response curve severely lacks any resolution, i.e. has been plotted showing 10 points over the complete frequency range, check your AC sweep simulation settings for the correct AC sweep type, linear of logarithmic. Another common mistake is when specifying megahertz, to use MHz, which is the same as mHz (millihertz) as PSpice is not case sensitive. For megahertz, use megHz or MEGHz or 10e6 Hz. You do not have to enter the units (Hz), for example 100 meg.

4.2. AC MARKERS

AC markers can be found under the **PSpice > Markers > Advanced** menu as shown in Fig. 4.5. These markers can be used to display dB magnitude, phase, group delay, and the real and imaginary parts of voltage and current. For example, a combination of these markers can be used for Bode and Nyquist plots.

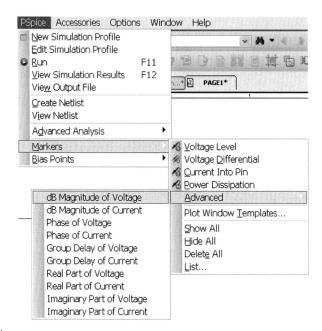

Fig. 4.5 AC Markers menu.

4.3. EXERCISES

Exercise 1

Fig. 4.6 shows a passive twin T notch filter. You will create an AC sweep and plot the frequency response of the circuit.

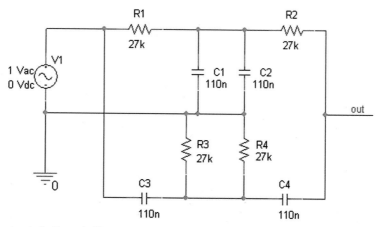

Fig. 4.6 Passive twin T notch filter.

1. Draw the circuit of the notch filter in Fig. 4.6. The V_{AC} source can be found in the **source** library.

2. Create a PSpice simulation profile to perform a logarithmic sweep from 1 Hz to 100 kHz in steps of 100 points per decade (Fig. 4.7).

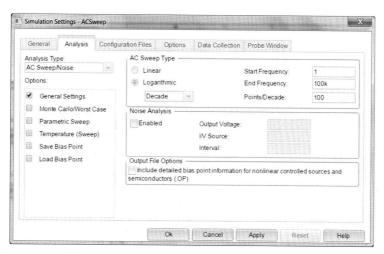

Fig. 4.7 AC sweep simulation settings.

3. Place a V_{dB} voltage marker on the output node, "out." This will automatically calculate the output voltage in dB: **PSpice > Markers > Advanced > dB Magnitude of Voltage** (Fig. 4.8).

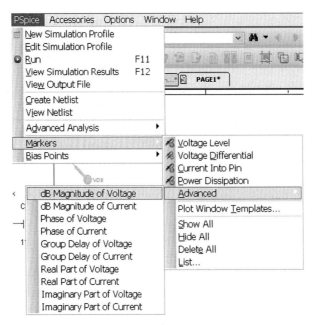

Fig. 4.8 Adding a dB voltage marker.

4. Run the simulation and you should see the notch filter response (Fig. 4.9). What we
need to do now is to determine the frequency at the deepest part of the notch.

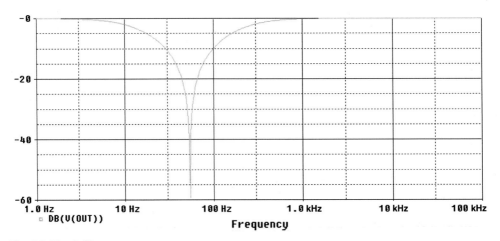

Fig. 4.9 Notch filter response.

5. Turn on the cursor, **Trace > Cursor > Display** ⌖ ▷ and place the cursor at the
bottom of the notch. For a more accurate reading, zoom in towards the bottom of
the trace, **View > Zoom > Area** or use the icons 🔍 🔍 🔍 🔍

    Alternatively, you can use one of the cursor functions, **Trace > Cursor > Min**
↯ or ↯.

The **Probe Cursor** box will display a notch frequency 53.703 Hz with an attenuation of -59.348 dB as seen in Fig. 4.10, depending on which software version you are using. In both software versions, the **Probe Cursor** box can be positioned anywhere in the Probe screen.

```
Probe Cursor
A1  =    53.703,    -59.348
A2  =    1.0000,    -24.147m
dif=     52.703,    -59.324
```
(A)

Trace Color	Trace Name	Y1	Y2
	X Values	53.703	1.0000
CURSOR 1,2	DB(V(OUT))	−59.348	−24.148m

(B)

Fig. 4.10 Cursor data: (A) version 16.2 and (B) version 16.3.

6. Restore the Probe display to its original size, **View > Zoom > Fit** 🔍
   Select the trace name at the bottom of the display and press delete on the keyboard or select **Trace > Delete all Traces**. Now we are going to manually add the trace for the output voltage V(out).
7. Select **Trace > Add Trace** 📈 and the **Add Traces** window will appear as shown in Fig. 4.11.

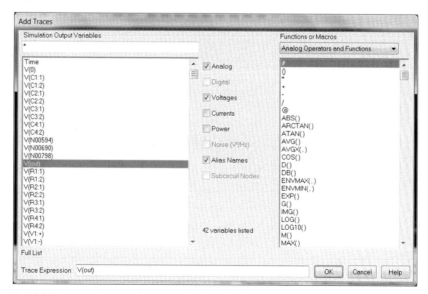

Fig. 4.11 Add Traces window showing the list of output variables and Analog Operators and Functions.

8. The **Add Traces** window displays all the data for all the nodes and devices in the circuit. Uncheck the boxes for **Currents** and **Power** and in the list of **Simulation Output Variables**, scroll down and select the V(out) variable.
   Click on OK.
9. The V(out) trace will be displayed in the Probe trace window. However, what we want is the voltage in dB. Select the trace name V(out) and press the delete key to remove the trace.
10. Select **Trace > Add Trace** and in the right-hand side of **Add Traces** under **Analog Operators and Functions**, select DB().
    Then, as before, select V(out) from the list of **Simulation Output Variables**. At the bottom of the window in the **Trace Expression** box, you should see DB(V (out)). The DB function automatically calculates the DB of V(out). See Fig. 4.12.

Trace expression: DB(V(out))

Fig. 4.12 Conversion of V(out) to dB.

Click on OK and you should see the trace shown in Fig. 4.13.

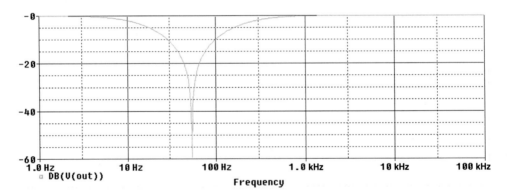

Fig. 4.13 Notch filter response.

11. Repeat Step 5 to determine the depth of notch attenuation in dB.

4.3.1 Twin T notch filter

Fig. 4.14 shows the implementation of a notch filter which has a notch frequency given by

$$f_o = \frac{1}{2\pi RC}$$

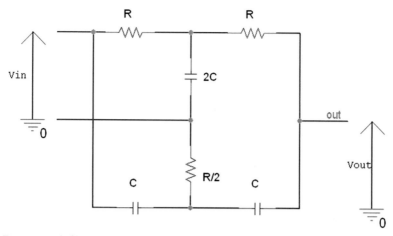

Fig. 4.14 Twin T notch filter.

Using only one value of resistor and one value of capacitor, the circuit in Fig. 4.14 can be implemented as shown in Fig. 4.15.

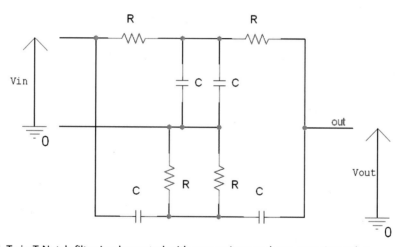

Fig. 4.15 Twin T Notch filter implemented with one resistor and one capacitor value.

The parallel combination of the two capacitors is given by:

$$C_p = C + C = 2C$$

The parallel combination of the resistors is given by:

$$R_p = \frac{R \times R}{R + R} = \frac{R}{2}$$

which conforms to the circuit implementation shown in Fig. 4.14.

For the notch filter in Fig. 4.6, $R = 27\,k\Omega$ and $C = 110n$. Therefore, using Eq. (4.1), the notch frequency is given by:

$$f_o = \frac{1}{2\pi RC}$$

$$f_o = \frac{1}{2\pi \times 27 \times 10^3 \times 110 \times 10^{-9}} = 53.6 \text{ Hz}$$

CHAPTER 5

Parametric Sweep

5.1. Property Editor   79
5.2. Exercises   83
    Exercise 1   83
    Exercise 2   89
    Exercise 3   92

A parametric sweep allows for a parameter to be swept through a range of values and can be performed when running a transient, AC or DC sweep analysis. Parameters that can be varied include a voltage or current source, temperature, a global parameter or a model parameter. A global parameter can represent a mathematical expression as well as a variable and is defined using the **PARAM** part from the **Special** library. To define the global variable, you have to add a new property to the **PARAM** part by editing its properties via the **Property Editor**. For example, in the resistor circuit shown in Fig. 5.1, the value of resistor R2 has been replaced with a variable called **{rvariable}**; you can name the variable anything you like. The braces otherwise known as curly brackets { } are required in PSpice to define global parameters.

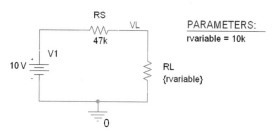

Fig. 5.1 Defining a global parameter and a default value.

The **PARAM** part has a heading called **PARAMETERS:** and contains a list of defined variables and their default values. In this case, RL is defined to have a default value of 10 kΩ if no parametric sweep is performed.

5.1. PROPERTY EDITOR

As mentioned above, the global parameter variable name and default value must be added to the **PARAM** part as a new property in the **Property Editor**. The **PARAM** part is

Analog Design and Simulation Using OrCAD Capture and PSpice
https://doi.org/10.1016/B978-0-08-102505-5.00005-7

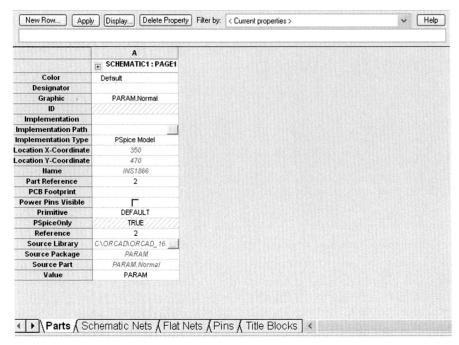

Fig. 5.2 Property Editor.

found in the **special** library and is placed anywhere in the schematic page. By double clicking on the **PARAM** part, the **Property Editor** (Fig. 5.2) will open.

The **Property Editor** is a spreadsheet that displays all the properties attached to a part. For example, a resistor will have defined properties such as a footprint, resistor value, power rating, tolerance, manufacturer's part number and PSpice model. Properties can be added to parts in the Property Editor and in the case of the **PARAM** part, we need to add a property to define the global variable that needs to be swept.

When first selected, the Property Editor will open up in one of two modes, displaying the properties in either rows (Fig. 5.3A) or columns (Fig. 5.3B).

It is easier to view all the properties listed in rows, as shown in Fig. 5.3A. If the properties are displayed in columns (Fig. 5.3B), you will have to use the scrollbar at the bottom of the Property Editor to scroll along to view all the other properties. The viewed mode can be changed, for example, from columns to rows by placing the cursor in the blank cell to the left of the **Color** property (Fig. 5.3B) and when the cursor changes to an arrow, **rmb > Pivot**. The properties will now be displayed as rows.

To change the view from rows to columns, place the cursor in the blank cell above the **Color** property (Fig. 5.3A). When the cursor changes to an arrow, **rmb > Pivot**.

To add a property which allows a global parameter to be swept, select **New Row** as shown in Fig. 5.3A and the **Add New Row** dialog box will appear (Fig. 5.4). Add the **Name** of the variable and the default **Value** of the variable. Fig. 5.4 shows a parameter

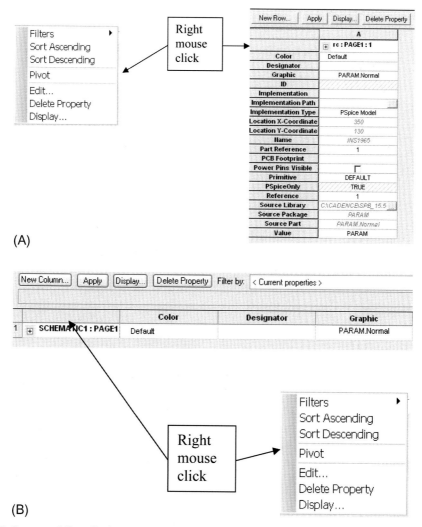

Fig. 5.3 Property Editor displaying PARAM part properties in either (A) rows or (B) columns.

called **rvariable** being defined with a default value of 10k and Fig. 5.5 shows the new row added in the Property Editor.

The properties listed in the Property Editor have a name and a value. For example, a transistor may have a standard TO5 footprint. Footprint is the property name and its value is TO5. Another example is a resistor with a 1% tolerance. Tolerance is the property name and 1% is the value of the property. Table 5.1 gives some examples of property name and value.

By default, new property names and values are not displayed on the part in the schematic diagram so they have to be made visible. This is done by highlighting the property cell in the Property Editor and selecting **Display** (or **rmb > Display**). This will open up the **Display Properties** dialog box as shown in Fig. 5.6.

Add New Row

Name:

rvariable

Value:

10k

Enter a name and click Apply or OK to add a column/row to the property editor and optionally the current filter (but not the <Current properties> filter).

No properties will be added to selected objects until you enter a value here or in the newly created cells in the property editor spreadsheet.

☐ Always show this column/row in this filter

| Apply | OK | Cancel | Help |

Fig. 5.4 Adding a new property to the PARAM part.

| New Row... | Apply | Display... | Delete Property |

10k

	A
	⊞ SCHEMATIC1 : PAGE1
Color	Default
Designator	
Graphic	PARAM.Normal
ID	
Implementation	
Implementation Path	
Implementation Type	PSpice Model
Location X-Coordinate	360
Location Y-Coordinate	370
Name	INS1858
Part Reference	1
PCB Footprint	
Power Pins Visible	☐
Primitive	DEFAULT
PSpiceOnly	TRUE
Reference	1
rvariable	10k
Source Library	C:\ORCAD\ORCAD_16. ...
Source Package	PARAM
Source Part	PARAM.Normal
Value	PARAM

Fig. 5.5 The new parameter variable **rvariable** with a default value of 10k has been added to the PARAM part.

Table 5.1 Property examples

Property name	Property value
Footprint	TO5
Tolerance	1%
Part reference	R1

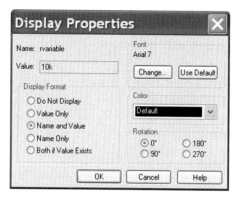

Fig. 5.6 Display Properties controls whether property names and values are displayed on the schematic.

It is recommended for Global parameters that both property **name** and **value** are both displayed when added to the **Param** part, as seen in the resistor circuit in Fig. 5.1.

Note

The Property Editor is closed by clicking on the lower right hand cross of the property window. Be careful not to select the upper top cross as this will close Capture

5.2. EXERCISES

Exercise 1

When you connect audio equipment together, for example the output of a microphone to the input of an amplifier, it is recommended that the output impedance of the microphone should match the input impedance of the amplifier. This is also true for video and radio frequency (RF) equipment. What happens is that the maximum power transfer of a signal between a source impedance and a load impedance occurs when the impedances match and the maximum power transfer that can be achieved is 50% of the source signal.

You will demonstrate simple resistance matching by plotting the load resistor power dissipation against load resistance for the circuit in Fig. 5.7.

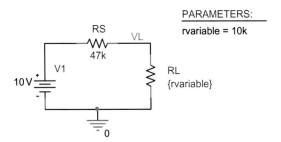

Fig. 5.7 Resistor network to demonstrate resistance matching.

1. Create a new PSpice project or use the resistor project from Chapter 1 as a starting point.
2. Place a V_{DC} source from the source library and set its value to 10 V.
   Place a resistor R from the analog library and name it RS and set its value to 47k. Place resistor RL and set its value to {rvariable}.
   Connect a 0 V symbol from the capsym library (Place > Ground).
   Name the net node connecting RS to RL as VL (Place Net > Alias) (Fig. 5.7).
3. Place a PARAM part from the special library anywhere on the schematic.
4. Double click on the Param part to open the Property Editor.
5. Depending on how the properties are displayed in the Property Editor (rows or columns), add a new property by either clicking on **New Row…** or **New Column…** Create a new property called **rvariable** with a value of 10k as shown in Fig. 5.8.
   The circuit is now set up with a global parameter, **rvariable**, with a default value of 10 kΩ, which will be the resistor value used for simulation if no parametric sweep is performed. Click on OK but do **not** exit the Property Editor.

Fig. 5.8 Creating a new global parameter.

6. Highlight the new property **rvariable** and select **Display**. In the **Display Properties** window (Fig. 5.9) select **Name and Value**. Close the Property Editor.

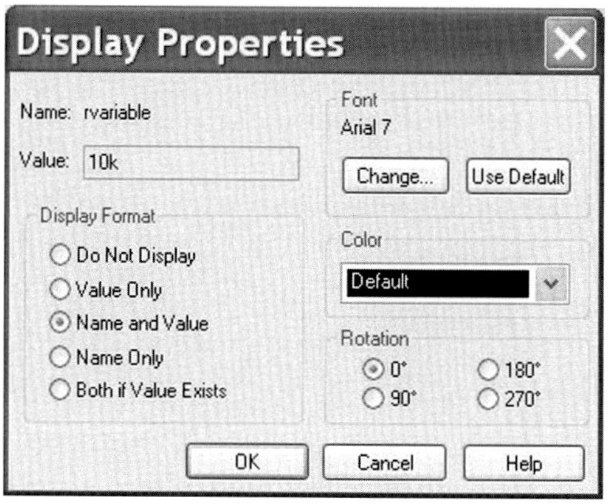

Fig. 5.9 Display Properties.

7. You will need to set up a DC sweep with a Global parameter named **rvariable** for a linear sweep from 500 Ω to 100 kΩ in steps of 500 Ω.

   Create a new simulation profile, **PSpice > New Simulation Profile**, and call it anything you like, for example, global sweep.

   Select **Analysis type** to **DC Sweep** and select the **Sweep variable** as a Global parameter with a **Parameter name: rvariable**. The **Sweep type** will be linear with a start value of 500 Ω, an end value of 100 kΩ and an increment value of 500 Ω (Fig. 5.10).

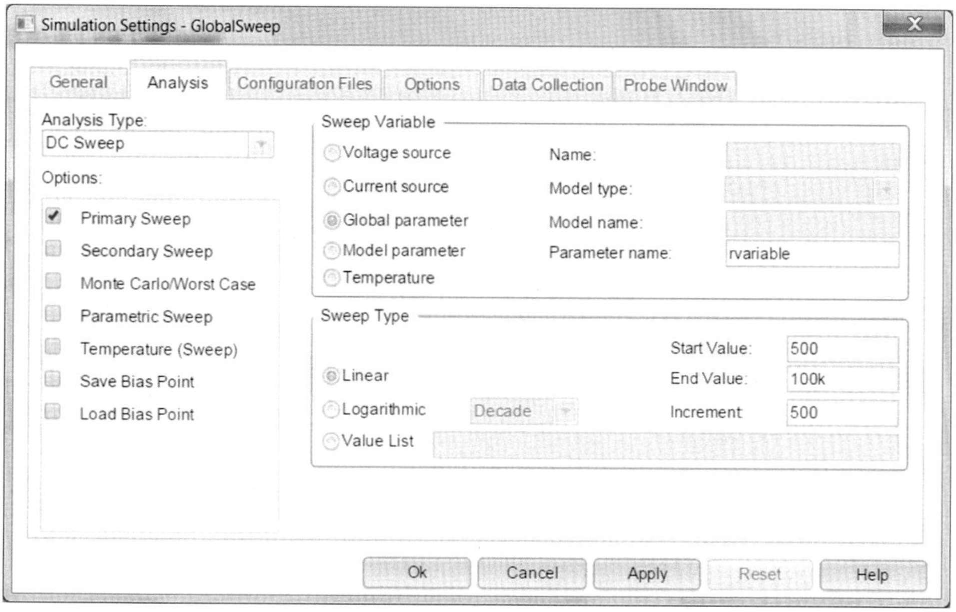

Fig. 5.10 Simulation settings for a global parameter sweep.

8. Place a power marker on the body of RL, i.e. in the middle of RL, by selecting **PSpice > Markers > Power Dissipation**, or select the icon or
   Your circuit should look the same as that in Fig. 5.7.

9. Run the simulation (**PSpice > Run**) .
   You should see the power dissipation curve in Fig. 5.11.

10. From the curve in Fig. 5.11 the cursors can be turned on to determine the value of load resistance for maximum load power (Figs. 5.12–5.14).

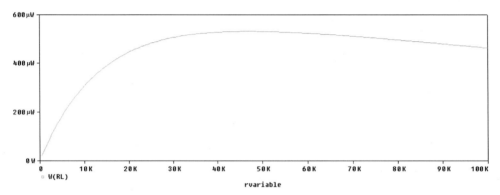

Fig. 5.11 Resistor load power dissipation curve.

Display the cursor, **Trace > Cursor > Display** or .

In Probe, there are two cursors and they follow either the left mouse button(lmb) or the right mouse button (rmb). When you first select **Cursor > Display**, a dashed white box will appear around the symbol used for the trace name (Fig. 5.12).

Fig. 5.12 Cursor activated for W(RL) trace.

The cursor will then follow the left mouse with the left mouse button held down. The cursor box will then appear and give a coordinate reading of the cursor. In release 16.2 and previous versions (Fig. 5.13a), A1 is the left mouse cursor with the first value the x-coordinate and the second value the y-coordinate. A2 is the second cursor, which follows the right mouse button. From release 16.3, the cursors are labeled as shown in Fig. 5.13b, where you can also add multiple cursor measurements from other traces and plots.

```
Probe Cursor
A1 =    47.000K,      5.3192u
A2 =   500.000,     221.607n
dif=    46.500K,      5.0975u
```
(A)

Trace Color	Trace Name	Y1	
	X Values	47.000K	500.000
CURSOR 1,2	W(RL)	531.915u	22.161u

(B)

Fig. 5.13 Cursor coordinates: (A) release 16.2 and (B) release 16.3.

11. Place the cursor on the maximum point of the curve and read off the value for the load resistor.

   Alternatively, there are predefined cursor functions (Fig. 5.14) which can be used to find the points on the curve such as maximum or minimum values. These can be accessed from **Trace > Cursor** or you can select the readily available icons on the top toolbar.

Fig. 5.14 Cursor icons.

   The function of each icon is shown by moving the cursor over the icons.

12. Select the **Cursor Max** function and you will see the cursor move to the maximum value on the curve. The cursor box then displays the maximum value as 47k, which is the load resistor value for maximum power transfer (Fig. 5.15).

Trace Color	Trace Name	Y1	
	X Values	47.000K	500.000
CURSOR 1,2	W(RL)	531.915u	43.403u

Fig. 5.15 Cursor max.

13. With the cursor placed at the maximum value, select **Plot > Label > Mark** or click on the icon [icon] or [icon]. The coordinates of the maximum point on the curve (47k, 531.915W) will now be marked and displayed (Fig. 5.16).

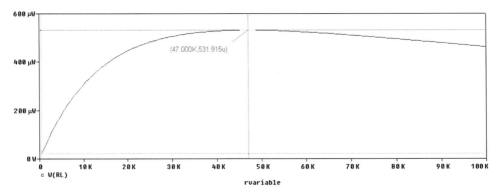

Fig. 5.16 Load resistor power versus load resistance.

Note
You can add arrows and text on the curves by selecting the **Plot > Label** menu.

Theory
In Fig. 5.17, the current in the circuit is given by:

$$I = \frac{Vs}{Rs + RL} \tag{5.1}$$

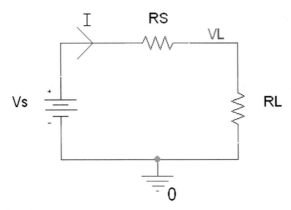

Fig. 5.17 Resistor network.

The power dissipated in the load resistor is given by:

$$P_{\mathrm{L}} = I^2 \mathrm{RL} \tag{5.2}$$

Substitute for I from Eq. (5.1) into Eq. (5.2):

$$P_{\mathrm{L}} = \left(\frac{\mathrm{Vs}}{\mathrm{Rs} + \mathrm{RL}}\right)^2 \mathrm{RL} \tag{5.3}$$

$$P_{\mathrm{L}} = \frac{\mathrm{Vs}^2}{\mathrm{Rs}^2 + 2\mathrm{RsRL} + \mathrm{RL}^2} \mathrm{RL} \tag{5.4}$$

Dividing by RL:

$$P_{\mathrm{L}} = \frac{\mathrm{Vs}^2}{\dfrac{\mathrm{Rs}}{\mathrm{RL}^2} + 2\mathrm{Rs} + \mathrm{RL}} \tag{5.5}$$

For maximum power in Eq. (5.5), the denominator must be a minimum. Rather than differentiating the whole equation, differentiating the denominator will give the same result.

$$\frac{dP_{\mathrm{L}}}{d\mathrm{RL}} = \frac{-\mathrm{Rs}^2}{\mathrm{RL}^2} + 1 \tag{5.6}$$

For a turning point,

$$\frac{dP_{\mathrm{L}}}{d\mathrm{RL}} = 0$$

Therefore,

$$0 = \frac{-\mathrm{Rs}^2}{\mathrm{RL}^2} + 1 \tag{5.7}$$

$$\mathrm{Rs} = \mathrm{RL}$$

It can be shown by differentiating Eq. (5.6) again, that the denominator is a minimum at the turning point and so when Rs = RL, the power is a maximum value.

Exercise 2

Notch Filter

You will globally sweep the resistor values in the notch filter circuit from Chapter 4 on AC analysis to see what effect this has on the circuit response. In the circuit shown in Fig. 5.18, the four resistor values have been replaced by the global parameter {Rvalue}, which has been defined by the Param symbol to have a default value of 27k if no parametric sweep is performed.

1. Place a **Param** part from the special library. Double click on the **Param** part and define the global variable as **Rvalue** with a default value of 27k. Display both the property name and value for **Rvalue**. The steps are the same as in Exercise 1.

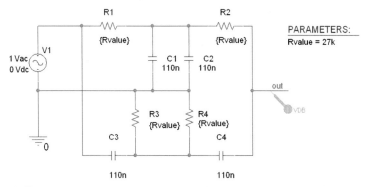

Fig. 5.18 Notch filter resistor values set to a global default parameter value of 27 kΩ.

2. Create a PSpice simulation profile for an AC sweep with the same settings for the passive notch filter in Exercise 1, performing a logarithmic sweep from 10 to 10 kHz with 100 **Points/Decade** (Fig. 5.19). Click on **Apply** but do exit.

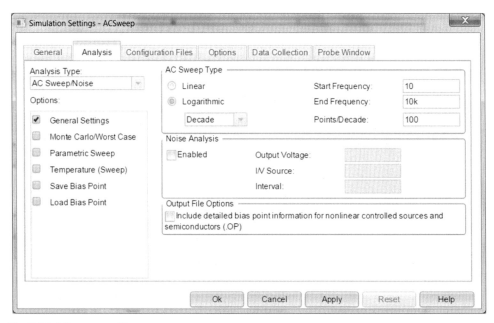

Fig. 5.19 AC sweep settings.

3. Select the **Parametric Sweep** in the **Options** box and set up a **Global** parametric **linear** sweep for the **Rvalue** variable starting at 24 to 30 kΩ in steps of 1 kΩ, which will perform a total of seven AC sweeps (Fig. 5.20). Click on OK.

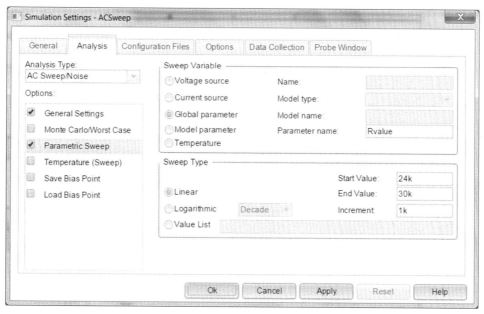

Fig. 5.20 Global parameter settings.

4. Place a V_{db} marker on the output node, "**out**". **PSpice > Markers > Advanced > dB Magnitude of Voltage**.

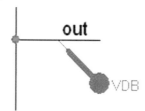

5. Run the simulation ▶.

The notch filter response is shown in Fig. 5.21. Here you can see that the notch frequency changes with a change in resistance **Rvalue**. However, the attenuation depth of the notch, also known as the Q of the filter, also changes with frequency.

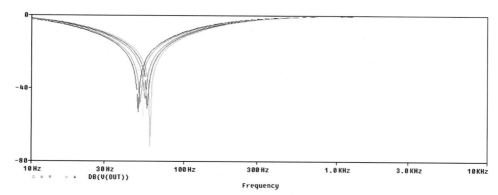

Fig. 5.21 Passive notch filter response.

Exercise 3

Active Notch Filter

Fig. 5.22 shows an active notch filter built upon the previous twin T notch filter. The notch frequency is set as before by the resistor and capacitors in the passive network but the potentiometer R5 is used to set the Q (the sharpness of the attenuation) of the notch, which is not dependent on frequency.

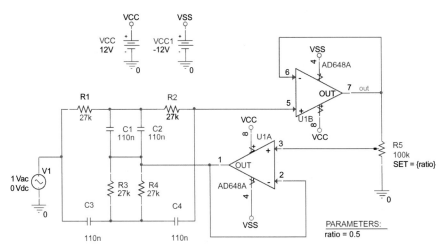

Fig. 5.22 Active notch filter.

The potentiometer, R5, has a **SET** parameter which effectively sets up the ratio between the two resistances on either side of the potentiometer wiper (pin 2). For example, if the ratio is set to 0.4, the resistance between pins 1 and 2 takes on a value of $0.4 \times 100\text{k} = 40\text{k}$ and the resistance between pins 2 and 3 takes on a value of $(1 - 0.4) \times 100\text{k} = 60\text{k}$. So by varying the SET parameter between 0 and 1.0, the pot is effectively being turned through its complete resistance range.

Now, in order to automatically sweep the SET parameter through its range of 0 to 1.0, a global parameter, **ratio**, has been set up which has a default value of 0.5 corresponding to the midrange of the potentiometer $(50\text{k}\Omega)$.

In order to draw the circuit of the active notch filter, the operational amplifiers (opamps) AD648A need to be found. Any opamps could be used, but this is a good exercise in how to search for specific parts.

Note

If you have the OrCAD Demo CD, search for the μA741 opamp in the eval library.

1. In the **Place Part** menu there is a **Search for Part** function which can help to locate the opamp library as shown in Fig. 5.23.

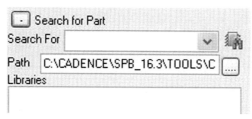

Fig. 5.23 Search for Part.

To change the search path, click on the Browse icon [....] to the right of the Search Path box, which will open the **Browse File** window (Fig. 5.24) showing the **[install path] > capture > library** folder selected on the right-hand side and the list of libraries in that folder shown on the left.

Browse File

Folders:
c:\...\capture\library

OK

Cancel

Amplifier.olb
Arithmetic.olb
ATOD.OLB
BusDriverTransceiver.
capsym.olb
CAPSYM.OLBlck
Connector.olb
Counter.olb

CAPTURE
library
fpga
iec
ieee
ieeelibs

List files of type:
Capture Libraries(*.olb)

Drives:
c: IBM_PRELOAD

Network...

Fig. 5.24 Browse File window displays the search part path.

In the **Folder** section, scroll down and double click on the **pspice** folder as shown in Fig. 5.25. On the left-hand side, you will see a list of the available PSpice libraries. Click on OK.

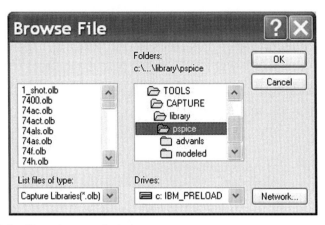

Fig. 5.25 The PSpice library is now selected.

Note

By default, the Search Path in Fig. 5.23 does not point to the PSpice libraries but to the Capture libraries, so running a search for the opamps will not return any devices. It is a common mistake when searching for parts to forget to change the **Search for Part** path.

---

2. Different manufacturers append their own specific numbers and letters to standard part numbers. For example, if you are searching for the BC337 transistor and type in BC337 in the **Search For** box, you will only see one result. However, if you type in a wildcard (*) after the transistor number, BC337*, you will see more results because the wildcard effectively ignores the manufacturer's extra characters after the transistor number when doing a search.

In the **Search for Part** enter the opamp number, AD648*, and either press return or click on the **Part Search** icon 🔍.

There should only be one instance of the AD648A from the opamp library, as shown in Fig. 5.26. Double click on the AD648A and you will see the opamp library added to the list of libraries and a Capture graphical representation of the opamp.

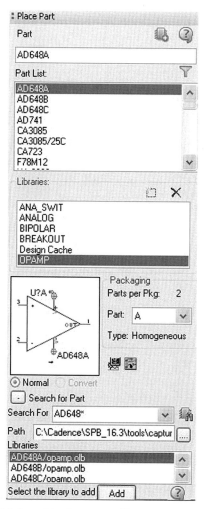

Fig. 5.26 The opamp AD648A is found in the opamp library.

Note
If you have the eval software version, search for the μA741* opamp.

In the Place Part menu alongside the graphical representation of the opamp, there is a **Packaging** section which shows that there are two Parts per package. As this is a dual opamp, there are two available sections, A and B. Fig. 5.26 shows Part A selected, while Fig. 5.27 shows Part B selected. Note the different pin numbers and different reference designators, U?A and U?B. Type: Homogeneous indicates that both sections in a part are the same, in contrast to, for example, a relay and coil, where both sections are different and are classed as heterogeneous.

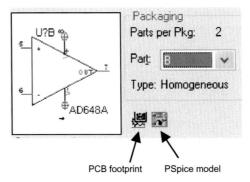

Fig. 5.27 Part B of the opamp is selected.

The two icons shown in Fig. 5.27 indicate that the AD648A has PCB footprint and a PSpice model attached and is therefore ready for simulation. It is important that the PSpice icon is displayed when selecting parts for simulation.

3. Place the A part of the opamp in the circuit. Highlight the opamp and **rmb > Mirror Vertically** or press V. The **Mirror** and **Rotate** operations can be used even if the opamp is connected in the circuit. You do not have to delete any wires.

4. Select the B part of the opamp and place it in the circuit.

5. With all the libraries selected (left mouse click at the top of library list and drag mouse pointer down to bottom of library list), type **pot**, in the **Part** box. The **pot** is found in the **breakout** library. Place the **pot** part and change its value to 100k. Double click on the SET property and change the default value of 0.5 to {ratio}. Do not forget the brackets.

6. We need to define **ratio** as a global parameter with a default value of 0.5. Place a Param part from the special library in the circuit.

7. Double click on the **param** part and enter a new row (or new column). Create a new property called **ratio** with a value of 0.5 and display the property name and value. The steps are the same as in Exercise 1.

8. Select **Place > Power** and scroll down to the VCC_CIRCLE symbol and in the **Name:** box, change the name to VCC and click on OK. Repeat for the VSS symbol.

9. Place and connect the remaining components as shown in Fig. 5.22.

10. The parametric sweep is run in conjunction with an AC analysis. In this example we are going to run a parametric sweep on the potentiometer ratio from 0.1 to 0.9 in steps of 0.1. The first thing to do is to set up the same AC analysis as before with the passive notch filter, sweeping the frequency logarithmically from 10 to 10 kHz with 100 points/decade (Fig. 5.28). Click on apply but do **not** exit.

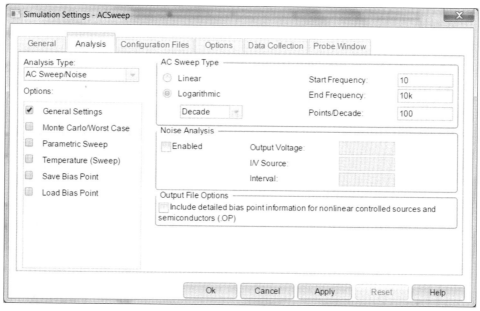

Fig. 5.28 AC sweep settings.

11. Select the **Parametric Sweep** in the **Options** box. The Sweep variable is a **Global parameter** and the **Parameter name** is **ratio**. The **Sweep type** is linear, the **Start value** 0.1, the **End value** 0.9 and the **Increment** 0.1 (Fig. 5.29). Click on OK.

Fig. 5.29 Global parameter settings.

12. Place a V_{db} marker on the output node "**out**". **PSpice > Markers > Advanced > dB Magnitude of Voltage.**

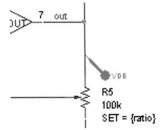

13. Run the simulation .

Fig. 5.30 shows that by varying R5, the Q of the circuit (the sharpness of the notch) can be varied without a change in the notch frequency.

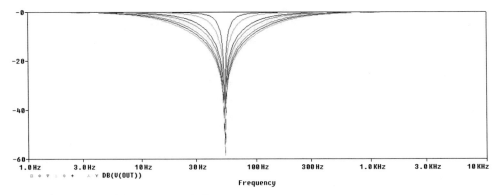

Fig. 5.30 Response of the active notch filter.

CHAPTER 6

Stimulus Editor

Chapter Outline

6.1. Stimulus Editor Transient Sources   100
    6.1.1 Exponential (Exp) Source   100
    6.1.2 Pulse Source   102
    6.1.3 VPWL   104
    6.1.4 SIN (Sinusoidal)   105
    6.1.5 SSFM (Single-Frequency FM)   105
6.2. User-Generated Time-Voltage Waveforms   106
6.3. Simulation Profiles   107
6.4. Exercise   107

The Stimulus Editor is a graphical tool to help you define transient analog and digital sources. The **sourcestm** library contains three source parts, shown in Fig. 6.1, each of which provides the interface with the defined stimulus in the Stimulus Editor.

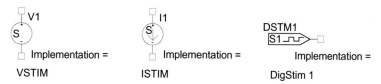

Fig. 6.1 Stimulus Editor, transient analog and digital sources.

When you first place one of the sources from the **sourcestm** library, the implementation property is displayed in the schematic. This property refers to the name of the stimulus which is defined in the Stimulus Editor. Either you can enter a name of the stimulus on the schematic to start with, or you will be prompted for the stimulus name in the Stimulus Editor when started.

To start the Stimulus Editor, highlight a **sourcestm** source, **rmb > Edit PSpice Stimulus**.

When the Stimulus Editor starts, the **New Stimulus** window will appear as shown in Fig. 6.2. Note that the stimulus file name in 16.3 has taken on the name of the PSpice simulation profile, in this case, transient.stl. In previous versions, the stimulus name takes on the name of the project name.

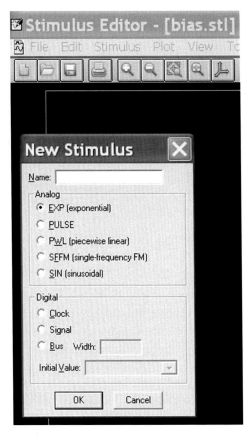

Fig. 6.2 Stimulus Editor started.

The New Stimulus window allows you to define analog and digital signals and prompts you to enter the stimulus name if you have not already defined the name in Capture.

6.1. STIMULUS EDITOR TRANSIENT SOURCES

6.1.1 Exponential (Exp) Source

Figs. 6.3 and 6.4 show the two possible exponential waveforms which can be defined for a voltage or a current using VSTIM or ISTIM sources, respectively.

Both exponential waveforms start after a time delay (td1) and then exponentially rise or fall, using a time constant (tc1) between two voltages V1 and V2 up to a time td2. The waveform then decays or rises after td2, using a time constant (tc2).

For example, in Fig. 6.3, the voltage is V1 (0 V) up to td1 (10 μs); then the voltage increases exponentially with a time constant given by tc1 (10 μs) towards V2 (10 V). The

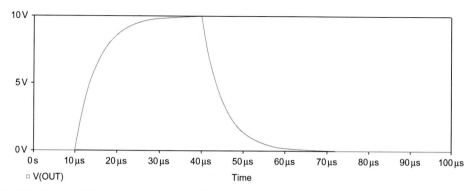

Fig. 6.3 Exponentially rising voltage waveform.

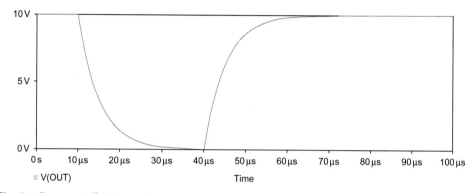

Fig. 6.4 Exponentially decreasing voltage waveform.

time for the exponential rise is defined by td2-td1 as 30 μs (40 − 10 μs), after which the voltage decreases exponentially with a time constant given by tc2 (5 μs) back towards V1.

Fig. 6.5 shows the attribute settings for the waveform in Fig. 6.3, where:

V1—initial starting value at time 0 s

V2—value that voltage rises or falls to

td1—start time (delay) of exponential rise (or fall)

tc1—time constant of rising (or falling) waveform

td2—start time (delay) of exponential fall (or rise)

tc2—time constant of falling (or rising) waveform

Fig. 6.3 was defined using V1 = 0 V, V2 = 10 V, td1 = 10 μs, tc1 = 5 μs, td2 = 40 μs and tc2 = 5 μs.

Fig. 6.4 was defined using V1 = 10 V, V2 = 0 V, td1 = 10 μs, tc1 = 5 μs, td2 = 40 μs and tc2 = 5 μs.

Now the exponential voltage is defined by:

$$v(t) = (V2 - V1)\left(1 - e^{\frac{-\text{time}}{\text{time constant}}}\right)$$

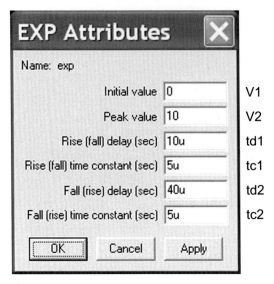

Fig. 6.5 Exponential attributes.

So between 0s and td1 the voltage is a constant:

$$v(t) = V1$$

Between td1 and td2:

$$v(t) = V1 + (V2 - V1)\left(1 - e^{\frac{-(\text{time}-\text{td1})}{\text{tc1}}}\right)$$

and for the time between td2 and the stop time, the voltage is given by:

$$v(t) = V1 + (V2 - V1)\left[\left(1 - e^{\frac{-(\text{time}-\text{td1})}{\text{tc1}}}\right) - \left(1 - e^{\frac{-(\text{time}-\text{td2})}{\text{tc2}}}\right)\right]$$

6.1.2 Pulse Source

Fig. 6.6 shows the definition for a voltage pulse waveform, where:

V1—low voltage
V2—high voltage
TD—the time delay before the pulse starts
TR—rise time specified in seconds, defined as the time difference between V1 and V2
TF—fall time specified in seconds, defined as the time difference between V1 and V2
PW—pulse width of the pulse
PER—period of the pulse, i.e. the pulse frequency
Similarly, current pulses can be defined using the ISTIM part as shown in Fig. 6.7.

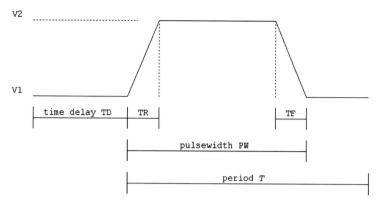

Fig. 6.6 Pulse waveform specification.

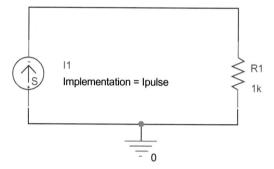

Fig. 6.7 Using ISTIM to define a current pulse waveform.

When you first place a VSTIM, ISTIM or DigSTIM part, the implementation property name and value are shown. You only need to display the **name** of the stimulus, so double click on the **Implementation =**, which will open up the Display Properties dialog box, and select **Value Only** (Fig. 6.8).

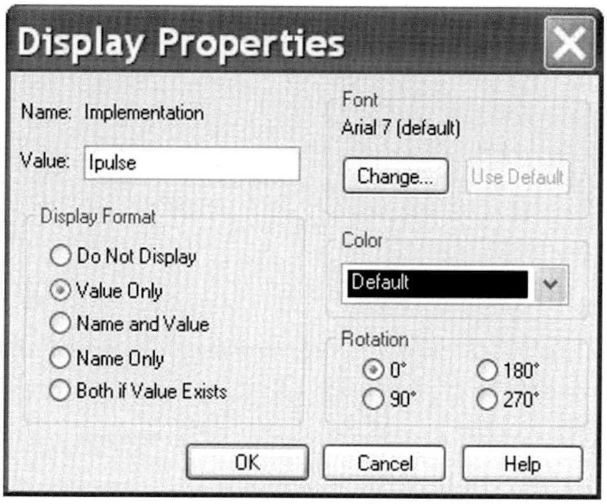

Fig. 6.8 Making the **Implementation** = invisible.

Fig. 6.9 shows the ISTIM part with the defined current stimulus, Ipulse displayed. As before, to start the Stimulus Editor, **rmb** and select **Stimulus Editor** and then select PULSE for **New Stimulus**. Fig. 6.10 shows the pulse attributes defined for a current pulse using the ISTIM part. The resulting current waveform is shown in Fig. 6.11.

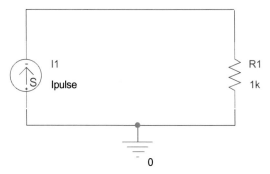

Fig. 6.9 ISTIM part with defined current stimulus, Ipulse displayed.

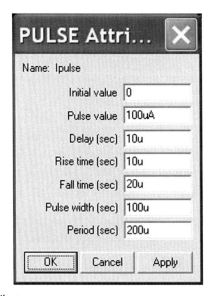

Fig. 6.10 Current pulse attributes.

6.1.3 VPWL

Piecewise linear (PWL) is where you actually draw the voltage or current waveform. You define the time and voltage (or current) axis and then use a cursor to draw the waveform. An example is given in the exercise at the end of the chapter.

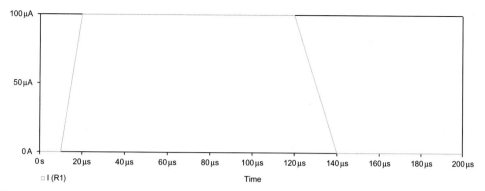

Fig. 6.11 Current pulse waveform using the attributes in Fig. 6.10.

6.1.4 SIN (Sinusoidal)

Fig. 6.12 shows the attributes for a sinewave. The complete definition includes attributes for a damped sinewave, with a phase angle and an offset value. Offset value is the initial voltage or current at time 0s, Amplitude is the maximum voltage or current, Frequency (Hz) is the number of cycles per second, Time delay (s) is the start delay, Damping factor (1/s) is the exponential decay, and Phase angle (degrees) is the phase angle (Fig. 6.13).

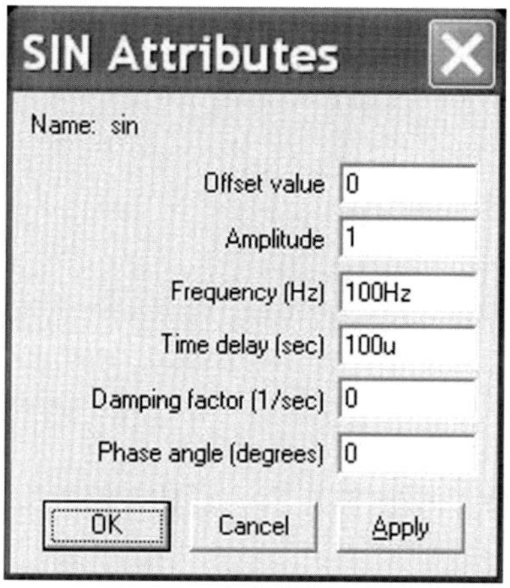

Fig. 6.12 Sinewave attributes.

6.1.5 SSFM (Single-Frequency FM)

This source generates frequency-modulated sinewaves as shown in Fig. 6.14, which shows the modulation of a carrier frequency. The sinewave is given by:

$$v(t) = V_{off} + V_{ampl} \times \sin\left[\left(2\pi f_c t + \left(\text{mod} \times \sin\left(2\pi f_m \text{time}\right)\right)\right)\right]$$

SFFM Attributes ✕

Name: sffm

Offset value `2`

Amplitude `1`

Carrier frequency (Hz) `10Hz`

Modulation index `5`

Modulation frequency (Hz) `1Hz`

[OK]   [Cancel]   [Apply]

Fig. 6.13 Frequency modulated sinewave.

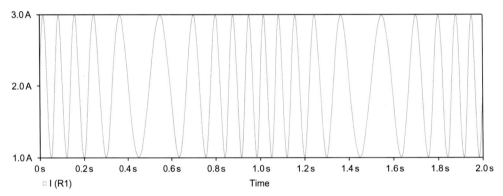

Fig. 6.14 Single-frequency FM attributes.

where V_{off} is the offset voltage, V_{ampl} is the maximum value of voltage, mod is the modulation index, f_c is the carrier frequency, and f_m is the modulation frequency.

6.2. USER-GENERATED TIME-VOLTAGE WAVEFORMS

You can also use the waveforms generated by a transient analysis in Probe as a time-voltage source. In Probe, select **File > Export**, which gives the options shown in Fig. 6.15.

Export ▶	Probe Data (.dat file)
Import	Stimulus Library (.stl file)
Open File Location	Text (.txt file)
Page Setup…	Comma Separated File (.csv file)

Fig. 6.15 Exporting time-voltage data.

An alternative method to create a time-voltage text file is to select the trace name in Probe, select **copy** and **paste** the data into a text file.

6.3. SIMULATION PROFILES

Prior to version 16.3, when you launch the Stimulus Editor, a stimulus file with the name of the project is created; for example a project named stimulus will create a stimulus.stl. All the stimuli you create are saved in the stimulus.stl file and so in order to select a different stimulus, all you need to do in the schematic is to change the name shown on the VSTIM, ISTIM or DigSTIM source.

From version 16.3 onwards, the stimulus file is associated with the current active simulation profile and can be accessed via the simulation profile under the Configuration Files tab. In previous versions, there were separate tabs for Stimulus, Library and Include options.

Under configured files you will see the stimulus.stl file (see Fig. 6.23). If you do not see the stimulus file then you can browse for the Filename. You can then add the stimulus file to the profile (Add to Profile). However, there are other options:

• Add as Global: all designs will have access to the stimulus file.
• Add to Design: only the current design will have access.

Adding the stimulus file as Global is useful if you have created a standard set of stimuli to test all your circuits. You can add several stimulus files and arrange the order by clicking on the up and down arrows. The red cross deletes the selected file.

6.4. EXERCISE

Note

From release 16.3 onwards, there are some differences compared to previous versions. When you first create a new project in any release, a PSpice bias simulation profile is created by default. This can be seen in the Project Manager under **PSpice Resource > Simulation Profiles**. In order to keep compatibility between releases, delete the bias simulation profile in the Project Manager.

---

1. Create a project called stimulus.
2. In the Project Manager, expand **PSpice Resources > Simulation Profiles** as shown in Fig. 6.16 and delete the SCHEMATIC1-Bias profile.
3. Draw the circuit diagram in Fig. 6.17. The VSTIM source (V1) is from the Sourcestm library.

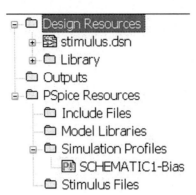

Fig. 6.16 Bias simulation profile in Project Manager.

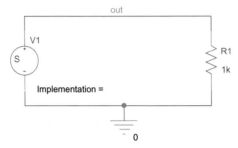

Fig. 6.17 VSTIM 100 Hz sinewave generation.

4. Highlight VSTIM and **rmb > Edit PSpice Stimulus**. In the **New Stimulus** window, name the source as sin100Hz and select a **SIN** source (Fig. 6.18).

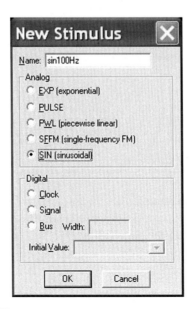

Fig. 6.18 New sinewave source.

5. Create a 100 Hz sinewave with no offset and amplitude of 1 V. Leave all the other values at their default value of 0 (Fig. 6.19). Click on OK and save the stimulus source and **Update Schematic** when prompted and exit the Stimulus Editor.

Fig. 6.19 Creation of a 100 Hz sinewave source.

6. In Capture, the name of the stimulus is now shown on V1. Double click on **Implementation= sin100Hz** and select **Value Only** (Fig. 6.20).

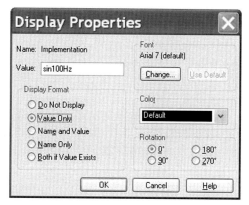

Fig. 6.20 Making the implementation name invisible.

Only the name of the stimulus will be displayed, as seen in Fig. 6.21.

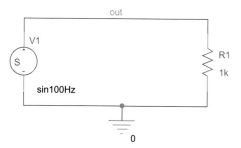

Fig. 6.21 Displaying only the stimulus name.

7. Create a PSpice simulation profile (PSpice > New Simulation Profile) and call it **transient**. In the **Analysis type:** pull-down menu, select **Time Domain (Transient)** and set the **Run to time:** to 20 ms (Fig. 6.22). Click on **Apply** but do **not** exit the profile.

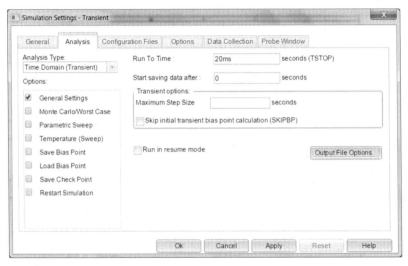

Fig. 6.22 Simulation settings.

8. Select **Configuration Files > Category > Stimulus** and you should see the stimulus.stl file listed (Fig. 6.23).

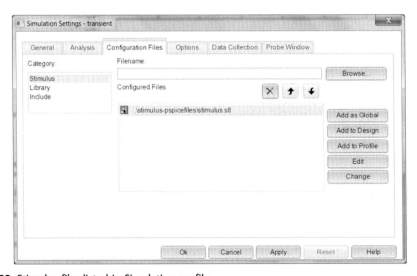

Fig. 6.23 Stimulus files listed in Simulation profile.

Highlight the stimulus.stl file and click on **Edit**. The Stimulus Editor will launch showing the sin100Hz stimulus. This is a quick way to see what stimuli are available in stimulus files. Close the simulation profile.

9. Place a voltage marker on node **out** (Fig. 6.24), run the simulation (**PSpice > Run**) and confirm that a sinewave voltage appears across the resistor.

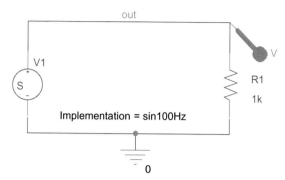

Fig. 6.24 Placing a voltage marker.

Note

If you see a flat voltage line in the Probe window, then you may have not deleted the default bias.stl file. See note at the beginning of the exercise. The Stimulus Editor was then invoked with the default bias.stl as the active simulation profile. If you do not see the stimulus file in the simulation profile (Fig. 6.23) then **Browse** for the file, which can be found in the bias folder, and then select **Add to Design**.

10. In the Project Manager, expand **PSpice > Resources > Stimulus Files** and you should see the **transient.stl** stimulus file as shown in Fig. 6.25. You can double click on the file here to open the Stimulus Editor to check the stimulus.

Fig. 6.25 Location of stimulus file, transient.stl.

11. Highlight VSTIM in Capture and launch the Stimulus Editor. In the Stimulus Editor the previous SIN Attributes will be displayed. Click on Cancel.

12. Create a pulse source named **Vpulse (Stimulus > New)** with an initial value of 0 V, an amplitude of 1 V, no initial delay, a rise time of 500 μs, a fall time of 1 ms, a pulse width of 2 ms and a period of 10 ms (Fig. 6.26). Click on OK and save the stimulus but **do not Update Schematic** when prompted. Exit the Stimulus Editor.

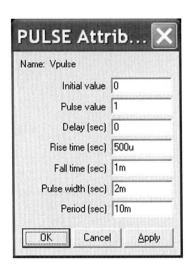

Fig. 6.26 Vpulse attributes.

Note

When entering attribute values, press the TAB button on the keyboard to move down to the next attribute box.

Note

The stimulus is already named as sin100Hz in Capture and cannot be updated from the Stimulus Editor. In previous software releases, if you say yes to **Update Schematic**, the cursor will change to an hourglass and just sit there. You will need to switch to Capture, where you will see the dialog box in Fig. 6.27. Just click on OK.

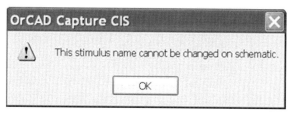

Fig. 6.27 Stimulus name change warning.

---

13. In Capture, double click on the stimulus name, sin100Hz, and change it to Vpulse as shown in Fig. 6.28.

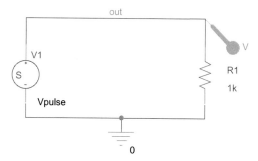

Fig. 6.28 Using the Vpulse source.

14. Run the simulation. **PSpice > Run** or click on the blue play button ⏵.
15. Confirm that a voltage pulse waveform appears across the resistor.
16. In Capture, highlight VSTIM and **rmb > EditStimulus Editor**. In the Stimulus Editor the previous PULSE Attributes will be displayed. Click on Cancel.
17. Create a new stimulus, **Stimulus > New**. Name the stimulus Vin and select PWL (piecewise-linear).

Note
You may be asked if you want to change the axis settings.

---

18. From the top toolbar, select **Plot > Axis Setting**. Set the waveform drawing resolution to that shown in Fig. 6.29.

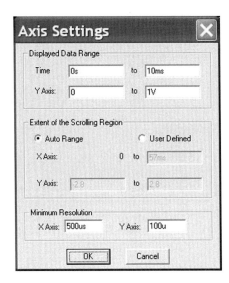

Fig. 6.29 Axis settings.

19. A pen cursor will appear. Draw a corresponding piecewise linear graph approximating the PWL voltage shown in Fig. 6.30. The first point at (0,0) has already been selected. The accuracy does not matter as long as there are three peaks defined. This stimulus will be used in Chapter 7 on transient analysis. Press escape to exit draw mode.

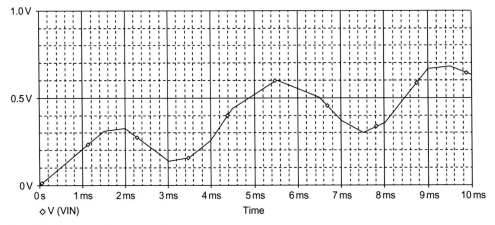

Fig. 6.30 Piecewise linear waveform.

20. If you want to delete or move a point, press escape out of draw mode and place the cursor on a point, which will turn red, and then delete or move the point. To return to draw mode, select **Edit > Add** or select the icon or .

Note

Prior to release 16.3, you can only place points forward in time; you cannot go backwards. If you want to delete or move a point, press escape out of draw mode and place the cursor on a point, which will turn red, and then delete or move the point.

---

21. Save the stimulus file and exit the Stimulus Editor but **do not Update Schematic**.
22. Change the name of the stimulus from Vpulse to Vin and simulate, and confirm that the piecewise voltage waveform appears across the resistor.

Note

The Vin source will be used in Chapter 7.

CHAPTER 7

Transient Analysis

Chapter Outline

7.1. Simulation Settings 118
7.2. Scheduling 118
7.3. Check Points 119
7.4. Defining a Time-Voltage Stimulus Using Text Files 120
7.5. Exercises 122
    Exercise 1 122
    Exercise 2 125

Transient analysis calculates a circuit's response over a period of time defined by the user. The accuracy of the transient analysis is dependent on the size of internal time steps, which together make up the complete simulation time known as the **Run to time** or **Stop time**. However, as mentioned in Chapter 2, a DC bias point analysis is performed first to establish the starting DC operating point for the circuit at time $t=0$s. The time is then incremented by one predetermined time step at which node voltages and current are calculated based on the initial calculated values at time $t=0$. For every time step, the node voltages and currents are calculated and compared to the previous time step DC solution. Only when the difference between two DC solutions falls within a specified tolerance (accuracy) will the analysis move on to the next internal time step. The time step is dynamically adjusted until a solution within tolerance is found.

For example, for slowly changing signals, the time step will increase without a significant reduction in the accuracy of the calculation, whereas for quickly changing signals, as in the case of a pulse waveform with a fast leading edge rise time, the time step will decrease to provide the required accuracy. The value for the maximum internal time step can be defined by the user.

If no solution is found, the analysis has failed to converge to a solution and will be reported as such. These convergence problems and solutions will be discussed in more detail in Chapter 8.

There are some circuits where a DC solution cannot be found, as in the case of oscillators. For these circuits, there is an option in the simulation profile to skip over the initial DC bias point analysis. If you add an initial condition to the circuit, the transient analysis will use the initial condition as its starting DC bias point.

Analog Design and Simulation Using OrCAD Capture and PSpice
https://doi.org/10.1016/B978-0-08-102505-5.00007-0

7.1. SIMULATION SETTINGS

Fig. 7.1 shows the PSpice simulation profile for a transient (time domain) analysis. In this example, the simulation time has been set to 5 μs. The **Start saving data after:** specifies the time after which data are collected to plot the resulting waveform in Probe in order to reduce the size of the data file.

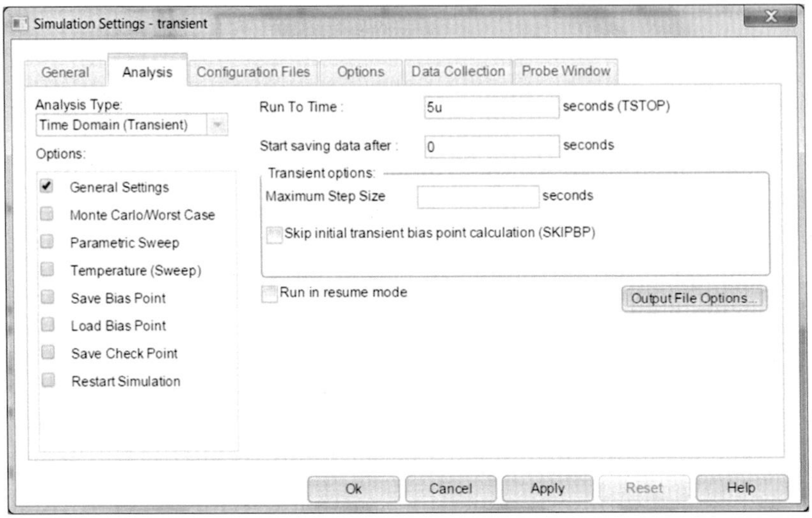

Fig. 7.1 Transient analysis simulation profile.

Maximum step size: defines the maximum internal step size, which is dependent on the specified **run to time** but is nominally set at the **run to time** divided by 50.

Skip the initial transient bias point calculation will disable the bias point calculation for a transient analysis.

7.2. SCHEDULING

Scheduling allows you to dynamically alter a simulation setting for a transient analysis; for example, you may want to use a smaller step size during periods that require greater accuracy and relax the accuracy for periods of less activity. Scheduling can also be applied to the simulation settings runtime parameters, RELTOL, ABSTOL, VNTOL, GMIN and ITL, which can be found in PSpice > Simulation Profile > Options. You replace the parameter value with the scheduling command, which is defined by:

{SCHEDULE(t1,v1,t2,v2…tn,tn)}

Note that t1 always starts from 0.

For example, it may be more efficient to reduce the relative accuracy of simulation from 0.001% to 0.1%, RELTOL, during periods of less activity by specifying a change in accuracy every millisecond. The format will be defined as:

{schedule(0,0, 1m,0.1, 2m,0.001, 3m,0.1, 4m,0.001)}
The simulation settings will be discussed in more detail in Chapter 8.

7.3. CHECK POINTS

Check points were introduced in version 16.2 to allow you to effectively mark and save the state of a transient simulation at a check point and to restart transient simulations from defined check points. This allows you to run simulations over selective periods. This is useful if you have convergence problems in that you can run the simulation from a defined check point marked in time before the simulation error, rather than having to run the whole simulation from the beginning.

Check points are only available for a transient simulation and are selected in the simulation profile in **Analysis > Options** box (Fig. 7.2) as **Save Check Points** and **Restart Simulation**. Check points are defined by specifying the time interval between check points. The simulation time interval is measured in seconds and the real time interval is measured in minutes (default) or hours. The time points are the specific points when the check points were created.

Before you restart a simulation from a saved check point, you can change component values, parameter values, simulation setting options, check point restart and data save options. Fig. 7.3 shows the **Restart Simulation** option selected.

The saved check point data are set to simulation time in seconds such that **Restart At** shows 4 ms, which was specified in the saved check point data file. The simulation will then start at 4 ms using the saved state of the transient simulation.

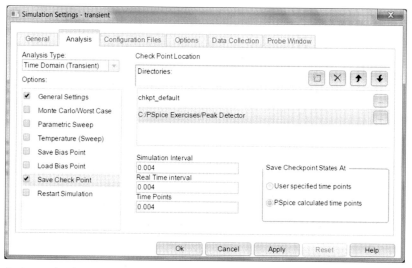

Fig. 7.2 Saving a check point.

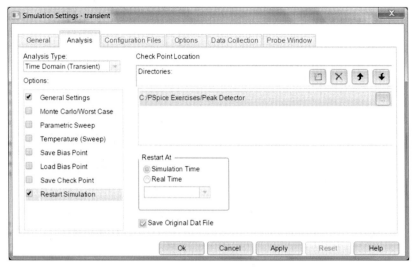

Fig. 7.3 Restarting a simulation using a saved check point.

7.4. DEFINING A TIME-VOLTAGE STIMULUS USING TEXT FILES

The piecewise linear stimulus was introduced in Chapter 6, where a graphically drawn voltage waveform was used as an input waveform to a circuit. Input waveforms can also be defined using pairs of time-voltage coordinates, which can be entered in the Property Editor or read from an external text file.

Fig. 7.4 shows the voltage VPWL and current IPWL sources and the corresponding time and voltage properties (Fig. 7.5) in the Property Editor. By default, eight time-voltage pairs are displayed in the Property Editor for the VPWL and IPWL parts, but, as seen in Fig. 7.5, more time-voltage pairs have been added. It is more efficient and easier to define a large number of time-voltage pairs in a text file.

Fig. 7.6 shows the VPWL_FILE part referencing a text file which contains time-voltage pairs as shown in Fig. 7.7. For example, at 1 ms the voltage is 0.2055 V, at

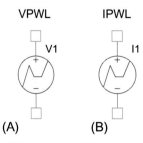

Fig. 7.4 Piecewise linear sources for (A) voltage and (B) current.

T1	0
T2	1 ms
T3	2 ms
T4	3 ms
T5	4 ms
T6	5 ms
T7	6 ms
T8	7 ms
T9	8 ms
T10	9 ms
T11	10 ms
V1	0
V2	0.2055
V3	0.3273
V4	0.1382
V5	0.2852
V6	0.5182
V7	0.5527
V8	0.3727
V9	0.3584
V10	0.6673
V11	0.6291
Value	VPWL

Fig. 7.5 VPWL and IPWL time-voltage properties displayed in the Property Editor.

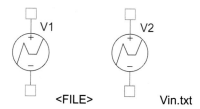

Fig. 7.6 Piecewise linear part VPWL_FILE referencing a file.

```
* Stimulus Vin
0, 0
0.001, 0.2055
0.0015, 0.3109
0.002, 0.3273
0.0025, 0.2345
0.003, 0.1382
0.0035, 0.1564
0.004, 0.2582
```

Fig. 7.7 Time-voltage data points describing the input voltage waveform, V_{in}.

2 ms 0.3273 V, and so on. It is always a good idea to make the first line a comment as PSpice normally ignores the first line.

When you reference a text file such as Vin.txt, you need to specify the location of the text file. You can use absolute addressing specifying the direct path to the file or relative addressing specifying the path location relative to the project location. Fig. 7.8 shows the hierarchy of a project showing the different folders in which the Vin.txt file can be placed and the corresponding <FILE> name for the referenced Vin.txt on the VPWL_FILE part.

Project Folder > PSpiceFiles > schematics > simulation profiles

..\..\Vin.txt   ..\Vin.txt   Vin.txt

Fig. 7.8 Referencing the Vin.txt time-voltage text file for VPWL_FILE.

For example, if you place the Vin.txt file in the same folder which contains the schematics, then you enter ..\Vin.txt in the <FILE> property of the VPWL_FILE.

Project Folder > PSpiceFiles > schematics > simulation profiles

..\..\Vin.txt ..\Vin.txt Vin.txt

You can also provide an absolute path to a text file. For example, if you had a folder named stimulus, then you enter C:\stimulus\Vin.txt.

In the **source** library there are other VPWL and IPWL parts which allow you to make a VPWL periodic for a number of cycles or repeat forever. These are given as:

VPWL_F_RE_FOREVER
VPWL_F_RE_N_TIMES
VPWL_RE_FOREVER
VPWL_RE_N_TIMES
IPWL_F_RE_FOREVER
IPWL_F_RE_N_TIMES
IPWL_RE_FOREVER
IPWL_RE_N_TIMES

The above source will be introduced in the exercises.

7.5. EXERCISES

Exercise 1

This exercise will demonstrate the effect that the maximum time step has on the resolution of a simulation and introduce the use of the scheduling command.

1. Draw the circuit in Fig. 7.9, which consists of a VSIN source from the **source** library, connected to a load resistor R1.

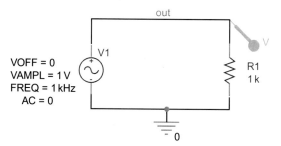

Fig. 7.9 Sinewave voltage applied to a load resistor.

2. Create a PSpice simulation profile called **transient** and select **Analysis type:** to **Time Domain (Transient)** and enter a **Run to time** of 10 ms, which will display 10 cycles of the sinewave (Fig. 7.10).

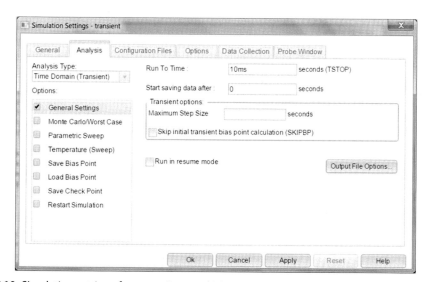

Fig. 7.10 Simulation settings for a transient analysis.

Place a voltage marker on node "**out**" and run the simulation. You should see the resultant waveform as shown in Fig. 7.11, which is lacking in resolution.

3. In Probe select **Tools > Options** and check the box for **Mark Data Points** or click on the icon 　. You will see the data points that make up the sinewave.

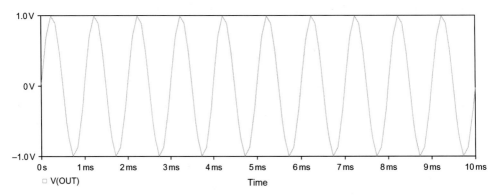

Fig. 7.11 Distorted resultant sinewave lacking resolution.

4. In the simulation profile, set up a schedule command to decrease the time step at set time points. You can enter the schedule command in the **Maximum step size** box, but because of the small field in which to type in the command it is recommended to type the schedule command in a text editor such as Notepad and cut and paste the following command into the box:

{schedule(0,0, 2m,0.05m, 4m,0.01m, 6m,0.005m, 8m,0.001m)}

Run the simulation. As the **Mark Data Points** is still on, you should see the resolution of the waveform improve with a decrease in the limit of the maximum step size (Fig. 7.12).

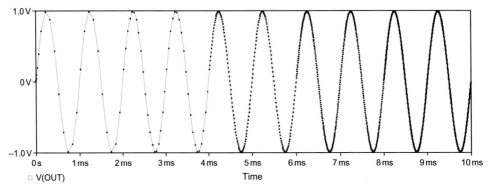

Fig. 7.12 Improved sinewave resolution with a successive decrease of the maximum time step using the schedule command.

Exercise 2

Fig. 7.13 shows a peak detector circuit, where the input stimulus can be either the Vin Sourcstm source created in the Stimulus Editor exercise or a file containing a time-voltage definition of the input waveform. Both implementations will be described.

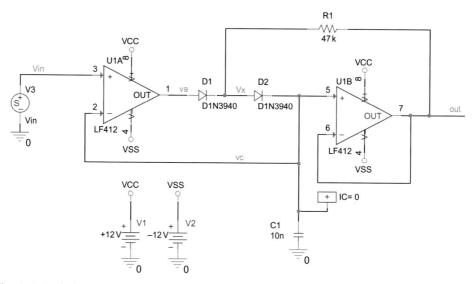

Fig. 7.13 Peak detector circuit.

1. Create a project called Peak Detector and draw the circuit in Fig. 7.13. If you are using the demo CD, use the uA741 opamps.
2. Rename the SCHEMATIC1 folder to Peak Detector.
3. You need to set up an initial condition (IC) on the capacitor, C1, by using an IC1 part from the special library. This ensures that at time $t=0$, the voltage on the capacitor is 0 V .

    Alternatively, you can double click on the capacitor, C1, and in the **Property Editor** enter a value of 0 for the IC property value (Fig. 7.14). This ensures that at time $t = 0$, the voltage on the capacitor is 0 V. If you change the capacitor, then you have to remember to set the initial condition, whereas an IC1 part will always be visible on the schematic.

Fig. 7.14 Setting an initial value of 0V on the capacitor.

4. Create a simulation profile, make sure you name it **transient** and set the run to time to 10 ms. Close the simulation profile.

   Two methods are described to define the input waveform V_{in} for the Peak Detector.

Using the Graphically Created Waveform in the Stimulus Editor

5. For the input stimulus, using the predefined V_{in} sourcestm in Chapter 6, edit the simulation profile and select the **Configuration Files** tab, select **Category** to **Stimulus** and **Browse** to the location of the **stimulus.stl** file. Click on **Add to Design** as shown in Fig. 7.15. An explanation of stimulus files added to the simulation profile was given in Chapter 6.

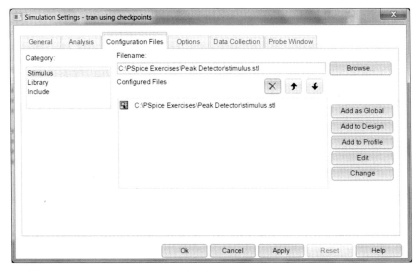

Fig. 7.15 Adding the stimulus.stl file to the simulation profile.

6. Check the stimulus by highlighting the stimulus name and click on **Edit**. This will launch the Stimulus Editor and display the V_{in} waveform. Close the simulation profile.
7. Go to Step 12.

Using a File With Time-Voltage Data Describing the Input Waveform

8. Enter the time-voltage data points in Fig. 7.16 in a text editor such as Notepad. By default, the simulator ignores the first line, so do **not** enter data on the first line. However, it is always a good idea to add a description or a comment to the data file using an asterisk * character to describe, for example, what the data is. The simulator will ignore any lines beginning with a * character.

```
* Stimulus Vin
0, 0
0.001, 0.2055
0.0015, 0.3109
0.002, 0.3273
0.0025, 0.2345
0.003, 0.1382
0.0035, 0.1564
0.004, 0.2582
0.0045, 0.44
0.005, 0.5182
0.0055, 0.6018
0.006, 0.5527
0.0065, 0.5018
0.007, 0.3727
0.0075, 0.3
0.008, 0.3564
0.0085, 0.5109
0.009, 0.6673
0.0095, 0.6782
0.01, 0.6291
```

Fig. 7.16 Time-voltage data points describing the input voltage waveform, V_{in}.

Peak detector   SCHEMATIC1   Vin.txt

Fig. 7.17 Place the V_{in} text file in the PSpiceFiles folder.

Name the file Vin and save the file as a text file in the PSpice folder for the Project, **Peak Detector > peak detector-PSpiceFiles** (Fig. 7.17).

Make sure the file has been saved with a .txt extension as Vin.txt.

9. Place a VPWL_FILE from the **source** library and rename the <FILE> shown in Fig. 7.18 as ..\..\Vin.txt

10. Your Peak Detector circuit will be as shown in Fig. 7.19.

11. Go to Step 12.

12. Place voltage markers on nodes **in** and **out** and run the simulation.

Fig. 7.20 shows the simulation response of the peak detector to the input voltage V_{in}.

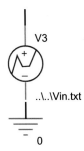

Fig. 7.18 Adding a VPWL_FILE.

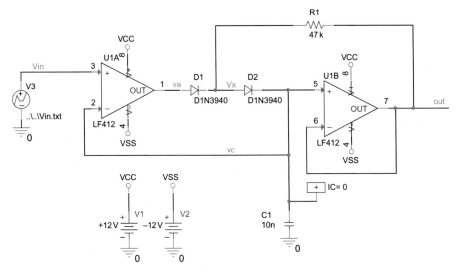

Fig. 7.19 Peak detector circuit using a text file.

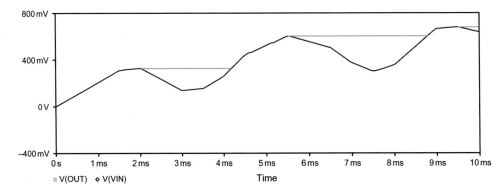

Fig. 7.20 Peak detector transient response.

Generating a Periodic V_in

13. Delete the VPWL_FILE source and replace it with a VPWL_F_RE_FOREVER from the source library. Double click on <FILE> and, as in Step 9, enter ..\..\Vin.txt

14. Edit the PSpice Simulation Profile, increase the simulation run to time to 50 ms and run the simulation. You should see the response as shown in Fig. 7.21, where V_{in} is now periodic (repeats forever).

15. Investigate the VPWL_F_RE_N_TIMES source.

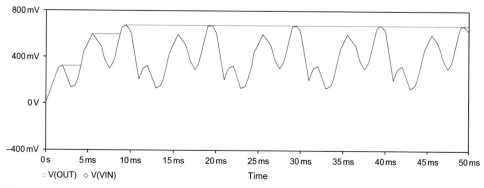

Fig. 7.21 The V_{in} signal is periodic.

CHAPTER 8

Convergence Problems and Error Messages

Chapter Outline

8.1. Common Error Messages   131
8.2. Establishing a Bias Point   132
8.3. Convergence Issues   133
8.4. Simulation Settings Options   134
8.5. Exercises   136
    Exercise 1   136
    Exercise 2   137
    Exercise 3   138
    Exercise 4   139

PSpice uses the Newton-Raphson iteration method to calculate the nodal voltages and currents for nonlinear circuit equations. The algorithm will start off with an initial "guess" to the solution and perform an iterative process until the voltages and currents converge to a consistent solution. As discussed in the chapter on Transient analysis, at every time step, the node voltages and currents are calculated and compared to the previous time step DC solution. Only when the difference between two DC solutions falls within a specified tolerance (accuracy), will the analysis move on to the next internal time step. The time step is dynamically adjusted until a solution within tolerance is found. However, if a solution cannot be found, PSpice will report that the simulation has failed due to a convergence problem. There are also occasions where the time step becomes too small for the iteration process to continue. This can occur if there is a fast moving signal in the circuit such as a pulse with a very short unrealistic rise time.

Simulations can also fail because of circuit errors and missing or incorrect parameters specified. Some of these common errors will be discussed in this chapter.

8.1. COMMON ERROR MESSAGES

Error—Node <name> is floating

There is no zero volt node "0" in the circuit. See Chapter 2, exercise 1 and Chapter 3, exercise 1.

Error—Missing DC path to ground

Analog Design and Simulation Using OrCAD Capture and PSpice
https://doi.org/10.1016/B978-0-08-102505-5.00008-2

There is no direct DC path to ground. Add a large value resistor either to the 0 node or to a DC ground path. See Chapter 3, exercise 1.

Error—Less than two connections at node <name>

There is no PSpice model attached to the Capture part. The Capture part is missing the required properties for PSpice simulation. This error can also happen if nets are left floating in that a wire is left "dangling."

Error—Voltage source or inductor loop

Voltage sources are modeled as ideal in that they have no internal resistance therefore connecting two voltage sources in parallel will result in an infinite current which will exceed the maximum current limit. Voltages and currents are limited to $\pm 1e10\,V$ and $\pm 1e10\,A$.

An inductor is essentially a time varying voltage source and if connected in parallel with another inductor or voltage source, will result in the same error message. Inductors are modeled as ideal in that they have zero series winding resistance.

8.2. ESTABLISHING A BIAS POINT

The Bias Point analysis is the starting point for a transient analysis and a DC Sweep. However, if PSpice cannot calculate the Bias Point for a circuit, the power supplies will be reduced from 100% toward zero where the nonlinearities of the circuit will effectively be linearized and hence improve the chances of a Bias Point solution to be found. The power supplies are then stepped back up to 100% in order to establish a Bias point upon which a DC Sweep or a transient analysis can be started.

When you run a transient analysis, PSpice launches and the simulation progress is shown in the **Simulation Status Window** and **Output Window**. Fig. 8.1 shows an example of the output window reporting that the Bias point was calculated, the transient analysis was finished, and that the simulation is complete.

```
Reading and checking circuit
Circuit read in and checked, no errors
Calculating bias point for Transient Analysis
Bias point calculated
Transient Analysis
Transient Analysis finished
Simulation complete
```

Fig. 8.1 Output window in PSpice.

The information displayed helps us to determine if the convergence failed during the Bias point analysis or the subsequent transient, DC, or AC analysis.

8.3. CONVERGENCE ISSUES

If there is a convergence issue, then the simulation pauses and the PSpice runtime settings window appears as shown in Fig. 8.2.

You can then change a simulation parameter and resume the simulation.

The output file will also be displayed and will report the last node voltages tried and which devices failed to converge which may relate to the nodes where the problem is.

There are no clearly defined rules for solving convergence problems. What you need to do is to try and localize the problem. For large circuits, methodically remove and simulate smaller portions of the circuit. This is where a hierarchical design, made up of blocks of circuitry, can help in solving convergence problems. Each block can be simulated separately and successively in the hierarchy, building up to the complete design. Hierarchical designs are covered in Chapter 20.

Another approach for large circuits is to replace parts of circuits with analog behaviorial models (ABM) which use mathematical expressions or tables to model components or circuit behavior. These devices simulate faster and can help localize which circuits are not converging by a process of replace and elimination. However, ABM's, if not used properly, can cause convergence problems in their own right especially if a mathematical expression contains a denominator variable which can under certain circuit conditions be set to equal zero resulting in large numbers exceeding the PSpice limits of $\pm 1e10 \, V$ and $\pm 1e10 \, A$.

Small circuits consisting of a few components can also cause convergence problems, even the humble diode if not modeled with a series resistance can cause currents and voltages to exceed the PSpice limits of $\pm 1e10 \, V$ and $\pm 1e10 \, A$.

Ideally the models used from semiconductor vendors are complete and have been tested for simulation and should not be the main cause of convergence problems. The

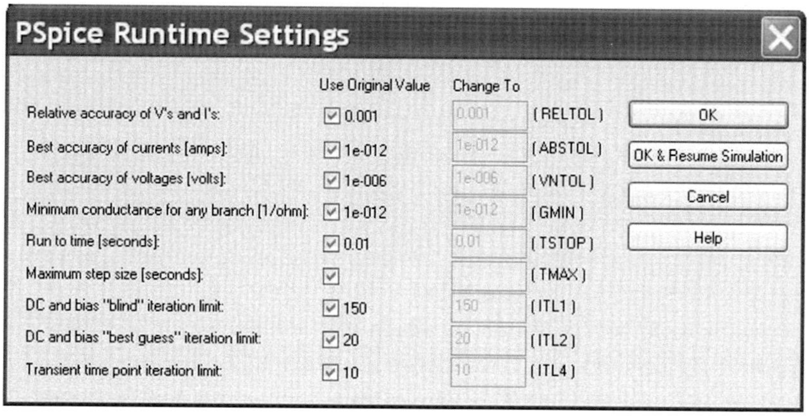

Fig. 8.2 PSpice runtime settings.

only issue is that some semiconductor models are represented by a subcircuit, especially power MOSFETS which can lead to convergence problems.

8.4. SIMULATION SETTINGS OPTIONS

The simulation settings can be accessed via the simulation profile and selecting the Options tab as shown in Fig. 8.3.

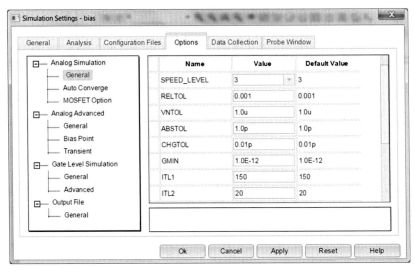

Fig. 8.3 Simulation settings options.

The calculated voltages and currents are based on previous calculated values such that the condition for convergence, for a node voltage is given by

$$|v(n-1) - v(n)| > \text{RELTOL}^* v(n) + \text{VNTOL} \tag{8.1}$$

Similarly, for a branch current:

$$|i(n-1) - i(n)| > \text{RELTOL}^* i(n) + \text{ABSTOL} \tag{8.2}$$

where RELTOL is the relative tolerance and VNTOL and ABSTOL are the absolute tolerances for voltage and currents, respectively. RELTOL has a default value of 0.001 which is equivalent to a 0.1% accuracy. Convergence will only occur when the relative difference between consecutive voltages and consecutive currents are calculated within the specified accuracy of simulation.

If you have a high voltage circuit in which the output voltage is 100 V, then using VNTOL which by default is 1 μV can be increased to 10 or 100 mV without affecting the resolution of accuracy and may help with subsequent convergence problems. The same applies to high current circuits where the default value of 1 pA for ABSTOL can be increased.

The number of times the simulator will try and reach convergence is set by the iteration limits, ITL1, ITL2, and ITL4. In some cases, just increasing the number of iteration limits will help a circuit to converge without reducing the accuracy of any of the simulation parameters. In Fig. 8.3, you can see that ITL and ITL2 are used in the calculation of the DC Bias Point and ITL4 is used in a transient analysis. When running a transient analysis it is always good idea to see if the convergence problem occurs with during the DC Bias Point calculation or the transient analysis so you can increase the appropriate iteration limit.

There is now a new feature in the simulation settings option window. Autoconverge, allows modified simulation settings to be used in a subsequent simulation run if there is a convergence problem. You enter the maximum **Relaxed limit** such that the simulator will automatically relax the simulation limit in an optimal manner subject to the maximum values set. The simulator will then automatically start again at time $t = 0$ using the modified values. Fig. 8.4A and 8.4B shows the AutoConverge settings for pre and post 17.2 software versions.

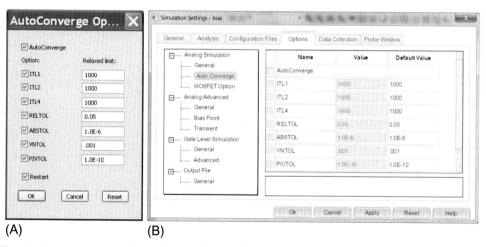

Fig. 8.4 Autoconverge settings: (A) pre 17.2 and (B) post 17.2.

Across each semiconductor there is a small conductance which provides a small conducting path for currents such that initial currents and voltages can be calculated for the initial DC bias point solution. This conductor is called GMIN and is globally available as one of the simulation setting options shown in Fig. 8.3. This is particularly useful when you have power MOSFETS or diodes with a large off resistance. By default, GMIN is $1.0E - 12$ Siemens but this can be increased up by a factor of 10 or 100. There is an option to automatically step GMIN, **Use GMIN stepping to improve convergence** which is very useful.

8.5. EXERCISES

Exercise 1

1. Draw the circuit in Fig. 8.5 and create a bias point simulation, **PSpice > New Simulation Profile**. In the **Analysis** type, select **Bias Point** and click on OK.

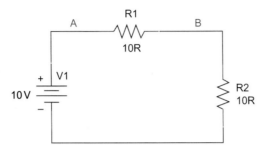

Fig. 8.5 Missing zero volt node.

2. Run the Simulation, **PSpice > Run** or click on the run button ▶.
You should see the warning message dialog box (Fig. 8.6) and a message will be displayed asking you to check the Session Log.

Fig. 8.6 Warning message.

The Session Log is normally open at the bottom of the screen, if not, the session log can be found from the top tool bar, **Window > Session Log**. The warning message will read:
    WARNING [NET0129] Your design does not contain a Ground (0) net.
    Click on OK and PSpice will launch.
3. In PSpice, the output file displays an error message as shown in Fig. 8.7.

```
V_V1            A N00514 10V
R_R1            A B    10R TC=0,0
R_R2            N00514 B   10R TC=0,0

**** RESUMING bias.cir ****
.END

ERROR -- Node A is floating
ERROR -- Node N00514 is floating
ERROR -- Node B is floating
```

Fig. 8.7 Floating node error message for missing ground..

All nodes are reported as floating as there is no reference to 0 V. Connecting a ground 0 V symbol will allow the circuit to simulate.

Exercise 2

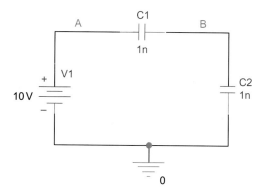

Fig. 8.8 Floating node.

1. Draw the circuit in Fig. 8.8 and create a bias point simulation, **PSpice > New Simulation Profile**. In the **Analysis** type, select **Bias Point** and click on OK.
2. Run the Simulation, **PSpice > Run** or click on the run button ▶.
3. In PSpice, the output file displays an error message as shown in Fig. 8.9.

This is because there is no DC path to ground at node B which will always be the case for two capacitors connected in series. Connecting a large value resistor across R1 or R2 will allow the circuit to simulate.

```
* source CAPACITORS
C_C1            A B   1n   TC=0,0
V_V1            A 0 10V
C_C2            0 B   1n   TC=0,0

**** RESUMING bias.cir ****
.END

ERROR -- Node B is floating
```

Fig. 8.9 Floating node error message for no DC path to ground.

Exercise 3

1. Draw the circuit in Fig. 8.10 and create a bias point simulation profile.

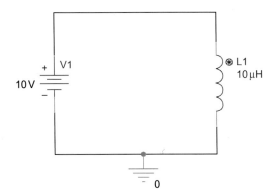

Fig. 8.10 Inductor loop.

2. Run the simulation ▶.
3. The PSpice output file will display an error message as shown in Fig. 8.11.
Inductors do not contain a series resistance so adding a series resistor will allow the circuit to simulate.

```
* source SINEWAVE
V_V1          N00502 0 10V
L_L1          N00502 0  10uH

**** RESUMING transient.cir ****
.END

ERROR -- Voltage source and/or inductor loop involving V_V1
You may break the loop by adding a series resistance▮
```

Fig. 8.11 Inductor loop error message.

Exercise 4

1. Draw the circuit in Fig. 8.12 using the transistor from the Capture **transistor** library.

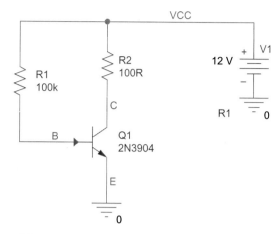

Fig. 8.12 No PSpice model.

2. Create a bias point simulation profile.
3. Run the simulation ⏵ .
4. The PSpice output file will display an error message as shown in Fig. 8.13.

```
* source SINEWAVE
R_R2          C VCC   100R TC=0,0
V_V1          VCC 0 12V
R_R1          B VCC   100k TC=0,0

**** RESUMING transient.cir ****
.END

ERROR -- Less than 2 connections at node C
ERROR -- Less than 2 connections at node B
```

Fig. 8.13 No Pspice model error message.

Note the transistor does not appear in the netlist.

5. In Capture a green circle will appear next to the transistor. Click on the circle and a Warning message will appear as shown in Fig. 8.14 stating that there is no PSpice template for Q1. The PSpice template is a required property for a Capture part to be simulated in PSpice. This will be covered in more detail in the Model Editor chapter.

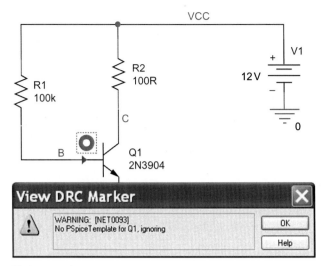

Fig. 8.14 Missing PSpice template.

CHAPTER 9

Transformers

Chapter Outline

9.1. Linear Transformer   141
9.2. Nonlinear Transformer   142
9.3. Predefined Transformers   144
9.4. Exercises   144
     Exercise 1   144
     Exercise 2   148

A transformer is implemented by magnetically coupling two or more coils (inductors) together. For air core transformers a K_Linear coupling device from the analog library is used whereas for nonlinear transformers a magnetic core model is referenced by the K coupling device. When creating a linear transformer the coils are specified in units of henry (H) whereas for nonlinear transformers you specify the number of turns for the inductors.

The PSpice magnetic cores model hysteresis effects and include a coupling coefficient which is used to define the proportion of flux linkage between the coils and has a value between 0 and 1. For coils wound on the same magnetic core, the coupling coefficient has a value almost equal to 1. For air cored coils, the flux linkage is smaller.

9.1. LINEAR TRANSFORMER

Fig. 9.1 shows a step-down linear transformer using inductors for the primary and secondary windings which are specified in Henry's. The two coils are magnetically coupled together using the K_Linear part, K1, which shows that L1 and L2 are coupled together. Ideally, the primary and secondary circuits are electrically isolated. However, for PSpice simulation, as mentioned in Chapter 2, there must be a DC path to ground for every node. This is achieved by using a large value resistor R4 connected from the secondary to 0 V, which will not have a significant effect on the accuracy of simulation.

Analog Design and Simulation Using OrCAD Capture and PSpice
https://doi.org/10.1016/B978-0-08-102505-5.00009-4

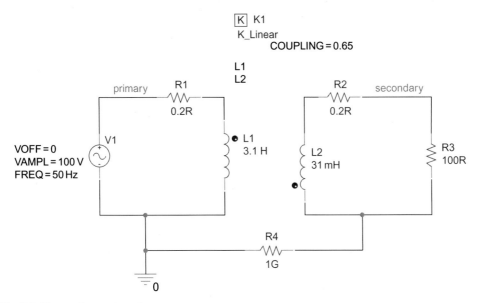

Fig. 9.1 Linear air core transformer.

Note

The dot convention has now been added to inductors to indicate the direction of current flow and subsequent voltage polarity, which is related to how the coils are wound relative to each other. In previous OrCAD versions, you have to display the pin numbers for the inductors, which can then be aligned relative to each other.

9.2. NONLINEAR TRANSFORMER

Fig. 9.2 shows a circuit for a nonlinear transformer created using three inductors for the coils, L1, L2 and L3. The reference designators, L1, L2 and L3, are added to the K device in the Property Editor and displayed on the schematic to indicate which coils make up the transformer. The standard K coupling devices allow for up to six inductors to be coupled together. Manufacturer's magnetic core models are found in the **magnetic** library. In this example, the E13_6_6_3C81 magnetic core is used.

The hysteresis curve for the magnetic cores can be displayed by selecting a K device and **rmb > Edit PSpice Model**, which will open the PSpice Model Editor. In the Model Editor the text description for the model will be displayed. By selecting **View > Extract Model** and clicking on **Yes** to the message window, the characteristic hysteresis curve will be plotted. At the bottom of the Model Editor, is a table of the magnetic core parameters. For each core, there is a gap parameter which can be specified.

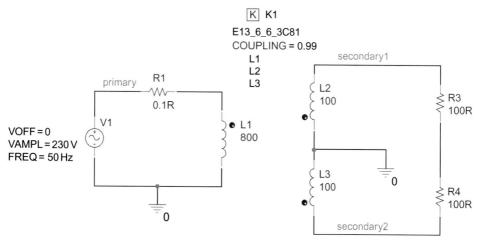

Fig. 9.2 Nonlinear center-tapped transformer.

There is also an entry table whereby you enter B–H curve data and extract your own model parameters. Fig. 9.3 shows the model parameters and hysteresis curve for the E13_6_6_3C81 magnetic core. The Model Editor will be covered in more detail in Chapter 16.

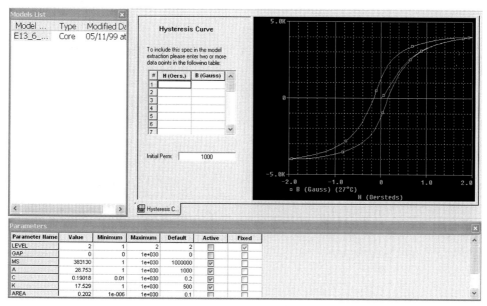

Fig. 9.3 Model parameters and hysteresis curve for the E13_6_6_3C81 magnetic core.

9.3. PREDEFINED TRANSFORMERS

A linear transformer, XFRM_LINEAR (Fig. 9.4), is available in the analog library. Non-linear transformers, which include center-tapped primary and secondary windings, can be found in the breakout library (Fig. 9.5). These transformers have properties that enable you to enter the inductance, coil resistances and number of turns. Double click on the transformers to access the properties in the Property Editor.

Fig. 9.4 Linear transformer XFRM_LINEAR.

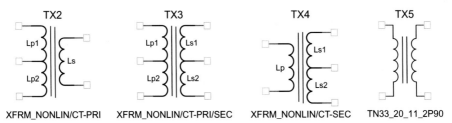

Fig. 9.5 Nonlinear transformers.

9.4. EXERCISES

Exercise 1

You will create a step-down transformer circuit with a primary inductance winding inductance of 3.1 H and a winding resistance of $0.2\,\Omega$. The secondary winding has an inductance of 31 mH and a winding resistance of $0.2\,\Omega$. The transformer provides a step-down ratio of 10 and is connected to a $100\,\Omega$ load resistance.

1. Create a new project called Linear Transformer and draw the circuit in Fig. 9.6. The inductors, resistors and K_Linear are all from the **analog** library. V1 is a VSIN source from the **source** library. Set the coupling coefficient to 0.65.

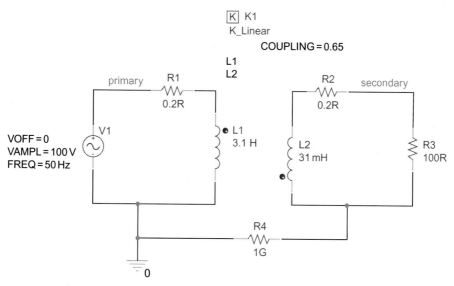

Fig. 9.6 Linear transformer.

2. If you have a previous version of OrCAD which does not have the dot convention shown on the inductors, then display pin 1 of the inductors and orientate the inductors accordingly.
3. Double click on the K_Linear device to open the Property Editor and add the inductor reference designators as shown in Fig. 9.7. Highlight L1 and L2, then **rmb >Display** and select **Value Only** (Fig. 9.8).

L1	L1
L2	L2
L3	
L4	
L5	
L6	

Fig. 9.7 Defining which coils are magnetically coupled together.

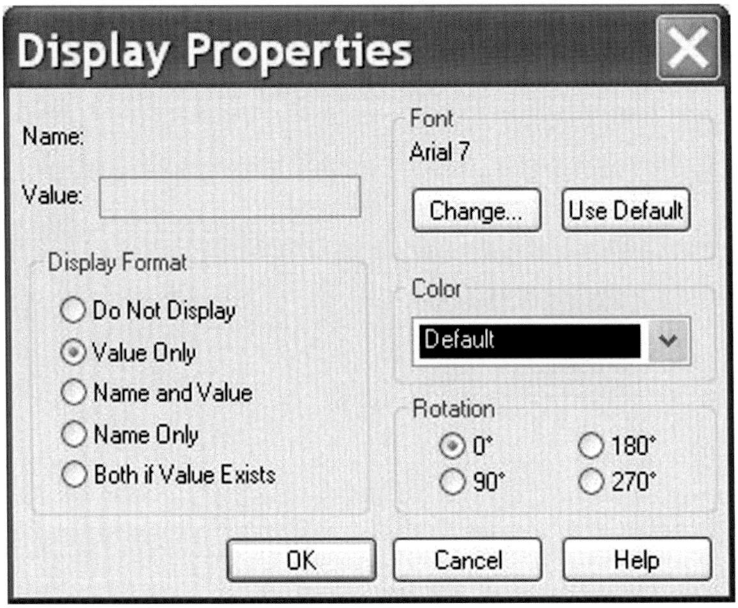

Fig. 9.8 Displaying the reference designators for the inductors.

Note

A common misunderstanding is that the coil reference designators must be entered as L1 to L6 in the Property Editor. If you have coils with reference designators L3, L4 and L5 in a circuit, then you enter the designators as shown in Fig. 9.9. It is recommended that you show the reference designators on the schematic for a K core device, especially if you annotate and change the coil numbering.

L1	L3
L2	L4
L3	L5

Fig. 9.9 Entering coil reference designators.

---

4. Set up a transient simulation profile with a **Run to time** of 50 ms (Fig. 9.10).

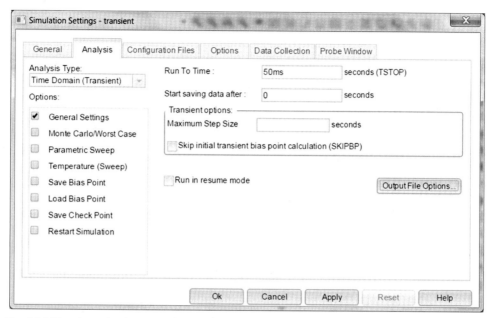

Fig. 9.10 Transient simulation profile.

5. Place voltage markers on the primary and secondary nets and run the simulation. You should see the response as shown in Fig. 9.11 with a secondary voltage of 6.4 V.

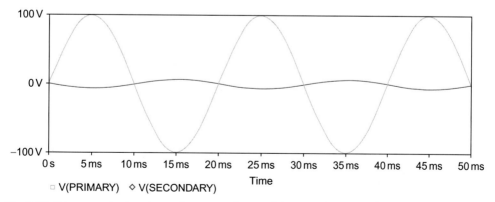

Fig. 9.11 Primary and secondary voltage waveforms of the step-down transformer.

If you have not placed markers on the circuit then when the PSpice window appears, select **Trace > Add Trace** and select from the left-hand side under **Simulation Output Variables**, **V(primary)** and **V(secondary)**.

6. The relationship between the primary and secondary voltages and the primary and secondary inductances for air cored transformers is given by:

$$\frac{V_s}{V_p} = \frac{\sqrt{L_s}}{L_p}$$

So, if you increased k to 1 you would see a secondary voltage of approximately 10 V. In practice, air cored transformers have a low coupling coefficient.

> **Tip**
> When running simulations with transformers, you may need to limit the maximum step size if you see distorted sinewave outputs.

Exercise 2

You will create a nonlinear, center-tapped transformer circuit.

1. Create a new project called Non Linear Transformer and draw the circuit in Fig. 9.12. Enter the inductor values as the number of turns as shown. The E13_6_6_3C81 core model is found in the **magnetic** library. Enter a value of 0.99 for the coupling coefficient.

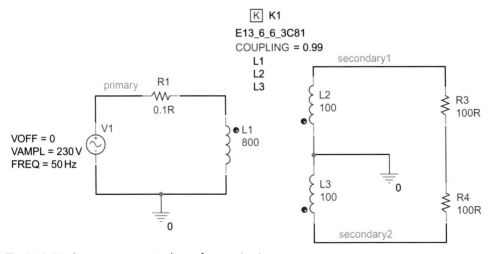

Fig. 9.12 Nonlinear, center-tapped transformer circuit.

2. As in Exercise 1, double click on the K core device and in the Property Editor, add and display the reference designators L1, L2 and L3.
3. Create a transient simulation profile with a run to time of 50 ms.
4. Place voltage markers on the nets, **primary**, **secondary1** and **secondary2** and run the simulation.
5. Fig. 9.13 shows the voltage waveforms for the nonlinear transformer. You should see 25.1 V on each secondary winding.

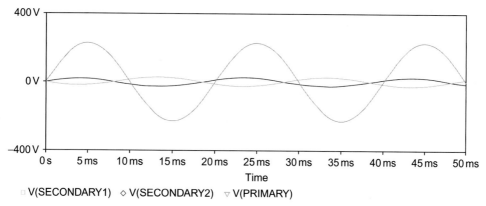

Fig. 9.13 Primary and secondary waveforms.

6. Investigate the other transformers in the analog and breakout libraries.

CHAPTER 10

Monte Carlo Analysis

Chapter Outline

10.1. Simulation Settings   152
    10.1.1 Output Variable   153
    10.1.2 Number of Runs   153
    10.1.3 Use Distribution   153
    10.1.4 Random Number Seed   154
    10.1.5 Save Data From   154
    10.1.6 MC Load/Save   154
    10.1.7 More Settings   154
10.2. Adding Tolerance Values   155
10.3. Exercises   156
    Exercise 1   156
    Exercise 2   161

Monte Carlo analysis is essentially a statistical analysis that calculates the response of a circuit when device model parameters are randomly varied between specified tolerance limits according to a specified statistical distribution. For example, all the circuits encountered so far have been simulated using fixed component values. However, discrete real components such as resistors, inductors and capacitors all have a specified tolerance, so that when you select, for example, a $10\,\mathrm{k}\Omega \pm 1\%$ resistor, you can expect the actual measured resistor value to be somewhere between 9900 and $10,100\,\Omega$. Other discrete components and semiconductors in a circuit will also have tolerances and so the combined effect of all the component tolerances may result in a significant deviation from the expected circuit response. This is especially the case in filter designs where applied component tolerances may result in a deviation from the required filter response.

What the Monte Carlo analysis does it to provide statistical data predicting the effect of randomly varying model parameters or component values (variance) within specified tolerance limits. The generated values follow a statistically defined distribution. The circuit analysis (DC, AC or transient) is repeated a number of specified times with each Monte Carlo run generating a new set of randomly derived component or model parameter values. The greater the number of runs, the greater the chances that every component value within its tolerance range will be used for simulation. It is not uncommon to perform hundreds or even thousands of Monte Carlo runs in order to cover as many possible component values within their tolerance limits. Monte Carlo, in effect, predicts the

Analog Design and Simulation Using OrCAD Capture and PSpice
https://doi.org/10.1016/B978-0-08-102505-5.00010-0

robustness or yield of a circuit by varying component or model parameter values up to their specified tolerance limits.

Although the results of a Monte Carlo analysis can be seen as a spread of waveforms in the PSpice waveform viewer (Probe), a **Performance Analysis** can be used to generate and display histograms for the statistical data together with a summary of the statistical data. This provides a more visual representation of the statistical results of a Monte Carlo analysis.

10.1. SIMULATION SETTINGS

A Monte Carlo analysis is run in conjunction with another analysis, AC, DC or transient analysis. Tolerances are applied to parts in the schematic via the **Property Editor** and the required analysis is created in the simulation profile. For the band pass filter in Fig. 10.1, component tolerances have been added to the resistors and capacitors and displayed on the circuit. The circuit response is analyzed by performing an AC sweep from 10 Hz to 100 kHz. Monte Carlo will run an initial analysis with all nominal values being used and then run subsequent analysis using randomly generated component values up to the number of Monte Carlo runs specified.

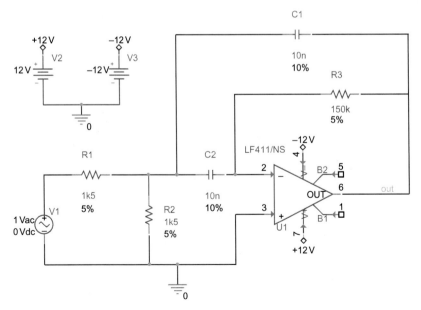

Fig. 10.1 1500 Hz band pass filter.

Fig. 10.2 shows the simulation profile for an AC sweep running a Monte Carlo analysis.

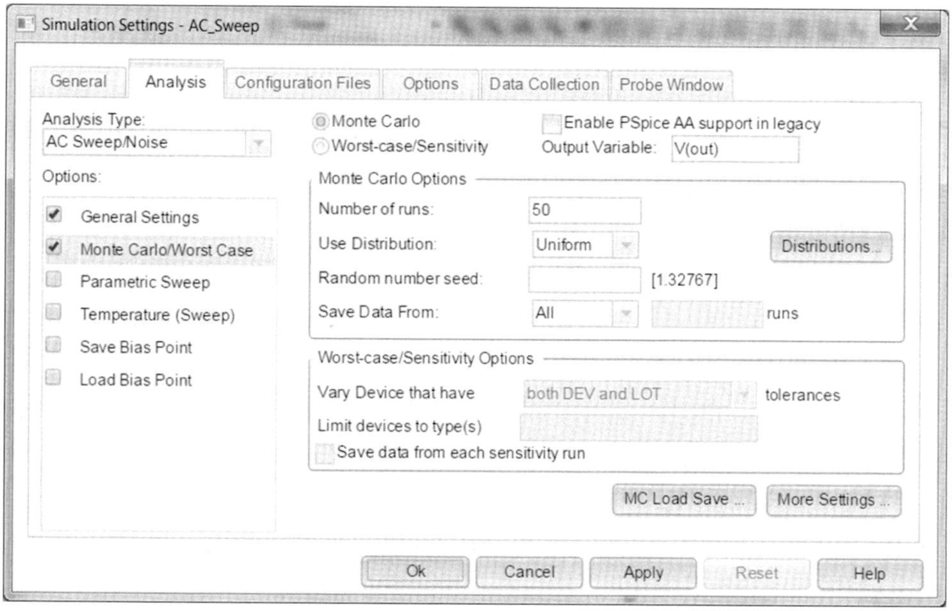

Fig. 10.2 Monte Carlo simulation settings.

10.1.1 Output Variable

You need to specify a node voltage or an independent current or voltage source. In this example, the output variable is V(out).

10.1.2 Number of Runs

This is the number of times the AC, DC or transient analysis is run. The maximum number of waveforms that can be displayed in Probe is 400. However, for printed results, the maximum number of runs has now been increased from 2000 to 10,000. The first run is the nominal run where no tolerances are applied.

10.1.3 Use Distribution

The model parameter deviations from the nominal values up to the tolerance limits are determined by a probability distribution curve. By default, the distribution curve is uniform; that is, each value has an equal chance of being used. The other option is the Gaussian distribution, which is the familiar bell-shaped curve commonly used in manufacturing. Component values are more likely to take on values found near the center of the distribution compared to the outer edges of the tolerance limits.

You can also define your own probability distribution curves using coordinate pairs specifying the deviation and associated probability, where the deviation is in the range -1 to $+1$ and the probability is in the range from 0 to 1. More information can be found in the PSpice A/D Reference Guide in <install dir> doc\pspcref\pspcref.pdf.

10.1.4 Random Number Seed

As with most random number generators, an initial seed value is required to generate a set of random numbers. This value must be an odd integer number from 1 to 32,767. If no seed number is specified, the default value of 17,533 is used.

10.1.5 Save Data From

This allows you to save data from selected runs. For example, if you just want to see the circuit response from the nominal run, select none. If you want to save all the runs, select **All.** If you want to select every third run from the nominal run, i.e. the fourth, seventh, tenth, etc., select **Every,** then enter 3 in the **runs** box. If you want to save data from the first three runs, select **First** and then enter 3 in the **runs** box. If you want to save data from the third, fifth, seventh and tenth run, select **Runs (list)** and enter 3, 5, 7, 10 in the **runs** box. Saved runs will be displayed in Probe.

10.1.6 MC Load/Save

PSpice now has history support which allows you to save randomly generated model parameters or component values from a Monte Carlo run in a file, which can be used for subsequent analysis.

10.1.7 More Settings

This option allows you to specify Collating Functions which are applied to the output waveforms returning a single resulting value. For example, the MAX function searches for and returns the maximum value of a waveform. The YMAX function returns the value corresponding to the greatest difference between the nominal run waveform and the current waveform. The RISE and FALL functions search for the first occurrence of a waveform crossing above or below a set threshold value. For each function, you can specify the range over which you want to apply these functions. Fig. 10.3 shows the available collating functions.

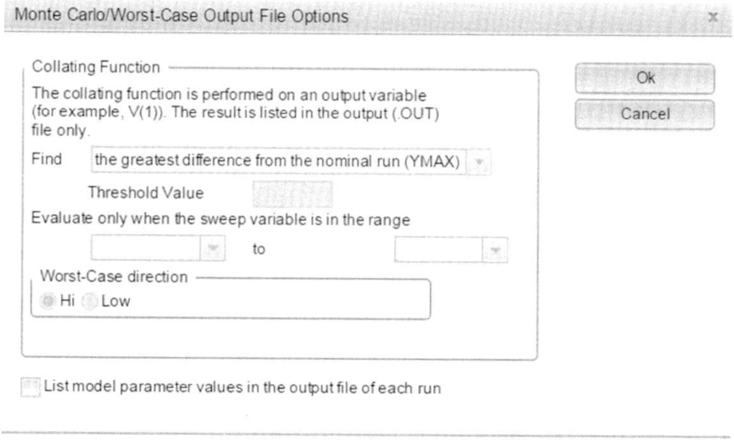

Fig. 10.3 Collating functions.

The collating functions are summarized as:

YMAX: Find the greatest difference from the nominal run

MAX: Find the maximum value

MIN: Find the minimum value

RISE_EDGE: Find the first rising threshold crossing

FALL_EDGE: Find the first falling threshold crossing

10.2. ADDING TOLERANCE VALUES

Tolerances can now be added to the discrete R, L and C parts in the Property Editor as these parts now include a tolerance property. You must make sure that you include the % symbol when entering the tolerance value i.e. 10%.

In previous OrCAD versions the only way to add tolerances to discrete parts was to use generic **Breakout** parts, which allowed you to edit the PSpice model. For example, to add a tolerance to a resistor, you had to use an Rbreak part from the **breakout** library, edit the model (rmb > Edit PSpice Model) and add tolerance statements to the PSpice model. The default model definition for an Rbreak is given by:

.model Rbreak RES R=1

where Rbreak is the model name that can be changed and appears on the schematic, RES is the PSpice model type and R is a resistance multiplier. There are two types of tolerance you can add, **dev** and **lot**, as shown below:

.model Rmc1 RES R=1 lot = 2% dev=5%

Here, the model name has been changed to Rmc1 and two tolerance types have been added to the model statement. The dev tolerance causes the tolerance values of devices that share the same model name (Rmc1 in this example) to vary independently from each other, while the lot tolerance will cause the tolerance values of devices, with the same model name, to track together.

Dev is the same as applying a standard 5% tolerance to all the resistors, where each resistor with the same model name will be assigned its own random resistance value independent from all the other resistors with the same model name.

Lot is where all the resistors with the same model name will be treated as one group and will track together by as much as $\pm$ 2%. This was mainly used in integrated circuit (IC) design where a change in temperature was equally applied to groups of components. The combined total tolerance for Rmc1 can therefore be as much as $\pm$ 7%.

One example in which you would use both lot and dev is for a single in-line (SIL) or dual in-line (DIL) resistor pack; each resistor will have a dev tolerance defined, and therefore each randomly generated resistance value will have a different value from the others. If there is a rise in temperature, the lot tolerance will ensure that all resistance values increase together by the same percentage.

As mentioned above, only discrete parts have an attached tolerance property which can be edited in the Property Editor. If you want to add for example a specified manufacturer's tolerance to the B_f of a transistor to see its effect on a circuit's performance, you will have to edit the transistor model and add, for example, the dev tolerance to the B_f model parameter as:

.model Q2N3906 PNP (Is$=$1.41f Xti$=$3 Eg$=$1.11 Vaf$=$18.7
Bf$=$180.7 dev$=$50% Ne$=$1.5 Ise$=$0

In this example, a dev tolerance of 50% with a default uniform distribution has been added to the B_f of the transistor. However, the distribution for the B_f is more likely to be Gaussian, so you would add dev/gauss$=$12.5% as PSpice limits a Gaussian distribution to $\pm$ 4σ:

.model Q2N3906 PNP (Is$=$1.41f Xti$=$3 Eg$=$1.11 Vaf$=$18.7
Bf$=$180.7 dev/gauss$=$12.5% Ne$=$1.5 Ise$=$0

10.3. EXERCISES

Exercise 1

Fig. 10.4 shows a Sallen and Key 1500 Hz band pass filter circuit. Tolerances will be added to the resistors and capacitors and a Monte Carlo analysis will be performed to predict the statistical variation of the band pass frequency.

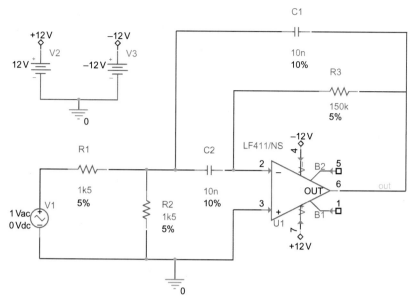

Fig. 10.4 1500 Hz band pass filter.

1. Draw the circuit in Fig. 10.4. If you are using the demo CD, use the ua741 opamps, which can be found in the **eval** library; otherwise, the LF411 opamp can be found in the **opamp** library. Performing a **Part Search** will result in the LF411 opamp being sourced from different manufacturers. It does not matter which one you use. V1 is a V_{AC} source from the **source** library, used for an AC analysis, and the power symbols are VCC_CIRCLE symbols from **Place > Power**, renamed +12 V and −12 V, respectively.

Note

When running a Part Search for the LF411, make sure the search part is pointing to the <install dir> \Tools\ Capture > Library > PSpice library (see Chapter 5, Exercise 3).

---

2. You need to add a 5% tolerance to all the resistors. Hold down the control key and select the resistors R1, R2 and R3, then **rmb > Edit Properties**. In the **Property Editor** highlight the entire **Tolerance** Row (or column) as shown in Fig. 10.5 and **rmb > Edit**. In the **Edit Property Values** box (Fig. 10.6), type in **5%** and click on OK.

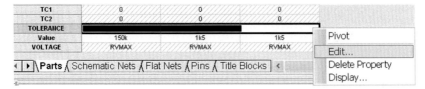

Fig. 10.5 Select entire TOLERANCE row and select **Edit**.

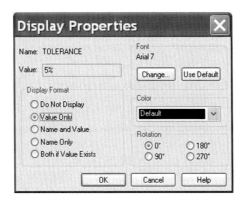

Fig. 10.6 Adding a 5% tolerance to the resistors.

3. To display the component tolerances on the schematic, select the entire TOLER-ANCE row as in Fig. 10.5, **rmb > Display** and in **Display Properties** (Fig. 10.7), select **Value Only** and click on OK.

Fig. 10.7 Displaying tolerance values.

4. Repeat Step 2 by assigning a 10% tolerance to capacitors C1 and C2.
5. Create a PSpice simulation profile called AC Sweep for an AC logarithmic sweep from 100 to 10 kHz with 50 points per decade (Fig. 10.8). Select **Monte Carlo/ Worst Case** in the **Options** box.

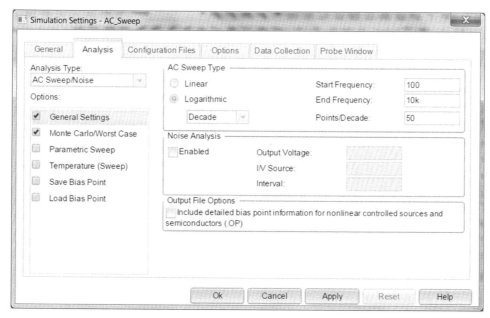

Fig. 10.8 Simulation profile for an AC sweep.

The output variable is **V(out)** and the number of runs is 50. The distribution used will be uniform and you will use the default random seed setting by leaving the Random number seed box empty. Your Monte Carlo simulation settings should be as shown in Fig. 10.9.

Fig. 10.9 Monte Carlo simulation settings.

6. Place a dB voltage marker on node **out** (**PSpice > Markers > Advanced > dB Magnitude of Voltage**) and run the simulation.

Note

Monte Carlo runs can produce large data files. Since we are only interested in the output node, the simulation can be set up to collect data for display waveforms only for markers placed on the schematic. In the simulation profile, select the **Probe Window** tab and make sure that **All markers on open schematic** is selected.

---

7. When PSpice launches, a list of available runs is shown (Fig. 10.10). Select **All** and click on OK.

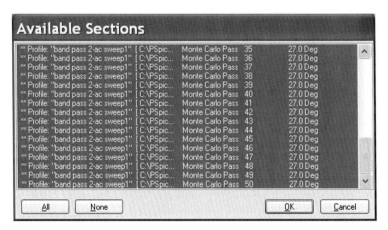

Fig. 10.10 Available sections.

8. Fig. 10.11 shows the band pass filter response showing a spread in the center frequency. Open the output file, **View > Output File{ XE "Output File" }** and scroll down to see the results for each Monte Carlo run. At the bottom of the file, you will see a summary of the statistical results and the deviation from the nominal for each run.

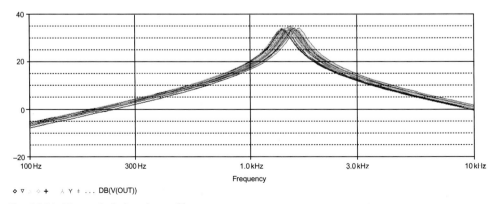

Fig. 10.11 Monte Carlo band pass filter response.

However, to get a better picture of the spread of the center frequency, you can use a **Performance Analysis** to display histograms representing the statistically generated data. Performance analysis will be covered in more detail in Chapter 12, but for now, the steps required for a performance analysis for the bandpass filter are introduced in Exercise 2.

Exercise 2

1. In Probe, from the top toolbar, select **Trace > Performance Analysis.** At the bottom of the **Performance Analysis** window, click on **Wizard,** then in the next window click on **Next.** You will be presented with a list of measurements as shown in Fig. 10.12.

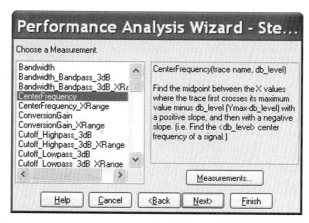

Fig. 10.12 Available measurements.

2. Select the **CenterFrequency** measurement and click on **Next**.
3. In the **Measurement Expression** window (Fig. 10.13), enter the name of the trace to search as **V(out)**. Alternatively, you can click on the **Name of trace to search** icon and search for the trace name, V(out).

Performance Analysis Wizard ...

Measurement Expression

CenterFrequency(V(out),)

Now you need to fill in the Measurement arguments. That is, you need
to tell the Measurement which trace(s) to look at, and if necessary,
the other numbers the Measurement needs to work.

The Measurement 'CenterFrequency' has 2 arguments.
Please fill them in now.

Name of trace to search [M] V(out)

db level down for measurement   3

[Help] [Cancel] [<Back] [Next>] [Finish]

Fig. 10.13 V(out) is selected as the trace variable with a −3 dB measurement down either side.

Enter a value of 3 for the **db level down for measurement** as shown. The center
frequency of the filter will be measured 3 dB down either side of the maximum value
for the Monte Carlo nominal run (no component tolerances applied). Click on **Next**
in the **Performance Analysis Wizard**.

4. The nominal trace waveform of the band pass circuit will be displayed (Fig. 10.14).
   This is the simulation response using the component nominal values. The trace shows
   two points where the measurements from the waveform will be taken. In this case, the
   center frequency is defined at 3 dB down from the maximum value. This is shown so
   that you can confirm whether you are seeing the correct circuit response and that the
   measurements are taken at the correct points on the waveform.

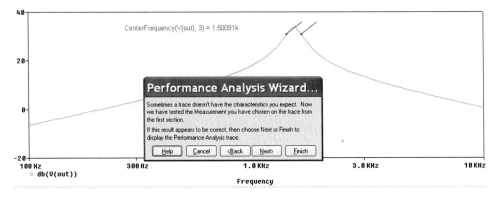

Fig. 10.14 Nominal trace showing where the −3 dB measurement will be taken on the nominal circuit
response waveform.

5. Click on **Next** in the **Performance Analysis Wizard**.

6. Histograms will now be displayed (Fig. 10.15) representing the generated statistical data together with a summary of statistical data.

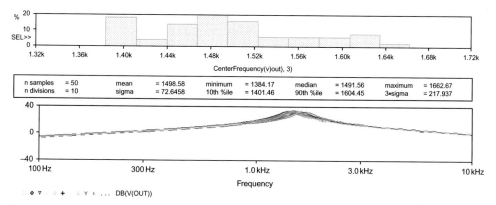

Fig. 10.15 Histograms shows the possible spread in center frequency.

7. Change the resistor tolerances to 1% and the capacitor tolerances to 5% and re-run the simulation. Run a performance analysis as before using the center frequency measurement as above. The results are shown in Fig. 10.16.

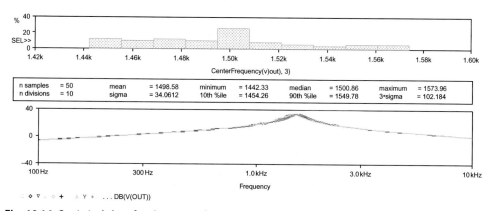

Fig. 10.16 Statistical data for the center frequency with tighter component tolerances.

Note

Because the tolerances are displayed on the schematic, there is no need to use the Property Editor to change the tolerance values. Just double click on the tolerance values in the schematic and enter the new tolerance value.

---

8. Run another Performance analysis but this time, use the bandwidth measurement.

Note

The number of displayed histograms can be changed in Probe. Select **Tools > Options > Histogram Divixions**.

Filter Specifications

The gain of the filter is given by:

$$|G| = \frac{R3}{2R1}$$

$$|G| = \frac{150 \times 10^3}{2 \times 1.5 \times 10^3} = 50 \qquad (10.1)$$

or in dB:

$$|G|_{dB} = 20 \log_{10} 50 = 34 \, dB$$

The band pass center frequency is given by:

$$f = \frac{1}{2\pi C \sqrt{\left(\dfrac{R1R2}{R1+R2}\right)R3}} \qquad (10.2)$$

where $C1 = C2 = C$

$$f = \frac{1}{2\pi \times 10 \times 10^{-9} \sqrt{\left(\dfrac{1.5 \times 10^3 \times 1.5 \times 10^3}{1.5 \times 10^3 + 1.5 \times 10^3}\right) \times 150 \times 10^3}}$$

$$f = 1501 \, Hz$$

The $-3 \, dB$ bandwidth is given by:

$$f = \frac{1}{\pi CR3}$$

$$f = \frac{1}{\pi \times 10 \times 10^{-9} \times 150 \times 10^3} = 212.2 \, Hz \qquad (10.3)$$

CHAPTER 11

Worst Case Analysis

Chapter Outline

11.1. Sensitivity Analysis   166
11.2. Worst Case Analysis   167
11.3. Adding Tolerances   168
11.4. Collating Functions   168
11.5. Exercise   169

Worst Case analysis is used to identify the most critical components which will affect circuit performance. Initially a sensitivity analysis is run on each individual component which has a tolerance assigned. The component value is effectively pushed toward both of its tolerance limits by a small percentage of its value to see which limit would have the most effect on the worst case output. A Worst Case analysis is then performed by setting all the component values to their end tolerance limits which gave an indication of the worst case results. In order to reduce the number of simulation runs, collating functions can be used to detect differences from the nominal worst case output such as minimum, maximum, or threshold differences.

In Fig. 11.1, an equivalent inductor circuit known as a gyrator is implemented using U1B, R4, R5, and C2 which enables a high inductance value to be realized, 100H in this example. The series connection of the equivalent inductor and C1 forms a series tuned circuit which determines the frequency of the notch filter. Using ideal component values, the initial simulation results will show the required notch filter response. However, components have tolerances which may affect the notch frequency. Adding component tolerances, a Worst Case analysis will determine which components are critical to the circuit's performance. The notch frequency can be determined by detecting the minimum output voltage of the filter. Therefore a Worst Case analysis can be run using a Collating function on the output voltage which will only record the minimum output voltages.

Analog Design and Simulation Using OrCAD Capture and PSpice
https://doi.org/10.1016/B978-0-08-102505-5.00011-2
165

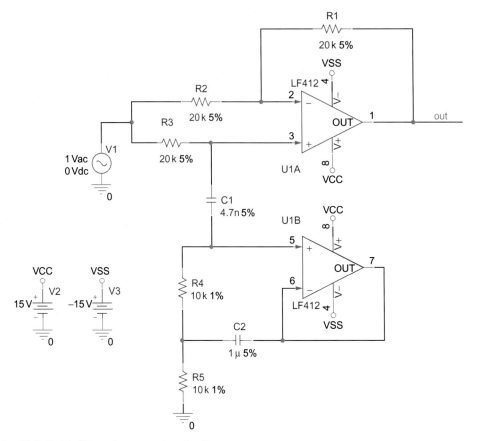

Fig. 11.1 Notch filter using a gyrator circuit.

11.1. SENSITIVITY ANALYSIS

What we are looking to investigate is the effect each component has on the notch frequency by recording the minimum output voltage of the filter and hence determine which components are critical to the notch frequency. A sensitivity analysis will be run on each component in turn with its sensitivity value given by

value = nominal value * (1 + RELTOL)

where RELTOL is the relative tolerance and can be found in the **PSpice Simulation Settings > Options** as shown in Fig. 11.2. By default, RELTOL is 0.001 (0.1%).

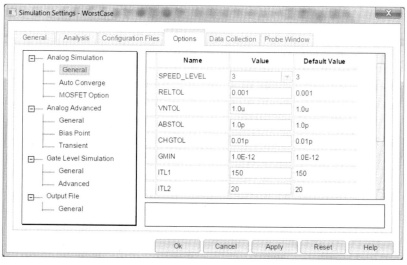

Fig. 11.2 Simulation settings options.

For example, R1 has a value of $20\,k\Omega$ and a tolerance of 5% and so has an expected value between $19\,k\Omega$ and $21\,k\Omega$. If RELTOL is set to 0.01 ($\pm1\%$), the resistor value will be increased to $20200\,\Omega$ and then decreased to $19800\,\Omega$. The direction in which the change in resistor value, gave the worst case result (minimum voltage), determines which tolerance limit (upper or lower) to use in the Worst Case analysis. If the minimum output voltage was recorded with a 1% decrease in the resistor value, then the lower tolerance limit of $19\,k\Omega$ will used for R1 in the Worst Case analysis.

The value of R1 is then reset to its nominal of $20\,k\Omega$ value and a sensitivity analysis is performed on R2, varying its value toward its tolerance limits and recording the "tolerance direction" which gave the minimum gain. The resistor values will then be set to their tolerance limit value which gave the worst case minimum gain and the gain of the amplifier will be calculated.

It must be remembered that the above assumes that the output voltage decreases continuously with a decrease in R1 and that there is no interdependencies on other components which may not be the case.

11.2. WORST CASE ANALYSIS

Based upon the results of the sensitivity analysis, Worst Case analysis sets the component values to one of their tolerance limits. If R1, $20\,k\Omega$ -1% gave the minimum voltage, then R2 will be set to $20\,k\Omega$ -5%.

A Worst Case analysis is run in conjunction with a DC, AC, or transient analysis and does not take into account the interdependence of the parameters. The results of the sensitivity and Worst Case analysis are written to the output file.

11.3. ADDING TOLERANCES

As with Monte Carlo analysis, tolerances can be added directly to the R, L, and C Capture parts in the schematic. Alternatively, breakout parts can be used and the following **dev** or **lot** statements are added as used with a Monte Carlo analysis.

.model Rwc1 RES R=1 dev=5% lot=2%

The dev statement causes the tolerance values of devices that share the same model name to vary independently from each other, while the lot statement will cause the tolerance values of devices to track together.

As with a Monte Carlo analysis, you have to define an output variable which can be a node voltage or an independent current or voltage source. In Fig. 11.3, the output variable is **V(out)**.

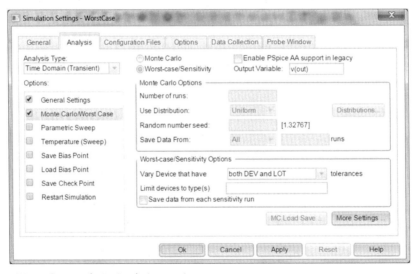

Fig. 11.3 Worst Case analysis simulation settings.

11.4. COLLATING FUNCTIONS

As with Monte Carlo, collating functions detect and compare the output response of the circuit with defined parameters. There are five functions which define the worst case results.

YMAX finds the greatest distance in the Y direction in each waveform from the nominal run.

MAX finds the maximum value of each waveform.

MIN finds the minimum value of each waveform.

RISE_EDGE(value) finds the first occurrence of the waveform crossing above the threshold (value). The function assumes that there will be at least one point that lies below the specified value followed by at least one above.

FALL_EDGE(value) finds the first occurrence of the waveform crossing below the threshold (value). The function assumes that there will be at least one point that lies above the specified value followed by at least one below.

11.5. EXERCISE

1. Draw the circuit in Fig. 11.4. The LF412 dual opamp is from the source library or you can use any opamps.

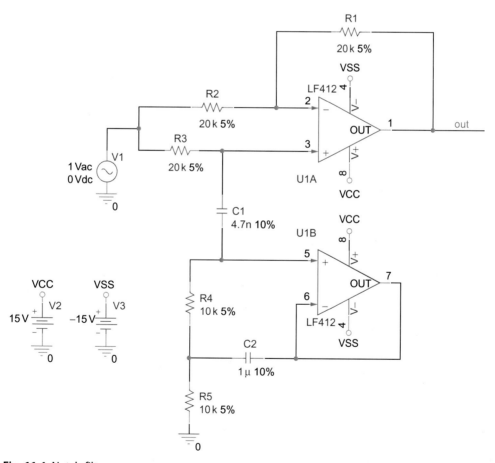

Fig. 11.4 Notch filter.

2. Select all the resistors by using holding down the control key and **right mouse button > Edit Properties**.

3. In the **Property Editor**, select and highlight the whole row (or column) for the **Tolerance** property and **rmb > Edit,** see Fig. 11.5.

Fig. 11.5 Selecting the tolerance property for all the resistors.

Add a 5% tolerance in the **Edit Property Values** window as shown in Fig. 11.6 and click on OK and then close the Property Editor.

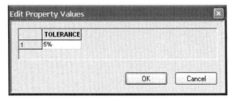

Fig. 11.6 Adding a 5% tolerance to the resistors.

4. Repeat steps 3 and 4 adding a 10% tolerance to the capacitors, C1 and C2.

5. Set up a simulation profile for an AC Sweep from 1 Hz to 10 kHz performing a logarithmic sweep with 100 points per decade, see Fig. 11.7. Click on **Apply** but do not exit.

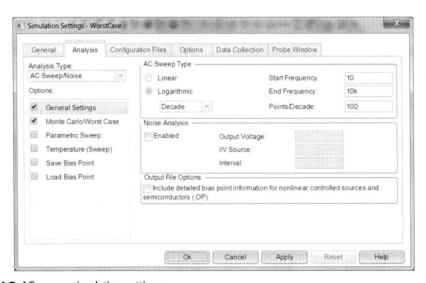

Fig. 11.7 AC sweep simulation settings.

6. In the simulation profile, in **Options** select **Monte Carlo/Worst Case** and enter **V (out)** for the **Output variable,** see Fig. 11.8. Click on **Apply** but do not exit.

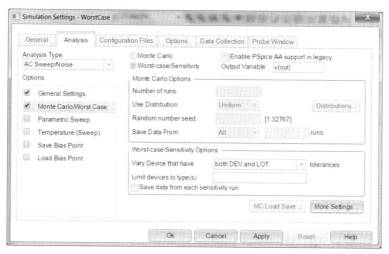

Fig. 11.8 Setting up a Worst Case analysis.

7. Click on **More Settings** and in **Find**, select the **minimum (MIN)** for the Collating Function and select **Low** for the **Worst Case Direction** as shown in Fig. 11.9. Click on OK but do not exit.

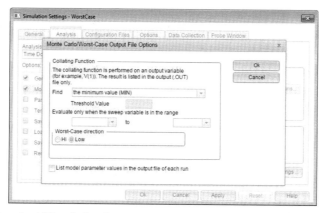

Fig. 11.9 Selecting the MIN collating function.

The Options window is different between pre and post 17.2 software versions. If you have pre 17.2 then follow steps 8.1 and 8.2. If you have post 17.2, then follow steps 9.1 and 9.2 to set the options.

8.1. Select the **Options** tab and select **Output File** and uncheck, **Bias Point Node Voltages** and **Model parameter values** as shown in Fig. 11.10. Click on OK but do not close the simulation profile window.

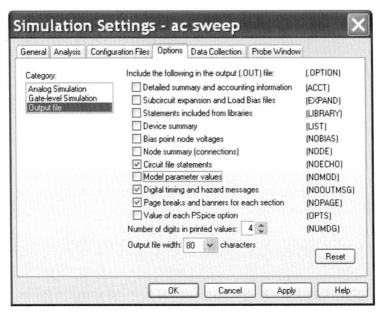

Fig. 11.10 Output File options (pre 17.2).

8.2. In the simulation profile, select **Options** and change RELTOL to 0.01, see Fig. 11.11. Click on OK to close the simulation settings profile.

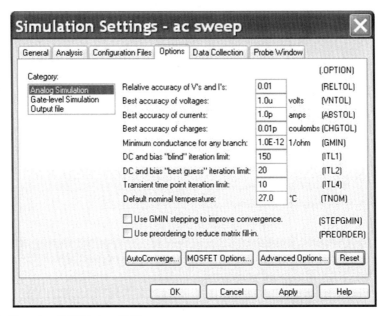

Fig. 11.11 Changing RELTOL (pre 17.2).

9.1. Select the **Options** tab and select **Output File** and uncheck, **Bias Point Node Voltages** and **Model parameter values** as shown in Fig. 11.12. Click on Apply but do not close the simulation profile window.

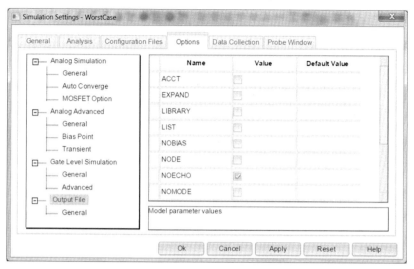

Fig. 11.12 Output File options (post 17.2).

9.2. In the simulation profile select **Options > Analog Simulation > General** and change RELTOL to 0.01, see Fig. 11.13. Click on OK to close the simulation settings profile.

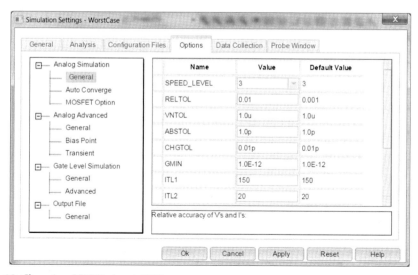

Fig. 11.13 Changing RELTOL (post 17.2).

10. Place a Vdb marker on node **out**, (**PSpice > Markers > Advanced > dB Magnitude of Voltage**).

11. Run the simulation.

12. In Probe you will see two output notch responses. The nominal response is at 234 Hz and the Worst Case response is at 199 Hz, see Fig. 11.14.

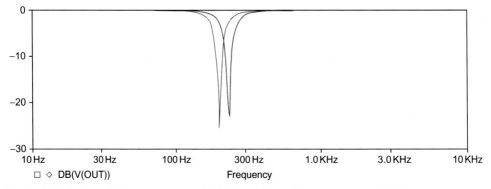

Fig. 11.14   Ideal notch filter response at 234 Hz and Worst Case response at 199 Hz.

13. Open the **Output File** and scroll down to the list of Minimum values as shown in Fig. 11.15. Here as expected R4, R5, C1, and C2 have a greater effect on the notch output voltage compared to R1, R2, and R3.

```
RUN                    MINIMUM VALUE

R_R2 R_R2 R            .0725 at F =  234.42
                     (     .2904% change per 1% change in Model Parameter)

R_R1 R_R1 R            .0725 at F =  234.42
                     (     .1786% change per 1% change in Model Parameter)

R_R3 R_R3 R            .0721 at F =  234.42
                     (    -.3169% change per 1% change in Model Parameter)

C_C2 C_C2 C            .0593 at F =  229.09
                     ( -37.543% change per 1% change in Model Parameter)

R_R4 R_R4 R            .0593 at F =  229.09
                     ( -37.601% change per 1% change in Model Parameter)

R_R5 R_R5 R            .0593 at F =  229.09
                     ( -37.616% change per 1% change in Model Parameter)

C_C1 C_C1 C            .0588 at F =  229.09
                     ( -38.143% change per 1% change in Model Parameter)
```

Fig. 11.15 Minimum value.

Scroll down to Worst Case All Devices as shown in Fig. 11.16 where you can see as the result of the sensitivity analysis, the tolerance direction in which the component values were set for the Worst Case analysis.

```
Device      MODEL      PARAMETER    NEW VALUE
C_C2        C_C2       C            1.1          (Increased)
C_C1        C_C1       C            1.1          (Increased)
R_R5        R_R5       R            1.05         (Increased)
R_R4        R_R4       R            1.05         (Increased)
R_R1        R_R1       R             .95         (Decreased)
R_R2        R_R2       R             .95         (Decreased)
R_R3        R_R3       R            1.05         (Increased)
```

Fig. 11.16 Table showing which tolerance limits were used.

14. At the bottom of the Output File is the Worst Case summary showing that the worst case scenario is for the notch filter frequency at 199 Hz which is a deviation of 7% from the nominal (Fig. 11.17).

```
RUN                        MINIMUM VALUE
WORST CASE ALL DEVICES
                          .0522 at F =  199.53
                          (   7.0339% of Nominal)
```

Fig. 11.17 Worst Case summary.

15. Change the tolerance of the capacitors to 5% and resistors R4 and R5 to 1% and run the simulation to see if this improves the predicted Worst Case response.

16. Fig. 11.18 shows an improvement in the Worst Case notch frequency that is given in the summary as 218 Hz (Fig. 11.19) being closer to the nominal frequency of 234 Hz.

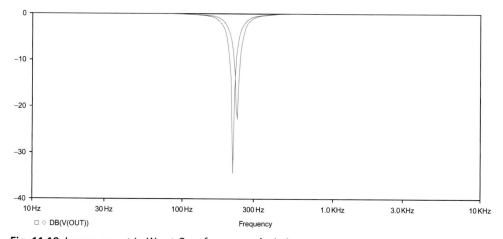

Fig. 11.18 Improvement in Worst Case frequency deviation.

```
RUN                        MINIMUM VALUE
WORST CASE ALL DEVICES
                          .019  at F =  218.78
                          (   4.8075% of Nominal)
```

Fig. 11.19 Improvement in Worst Case frequency.

CHAPTER 12

Performance Analysis

Chapter Outline

12.1. Measurement Functions   177
12.2. Measurement Definitions   177
12.3. Exercises   180
     Exercise 1   180
     Exercise 2   183

Performance analysis uses measurement definitions to scan a family of curves in Probe and to return a series of values based on the measurement definition. For example, after sweeping a voltage source connected to a CR network, a series of capacitor voltage charging curves will be obtained. Running a performance analysis using the rise time measurement definition on the resulting waveforms will generate a series of rise time values plotted against the swept source voltage.

12.1. MEASUREMENT FUNCTIONS

PSpice includes over 50 measurement definitions, some of which are listed in Table 12.1. A full list is given in the Appendix. The standard CenterFrequency and Bandwidth definitions were used in Chapter 10, Exercise 2. However, the measurement definitions also give you an option to define the range over which you want the measurement to be made. For example, the CenterFrequency_XRange definition gives you the option to measure the waveform over a specified x-range, which would be the frequency range. You can also custom design your own measurement definition.

12.2. MEASUREMENT DEFINITIONS

The measurement definitions can be viewed in PSpice from **Trace > Measurements**, which displays all the available measurements as well as the various options to create, view, edit and evaluate measurements (Fig. 12.1).

Analog Design and Simulation Using OrCAD Capture and PSpice
https://doi.org/10.1016/B978-0-08-102505-5.00012-4

Table 12.1 Some of the measurement definitions available in PSpice

Definition	Description
Bandwidth	Bandwidth of a waveform (you choose dB level)
Bandwidth_Bandpass_3dB	Bandwidth (3 dB level) of a waveform
CenterFrequency	Center frequency (dB level) of a waveform
CenterFrequency_XRange	Center frequency (dB level) of a waveform over a specified X-range
ConversionGain	Ratio of the maximum value of the first waveform to the maximum value of the second waveform
Cutoff_Highpass_3dB	High pass bandwidth (for the given dB level)
Cutoff_Lowpass_3dB	Low pass bandwidth (for the given dB level)
DutyCycle	Duty cycle of the first pulse/period
Falltime_NoOvershoot	Fall time with no overshoot
Max	Maximum value of the waveform
Min	Minimum value of the waveform
NthPeak	Value of a waveform at its nth peak
Overshoot	Overshoot of a step response curve
Peak value	Peak value of a waveform at its nth peak
PhaseMargin	PhaseMargin
Pulsewidth	Width of the first pulse
Q_Bandpass	Calculates Q (center frequency/bandwidth) of a bandpass response at the specified dB point
Risetime_NoOvershoot	Rise time of a step response curve with no overshoot
Risetime_StepResponse	Rise time of a step response curve
SettlingTime	Time from \<begin_x\> to the time it takes a step response to settle within a specified band
SlewRate_Fall	Slew rate of a negative-going step response curve

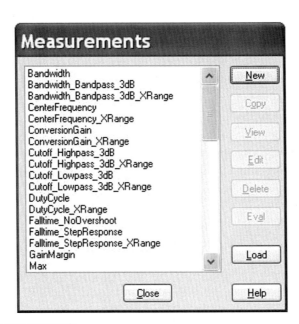

Fig. 12.1 Available measurements.

For example, Fig. 12.2 shows the measurement definition for Risetime_NoOvershoot.

Rise time, by definition, is the time difference between the value of the voltage or current at 10% and 90% of the maximum value. So two measurements are needed, one when the voltage (or current) curve is at 10% of the maximum value (x1) and the other measurement when the curve is at 90% of the maximum value (x2).

In order to find the points at 10% and 90% of the curve, search commands are used:

Search forward level(10%,p) !1

Search forward level(90%,p) !2

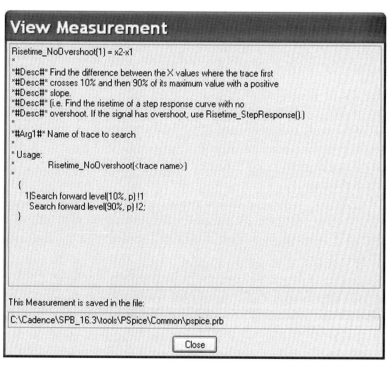

View Measurement

```
Risetime_NoOvershoot(1) = x2-x1
*
*#Desc#* Find the difference between the X values where the trace first
*#Desc#* crosses 10% and then 90% of its maximum value with a positive
*#Desc#* slope.
*#Desc#* (i.e. Find the risetime of a step response curve with no
*#Desc#* overshoot. If the signal has overshoot, use Risetime_StepResponse().)
*
*#Arg1#* Name of trace to search
*
* Usage:
*          Risetime_NoOvershoot(<trace name>)
*
  {
   1|Search forward level(10%, p) !1
     Search forward level(90%, p) !2;
  }
```

This Measurement is saved in the file:

`C:\Cadence\SPB_16.3\tools\PSpice\Common\pspice.prb`

[Close]

Fig. 12.2 Risetime_NoOvershoot measurement definition.

In this example, search in the positive going direction (p) for the 10% level and return the first data point (x1), and also search in the positive going direction (p) for the 90% level and return the second data point (x2).

The first line in Fig. 12.2 is called a marked point expression (has one value) and calculates the difference between x2 and x1 (x2 − x1). Any line starting with a # is a comment line and provides information to the user.

12.3. EXERCISES

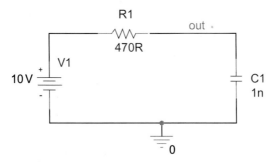

Fig. 12.3 Measuring rise time.

Exercise 1

1. Create a project called Risetime and draw the CR circuit in Fig. 12.3. Make sure you name the node **out** as shown.
2. Create a PSpice transient analysis with a run to time of 5 μs.
3. Place a voltage marker on node **out**.
4. Run the simulation.
5. What you will see is a flat line for the voltage (Fig. 12.4). This is because a DC bias point analysis has been run prior to the transient analysis such that the capacitor has reached a steady-state voltage of 10 V. So at time $t = 0$ s, the voltage across the capacitor is 10 V, as shown.

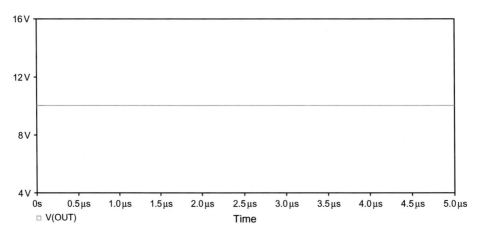

Fig. 12.4 Capacitor voltage has reached steady state.

6. You can place an initial condition, IC1 part from the special library to ensure that at time $t = 0\,$s, the voltage across the capacitor is $0\,$V, as in Chapter 7, Exercise 2. Alternatively, in this example, check the **Skip the initial transient bias point calculation** in the Simulation profile (Fig. 12.5) and run the simulation.

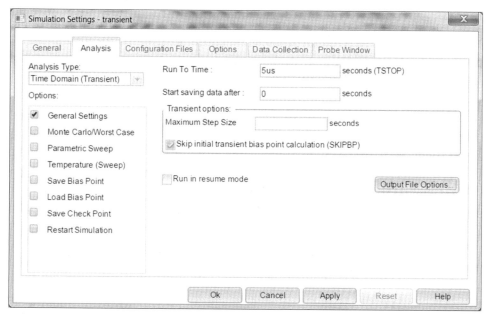

Fig. 12.5 Skipping the initial bias point calculation.

7. You should see the familiar exponential voltage rise across the capacitor (see Fig. 12.6).

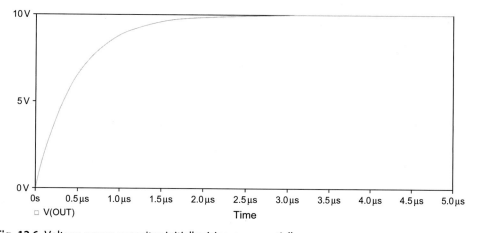

Fig. 12.6 Voltage across capacitor initially rising exponentially.

8. In PSpice, select **Trace > Evaluate Measurement**, scroll down on the right-hand side (Measurements) and select Risetime_NoOvershoot(1). The (1) means that the function is expecting one parameter.

   In the **Trace Expression** box, the cursor is automatically placed where the trace name will be entered. Select **V(out)** and the trace expression will be as shown in Fig. 12.7. Click on OK.

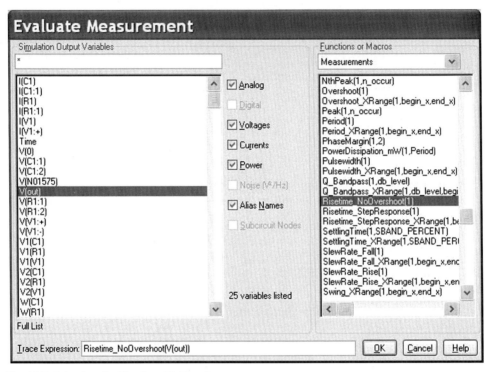

Fig. 12.7 Selecting the Risetime_NoOvershoot measurement.

9. The Trace measurement result will be displayed as seen in Fig. 12.8.

			Measurement Results	
	Evaluate	Measurement	Value	
▶	☑	Risetime_NoOvershoot(V(out))	1.03128u	
			Click here to evaluate a new measurement...	

Fig. 12.8 Trace measurement evaluation.

10. Turn on the cursors, measure the voltage at 1 V (10%) and 9 V (90%) and confirm that the rise time is correct or .

Exercise 2

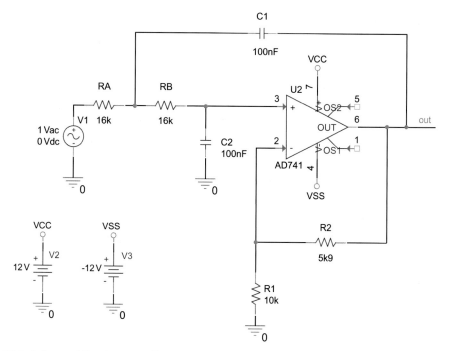

Fig. 12.9 Sallen and Key low pass filter.

You will measure the low-pass cut-off frequency for a Sallen and Key filter.
1. Draw the Sallen and Key filter in Fig. 12.9.
2. Set up an AC sweep for 1 Hz to 10 kHz, performing a logarithmic sweep with 20 points per decade.
3. Place a V_{dB} voltage marker on the output: **PSpice > Markers > Advanced > dB magnitude of Voltage**.
4. Run the simulation.
5. In PSpice, select **Trace > Evaluate Measurements > Cutoff_Lowpass_3dB()** and select V(out). The trace measurement should report a cut-off frequency of 99.6 Hz, as shown in Fig. 12.10.

	Evaluate	Measurement	Value	Measurement Results
▶	☑	Cutoff_Lowpass_3dB(V(out))	99.62219	
				Click here to evaluate a new measurement...

Fig. 12.10 Cut off frequency measured as 99.6 Hz.

6. The frequency response is shown in Fig. 12.11.

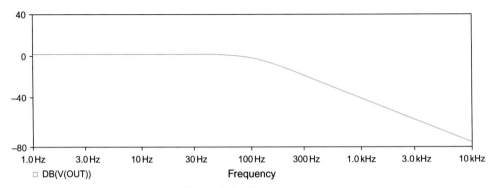

Fig. 12.11 Frequency response of Sallen and Key filter.

CHAPTER 13

Analog Behaviorial Models

Chapter Outline

13.1. ABM Devices 185
13.2. Exercises 192
    Exercise 1 192
    Exercise 2 193
    Exercise 3 193

Analog behaviorial models (ABM) devices are extended versions of the traditional Spice Voltage Controlled sources, the E device which is a voltage controlled voltage source (VCVS) and the G device, a Voltage Controlled Current Source (VCCS). They provide transfer functions, mathematical expressions, or look up tables to describe the behavior of an electronic device or circuit. ABM's can provide a systems approach to designing electronic circuits. The electronic system is represented by a block diagram with each block represented by an ABM device which will reduce the total simulation time. If the system meets the required specifications, then each block can be successively replaced by its final electronic circuit. Alternatively, working electronic circuits can be replaced by an equivalent ABM block.

There are two types of ABM devices, PSpice equivalent parts which have a differential input and double-ended output and control system parts which have a single input and output pin. The standard E, F, G, and H devices can be found in the analog library, whereas the ABM devices can be found in the ABM library.

13.1. ABM DEVICES

The extended sources provide five additional functions which are defined as:

Value	Mathematical expression
Table	Look up table
Freq	Frequency response
Chebyshev	Filter characteristics
Laplace	Laplace transform

Analog Design and Simulation Using OrCAD Capture and PSpice
https://doi.org/10.1016/B978-0-08-102505-5.00013-6
185

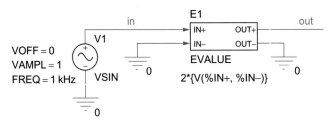

Fig. 13.1 Voltage amplitude doubler.

Fig. 13.1 shows a typical use of an ABM EValue device to implement a voltage doubler. The ABM has a differential input (IN+, IN−) and a double-ended output (OUT+, OUT−). When you first place the ABM, the default expression is given by
    V(%IN+, %IN−)
which calculates the difference between the voltages on the input pins IN+ and IN−. In order to multiply by a factor of 2, the expression is first enclosed in curly brackets (braces).
    2*{V(%IN+, %IN−)}
Fig. 13.2 shows the output voltage when a 1 V sinewave is applied to the input pins.

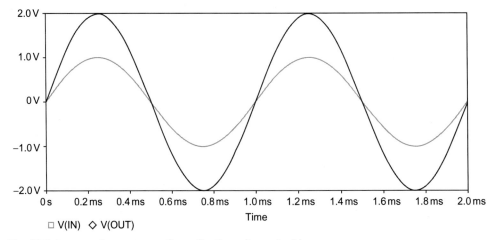

Fig. 13.2 Input and output waveforms for the voltage doubler.

Other mathematical functions that can be applied to ABM expressions are shown in Table 13.1.

Conditional statements can also be applied to ABM parts. For example, in Fig. 13.3, **if** the input voltage is greater than 4 V, **then** output 0 V **else** output 5 V. This effectively is a comparator. The resulting waveform is shown in Fig. 13.4.

Table 13.1 Mathematical functions

Function	Expression
ABS	Absolute value
SQRT	Square root $\sqrt{x}$
PWR	$|x|^{exp}$
PWRS	x^{exp}
LOG	Log base e $\ln(x)$
LOG10	Log base 10 $\log_{10}(x)$
EXP	e^x
SIN	sin
COS	cos
TAN	tan
ATAN	$\tan^{-1}$
ARCTAN	$\tan^{-1}$

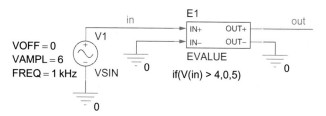

Fig. 13.3 ABM comparator.

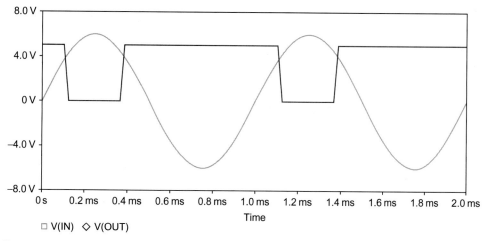

□ V(IN)   ◇ V(OUT)

Fig. 13.4 ABM comparator results.

The first-order 159 Hz low pass filter in Fig. 13.5 can be defined by the transfer function:

$$\frac{V_{out}}{V_{in}} = \frac{1}{1 + j\dfrac{\omega}{\omega_c}} \tag{13.1}$$

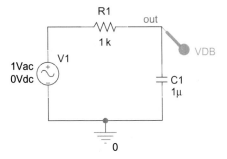

Fig. 13.5 Low pass filter.

where the cut off frequency is given by

$$\omega_c = \frac{1}{CR} = 2\pi f \tag{13.2}$$

$$f = \frac{1}{2\pi CR} = \frac{1}{2\pi \times 10^{-6} \times 10^3} = 159\,\text{Hz}$$

Fig. 13.6 shows the frequency response of the filter.

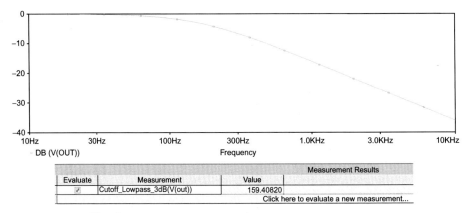

Evaluate	Measurement	Value
		Measurement Results
✓	Cutoff_Lowpass_3dB(V(out))	159.40820
		Click here to evaluate a new measurement...

Fig. 13.6 Low pass filter frequency response.

Laplace ABM's can be used to implement filter circuits where the transfer function is transformed to the s domain where $s = j\omega$ and $\omega = 2\pi f$ so the transfer function for the low pass filter is defined in the s domain by

$$\frac{V_{out}}{V_{in}} = \frac{1}{1 + s\tau} \qquad (13.3)$$

where

$$\tau = CR = 10^{-6} \times 10^{3}$$

$$\tau = 10^{-3} \text{ or } 0.001s$$

The filter circuit is redrawn in Fig. 13.7 with the transfer function given as:

$$\frac{V_{out}}{V_{in}} = \frac{1}{1 + 0.001 * s} \qquad (13.4)$$

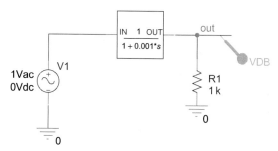

Fig. 13.7 Laplace low pass filter.

Note

When you enter the coefficient for s, make sure you enter the multiplier, i.e., $1 + 0.001 * s$ and not $1 + 0.001s$.

Fig. 13.8 shows the low pass frequency response for the filter.

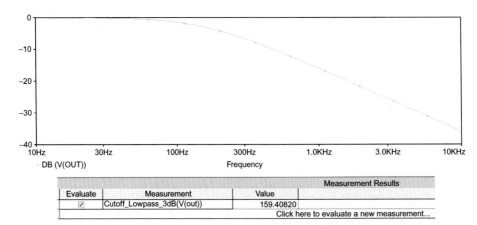

	Measurement Results		
Evaluate	Measurement	Value	
✓	Cutoff_Lowpass_3dB(V(out))	159.40820	
		Click here to evaluate a new measurement...	

Fig. 13.8 Low frequency response for the Laplace filter.

In the circuit of Fig. 13.9, both inputs are grounded but the ABM expression is referencing a net name called "source." This is useful to reduce the wiring complexity of a circuit especially if there are multiple ABM's being driven by a single source. Note that since a GValue ABM part is being used, the output is a current and therefore cannot be left unconnected. Hence the output resistor provides a DC path to 0 V.

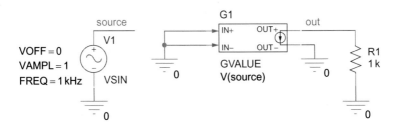

Fig. 13.9 Referencing the **source** node in the circuit.

In version 17.2, two new behavioral delay functions have been introduced, DelayT () and DelayT1(). Both these functions have improved simulation times and more efficient handling of convergence issues compared to the standard delay functions such as TLINE. DelayT() requires two parameters, delay time and max delay, whereas

DelayT1() only requires the delay value and also inverts the input signal. The delay functions are shown in Fig. 13.10 and can be found in the function.olb library in the advanced analysis folder.

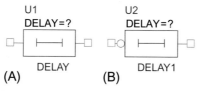

Fig. 13.10 Delay parts: (A) DelayT() and (B) DelayT1().

In later software versions you can search for the delay functions using:
Place > PSpice Component > Search
Then under **Categories,** select **Analog Behavioral Models > General Purpose** as shown in Fig. 13.11.

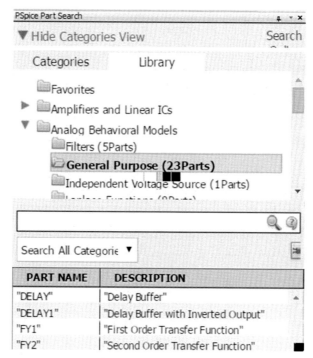

Fig. 13.11 PSpice component search for Delay functions.

13.2. EXERCISES

Exercise 1

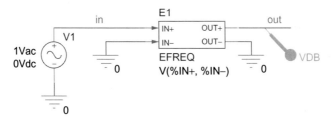

Fig. 13.12 Using a frequency table.

In Fig. 13.12, an EFreq ABM part is being used to model a bandpass filter using a table of parameters defining frequency, magnitude, and phase.

1. Draw the circuit in Fig. 3.12 using an EFreq part from the ABM library. The VAC source (V1) is from the source library.
2. Double click on EFreq to open the Property Editor and scroll to the Table property and enter:
   $(0.1, -40, 170)(1\,\text{k}, -40, 160)(2\,\text{k}, -20, 140)(3\,\text{k}, -0, 100)(6\,\text{k}, -0, -100)$
   $(10\,\text{k}, -20, -140)(20\,\text{k}, -40, -160)(30\,\text{k}, -40, -170)$.
3. Set up a simulation profile for an AC Sweep from 1 to 100 kHz. Place a VdB voltage marker. PSpice > Markers > Advanced > dB Magnitude of Voltage.
4. Run the simulation. You should see the bandpass response as shown in Fig. 13.13.

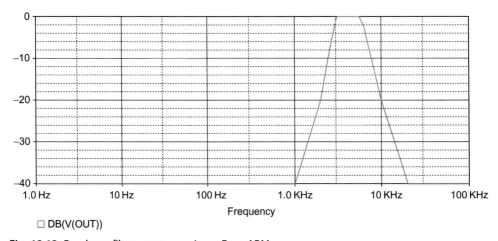

□ DB(V(OUT))

Fig. 13.13 Bandpass filter response using a Freq ABM.

Exercise 2

Use a GValue ABM to model to reference a net name and output a current.

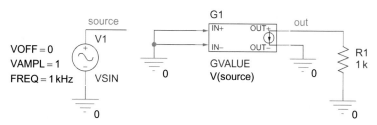

Fig. 13.14 Behavioral implementation of a rectifier.

1. Draw the ABM rectifier as shown in Fig. 13.14. The net name **source** is referenced inside the ABM expression brackets.
2. Set up a simulation profile for a transient sweep with a run to time of 4 ms.
3. Run the simulation.
4. Investigate the use of the EFREQ ABM device in modeling the 1500 Hz bandpass filter in Chapter 10.

Exercise 3

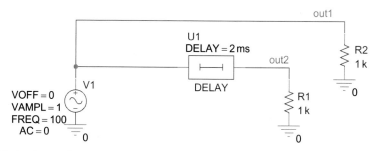

Fig. 13.15 Delay circuit diagram.

1. Draw the circuit in Fig. 13.15 using the **Delay** part. Either, **Place > PSpice Component > Search > Analog Behavioral Models > General Purpose (23Parts)** and select **Delay** or alternatively, use **Place > Part** and **Add Library > advanls > function** and select the Delay part.
2. Click on **DELAY =?** and add a delay value of 2 ms.

3. Place a 100 Hz, 1 V sinusoidal source from the source library. For later software versions, press shift-R to open the Independent Sources window as shown in Fig. 13.16.

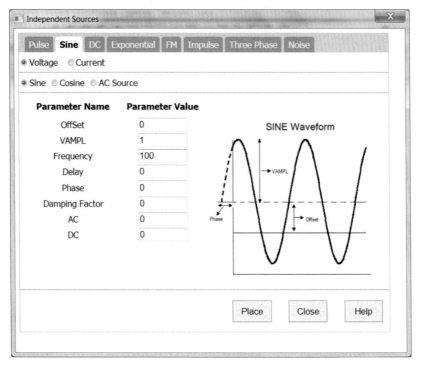

Fig. 13.16 Placing a sinusoidal source.

4. Name the output signals as out1 and out2 and add voltage probes as shown and set up a transient analysis for a Run-To Time of 20 ms.

5. Run the simulation. Note that **out2** has been delayed by 2 ms, see Fig. 13.17.

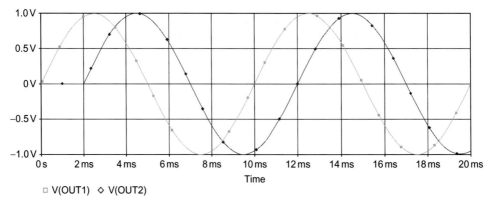

Fig. 13.17 Delayed output of Out2 by 2 ms.

6. Replace the Delay part with a DelayT1 part and add a delay of 2 ms.
7. Run the simulation. Note that **out2** has been inverted and delayed by 2 ms, see Fig. 13.18.

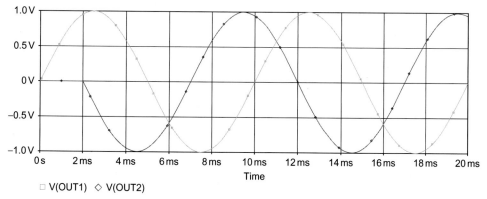

□ V(OUT1)   ◇ V(OUT2)

Fig. 13.18 Inverted and delayed output of Out2 by 2 ms.

CHAPTER 14

Noise Analysis

Chapter Outline

14.1. Noise Types   197
     14.1.1 Resistor Noise   197
     14.1.2 Semiconductor Noise   198
14.2. Total Noise Contributions   198
14.3. Running a Noise Analysis   199
14.4. Noise Definitions   200
14.5. Exercise   203

Noise analysis is run in conjunction with an AC analysis and calculates the output noise and equivalent input noise in a circuit. The output noise, at a specified output node, is the root mean square (RMS) sum of the noise generated by all the resistors and semiconductors in the circuit. If the circuit is considered as noiseless, then the equivalent input noise is defined as the noise required at the input in order to generate the same output noise. This is the same as dividing the output noise by the gain of the circuit in order to obtain the equivalent input noise.

14.1. NOISE TYPES

14.1.1 Resistor Noise

Johnson noise or thermal noise is due to the random thermal agitation of electrons in a conductor, which increases with frequency and temperature. In PSpice, the thermal noise contribution from a resistor is represented by a current source in parallel with a noiseless resistor. Because of its random nature, the noise current source is represented as a mean square value given by:

$$\overline{i^2} = \frac{4kT\Delta f}{R} \left(A^2/\mathrm{Hz}\right) \tag{14.1}$$

where:

   k—Boltzmann's constant: $1.38\mathrm{e}^{-23}$ $(\mathrm{J\ K^{-1}})$
   T—absolute temperature (K)
   R—resistance (Ω)
   Δf—frequency bandwidth (Hz)

Analog Design and Simulation Using OrCAD Capture and PSpice
https://doi.org/10.1016/B978-0-08-102505-5.00014-8

14.1.2 Semiconductor Noise

Semiconductor noise is generally made up of thermal, shot and flicker noise. Thermal noise is generated by the intrinsic parasitic resistances for the device. Shot noise, however, is a randomly fluctuating noise current generated when a current flows across a PN junction and is given by:

$$\overline{i^2} = 2qI \left(\text{A}^2/\text{Hz} \right) \tag{14.2}$$

where:

q—electron charge, 1.602×10^{-19} C

I—current through the device (A)

Flicker noise is a phenomenon not widely understood but has been attributed to imperfections in semiconductor channels and the generation and recombination of charge carriers. However, what is known is that flicker noise occurs at low frequencies and that the noise current decreases with frequency exhibiting a $1/f$ characteristic. Flicker noise current is given by:

$$\overline{i^2} = \frac{KF \times Id^{AF}}{\Delta f} \left(\text{A}^2/\text{Hz} \right) \tag{14.3}$$

where:

KF—flicker noise coefficient

Id—current through the device

AF—flicker noise exponent

Δf—frequency bandwidth

14.2. TOTAL NOISE CONTRIBUTIONS

After a noise analysis is run, the thermal, shot and flicker noise contributions for resistors and semiconductors are made available as trace variables in Probe. Table 14.1 shows the available noise variables for some of the devices.

Table 14.1 Noise output variables available in probe

Device	Output variable	Noise
Resistor	NTOT	Thermal noise
Diode	NRS	Parasitic thermal noise for RS
	NSID	Shot noise
	NFID	Flicker noise
	NTOT	Total of noise contributions

Table 14.1 Noise output variables available in probe—cont'd

Device	Output variable	Noise
Bipolar transistor	NRB	Parasitic thermal noise for RB
	NRC	Parasitic thermal noise for RC
	NRE	Parasitic thermal noise for RE
	NSIB	Shot noise for base current
	NSIC	Shot noise for collector current
	NFIB	Flicker noise
	NTOT	Total of all noise contributions
MOSFET	NRD	Parasitic thermal noise for RD
	NRG	Parasitic thermal noise for RG
	NRS	Parasitic thermal noise for RS
	NRB	Parasitic thermal noise for RB
	NSID	Shot noise
	NFID	Flicker noise
	NTOT	Total of all noise contributions

The noise spectral density for NTOT(device) is measured in units of V^2/Hz.

The total circuit noise is represented as either NTOT(ONOISE) in units of V^2/Hz or the RMS summed output, V(ONOISE) in units of $V/\sqrt{Hz}$.

The equivalent input noise is V(INOISE) and is calculated from $\left(\frac{V(ONOISE)}{gain\ of\ circuit}\right)$ in units of $V/\sqrt{Hz}$ or $A/\sqrt{Hz}$. If the input source is a current source, the units are $A/\sqrt{Hz}$. If the input source is a voltage source, the units are $V/\sqrt{Hz}$.

14.3. RUNNING A NOISE ANALYSIS

You have to run an AC analysis in order to run a noise analysis. In the simulation settings for an AC sweep, there is an option to enable the Noise Analysis as shown in Fig. 14.1. In this example, the output voltage node V(out) has been specified as the node at which the total output noise will be calculated. The **I/V Source** references the current or voltage source used as an input to the circuit, which is usually a V_{AC} or I_{AC} source. In this example, the reference designator is V1.

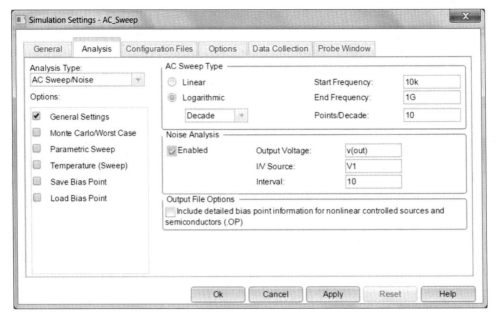

Fig. 14.1 AC sweep and noise analysis settings.

The **Interval** is an integer that specifies how often results are written to a table which is generated in the output file. Each table entry will be determined by the frequency **Interval**, which specifies the nth interval of the range set in the AC sweep. For example, in Fig. 14.1, the AC sweep is from 10 kHz to 1 GHz with 10 points per decade; the decade frequencies will be 10 kHz, 100 kHz, 1 MHz, 10 MHz, 100 MHz and 1 GHz. So the 10th frequency interval will be 100 kHz, the 20th interval 1 MHz, and so on. If the number of points/decade was set at 5, with the same interval integer of 10, then the frequencies written to the output file would be 10 kHz, 1 MHz and 100 MHz. If no interval number is entered, then no table will be generated in the output file. This is not to be confused with the frequency interval of the waveform data points in Probe, which is determined by the AC frequency sweep settings.

14.4. NOISE DEFINITIONS

The instantaneous value of noise voltage at any time t, is given by $v_n(t)$. As the noise voltage is statistical in nature, the root mean square (rms) value is given by:

$$E_n = \sqrt{\overline{v_n(t)^2}}\,(\mathrm{V}) \tag{14.4}$$

Similarly for noise current, the rms value is given by:

$$I_n = \sqrt{\overline{i_n(t)^2}}\,(\mathrm{A}) \tag{14.5}$$

The rms noise voltage across a resistor is given by:

$$E_n = \sqrt{4kTR\Delta f}\,(\text{V}) \tag{14.6}$$

and the rms noise current for a resistor is given by:

$$I_n = \frac{\sqrt{4kTR\Delta f}}{R}\,(\text{A}) \tag{14.7}$$

where:
k—Boltzman's constant 1.38×10^{-23} $(\text{J}\,\text{K}^{-1})$
T—absolute temperature (K)
R—resistance (Ω)
Δf—frequency bandwidth (Hz)

In PSpice, noise calculations are made assuming a unity gain bandwidth, i.e. $\Delta f = 1\,\text{Hz}$.

The rms noise power for a resistor is given by:

$$P_n = \frac{E_n^2}{R} = I_n^2 R\,(\text{W}) \tag{14.8}$$

The noise power spectral density, S, is given by:

$$S = \frac{P_n}{\Delta f}\,(\text{W/Hz}) \tag{14.9}$$

Substituting for P_n from Eq. (14.8) into Eq. (14.9):

$$S = \frac{\left(\dfrac{E_n^2}{R}\right)}{\Delta f} = \frac{\left(\dfrac{4kTR}{R}\right)}{\Delta f} \tag{14.10}$$

$$S = 4kT\,(\text{W/Hz})$$

Noise voltage spectral density, e_n, is given by:

$$e_n = \frac{E_n}{\Delta f} = \frac{\sqrt{4kTR\Delta f}}{\Delta f}$$

$$e_n = \frac{\sqrt{4kTR}\,\sqrt{\Delta f}}{\Delta f}$$

$$e_n = \frac{\sqrt{4kTR}\,\sqrt{\Delta f}}{\Delta f}\frac{\sqrt{\Delta f}}{\sqrt{\Delta f}} \tag{14.11}$$

$$e_n = \frac{\sqrt{4kTR}}{\sqrt{\Delta f}}\,\left(\text{V}/\sqrt{\text{Hz}}\right)$$

Similarly the noise current spectral density, i_n, is given by:

$$i_n = \frac{\sqrt{4kT/R}}{\sqrt{\Delta f}} \left(A/\sqrt{Hz}\right) \tag{14.12}$$

If the rms quantities, E_n and I_n, are uncorrelated (independent of each other), then noise sources can be added together such that:

$$E_n^2 = E_{n1}^2 + E_{n2}^2$$

or $\tag{14.13}$

$$E_n = \sqrt{E_{n1}^2 + E_{n2}^2}$$

In PSpice, the noise contributions for a bipolar transistor are represented by the thermal noise sources for the intrinsic base, emitter and collector resistances and the shot and flicker noise contributions for the base and collector currents. Each noise source is represented by the following spectral power densities assuming a unit frequency bandwidth ($\Delta f = 1\,Hz$).

Collector parasitic resistance thermal noise:

$$Ic^2 = \frac{4kT}{\left(\dfrac{RC}{AREA}\right)} \left(A^2/Hz\right) \tag{14.14}$$

Base parasitic resistance thermal noise:

$$Ib^2 = \frac{4kT}{RB} \left(A^2/Hz\right) \tag{14.15}$$

Emitter parasitic resistance thermal noise:

$$Ie^2 = \frac{4kT}{\left(\dfrac{RE}{AREA}\right)} \left(A^2/Hz\right) \tag{14.16}$$

Base shot and flicker noise currents:

$$Ib = 2qIb + \frac{KF \times Ib^{AF}}{\Delta f} \left(A/Hz\right) \tag{14.17}$$

Collector shot noise current:

$$Ic = 2qIc \left(A/Hz\right) \tag{14.18}$$

where:
    AREA—area scaling factor. By default this is set to 1
    AF—flicker noise exponent
    KF—flicker noise coefficient

14.5. EXERCISE

A simple transistor circuit is used to illustrate the contribution of component noise in a circuit. A large value resistor is used for RB in order to compare its noise contribution to the noise contribution from the transistor.

1. Draw the circuit in Fig. 14.2. The Q2N304 transistor is from the bipolar library and the V_{AC} source V1 is from the source library.

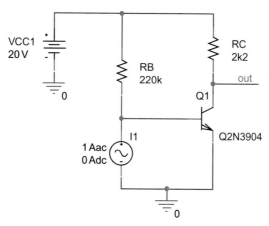

Fig. 14.2 A simple transistor amplifier.

2. Create a PSpice simulation profile and set up an AC sweep from 10 kHz to 1 GHz using a logarithmic sweep using 10 points/decade. Enable **Noise Analysis**, defining **V(out)** as the output node and I1 as the source (Fig. 14.3).

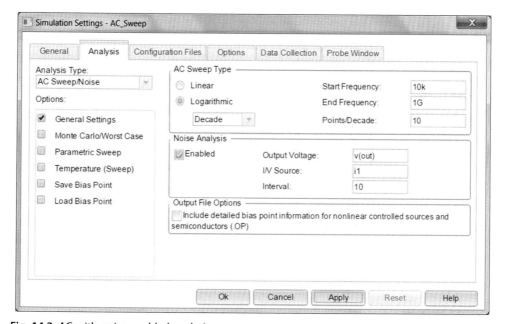

Fig. 14.3 AC with noise-enabled analysis.

3. Run the simulation.
4. In PSpice, add the trace (**Trace > Add**) for the noise contribution from the collector resistor, NTOT(RC) (Fig. 14.4). Units are V^2/Hz.

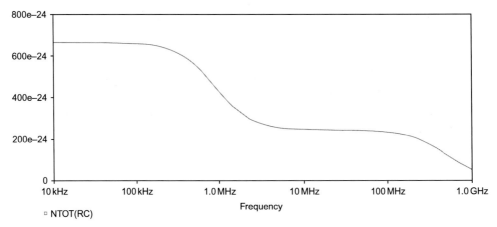

FIG. 14.4 Noise spectral density for the collector resistor.

5. Add the trace for the noise contribution from the base resistor, NTOT(RB). Fig. 14.5 shows that the larger base resistor contributes a larger noise contribution compared to the collector resistor.

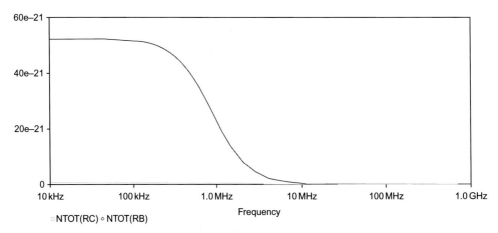

Fig. 14.5 A greater noise contribution is from the base resistor.

6. Delete the resistor traces and add the noise traces for Q1: NFIB(Q1), NRB(Q1), NRC(Q1), NRE(Q1), NSIB(Q1), NSIC(Q1) (Fig. 14.6). You should see that the greatest noise contribution is from the shot noise associated with the base current, NSIB(Q1), which exhibits a low pass frequency response.

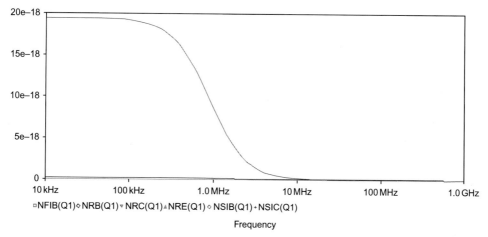

Fig. 14.6 Transistor noise traces.

Note

To delete a trace, select the trace name, which turns red, and press the delete key.

---

7. Delete the collector current shot noise trace NSIC(Q1) and you will see the noise contributions from the intrinsic transistor resistances.
8. Delete all the traces, **Trace > Delete All Traces**, and add the noise trace for the base resistor NTOT(RB) and the total noise trace for Q1, NTOT(Q1). As you can see in Fig. 14.7, the transistor contributes the greatest noise in the circuit.

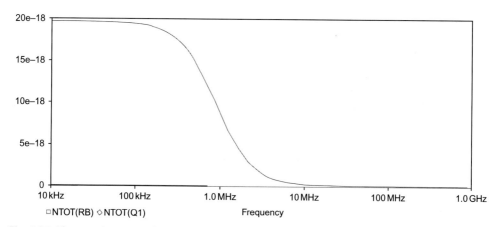

Fig. 14.7 The transistor contributes the greatest noise in the circuit.

9. Delete all the traces.

10. Add the NTOT(ONOISE) trace, which is the total output noise for the circuit in units of V^2/Hz as shown in Fig. 14.8.

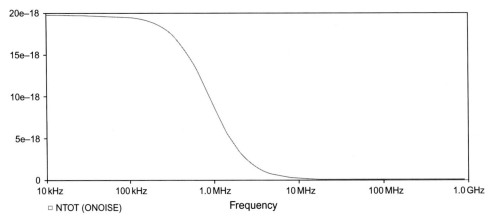

Fig. 14.8 Total output noise in the circuit.

11. Delete NTOT(ONOISE) and add the trace for the equivalent input noise I(NOISE). You should see the noise increase with an increase in frequency due to the current gain of the transistor being proportional to frequency (Fig. 14.9). As the input source is a current source, the units are $A/\sqrt{Hz}$.

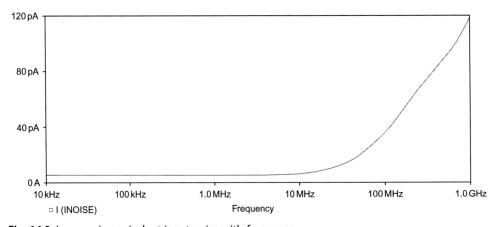

Fig. 14.9 Increase in equivalent input noise with frequency.

12. Delete I(NOISE) and add the trace for V(ONOISE) (Fig. 14.10). The units are $V/\sqrt{Hz}$.

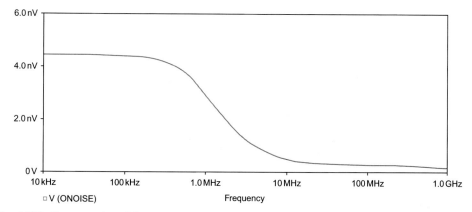

Fig. 14.10 Output noise of the circuit.

13. A table of results will be generated for the noise analysis and can be seen in the **Output File** for decade frequencies of 10 kHz, 100 kHz, 1 MHz, 10 MHz, 100 MHz and 1 GHz. Select **View > Output File** and scroll down to the start of the noise analysis, which is 10 kHz (Fig. 14.11). The first section of the Output File states the frequency used for the noise calculation. The calculated noise contributions from the transistor and the two resistors are shown in units of V^2/Hz. Next, the total circuit noise was calculated and shown in units of V^2/Hz for NTOT(ONOISE) and $V/\sqrt{Hz}$ for V(ONOISE).

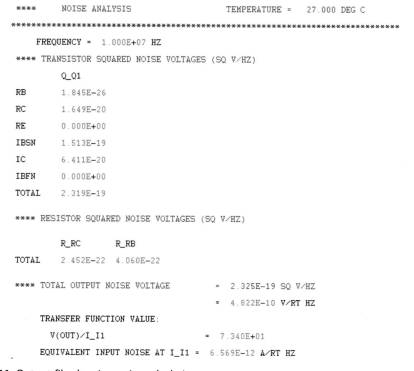

Fig. 14.11 Output file showing noise calculations.

The transfer function was then calculated giving, in the above case, a circuit gain of 832.9 at 10 kHz, from which the equivalent input noise I(NOISE) was calculated and shown in units of $A/\sqrt{Hz}$, as the input source was a current source.

CHAPTER 15

Temperature Analysis

15.1. Temperature Coefficients   209
15.2. Running a Temperature Analysis   210
15.3. Exercises   212
    Exercise 1   212
    Exercise 2   214

A change in temperature can affect the performance and characteristics of a circuit. The components most affected by a change in temperature include semiconductors, resistors, capacitors and inductors. All of these components have an inbuilt temperature dependence model parameter such that performing a temperature sweep will change component and subsequent circuit behavior.

15.1. TEMPERATURE COEFFICIENTS

For a resistor, the change in its nominal value due to a change in temperature is defined as:

$$R = R(\text{nom})*\left(1 + TC1*(T - T_{\text{nom}}) + TC2*(T - T_{\text{nom}})^2\right) \qquad (15.1)$$

where

TC1—linear temperature coefficient (ppm/°C)

TC2—quadratic temperature coefficient (ppm/°C$-^2$)

T—simulation temperature (°C)

T_{nom}—nominal temperature (°C), set by default to 27°C

There is a TCE—exponential coefficient which if specified gives the resistor value as:

$$R = R(\text{nom})*1.01^{TCE*(T-T_{\text{nom}})} \qquad (15.2)$$

Manufacturers normally specify the linear coefficient TC1 for resistors.

The temperature coefficients specified for resistors are given in parts per million per degree Celsius (ppm/°C). So for a 10 kΩ resistor with a linear temperature coefficient of +200 ppm/°C, TC1 = 0.0002 and with no TC2 specified, a 20°C rise in temperature will give:

$$R = 10,000 \times \left(1 + (0.0002*20)\right)$$

Therefore, for a 20°C rise in temperature, R = 10,040 Ω.

Analog Design and Simulation Using OrCAD Capture and PSpice
https://doi.org/10.1016/B978-0-08-102505-5.00015-X

Similarly, for inductors and capacitors the component values are given by:

$$L = L(\text{nom})*\left(1 + \text{TC1}*(T - T_{\text{nom}}) + \text{TC2}*(T - T_{\text{nom}})^2\right)$$

$$C = C(\text{nom})*\left(1 + \text{TC1}*(T - T_{\text{nom}}) + \text{TC2}*(T - T_{\text{nom}})^2\right)$$

There is no TCE exponential coefficient for the inductors and capacitors.

In previous versions of OrCAD, the temperature coefficients were not readily available on the Capture parts therefore to add TC1 and TC2. Breakout parts as used for Monte Carlo analysis are used where the temperature coefficients are added to the PSpice model definition.

For example, to add a linear temperature coefficient, TC1 = 0.02 (ppm/°C), place an Rbreak resistor from the Breakout library, **rmb > Edit PSpice Model** and edit the PSpice model from:

.model Rbreak RES R = 1

to

.model Rtemp RES R = 1 TC1 = 0.02

15.2. RUNNING A TEMPERATURE ANALYSIS

An AC, DC or transient analysis is normally run using the default nominal temperature (TNOM) of 27°C, which is set in the simulation profile under the **Options** tab (Fig. 15.1). TNOM is the default nominal temperature and is also the temperature at which model parameters were measured.

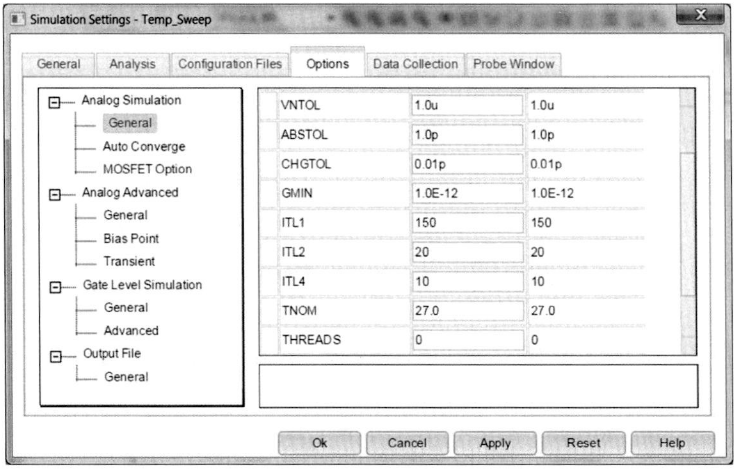

Fig. 15.1 Default simulation options.

If you want to run a transient analysis at a different temperature then you need to specify the simulation temperature by selecting **Temperature (Sweep)** in the simulation profile and then enter either a single simulation temperature or a list of temperatures values as shown in Fig. 15.2.

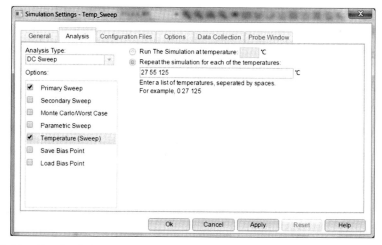

Fig. 15.2 Setting the simulation temperature.

In the example above, three transient analyses will be run at the specified temperatures of 27°C, 55°C, and 125°C and plotted, one for each temperature, on one graph in Probe, the PSpice waveform viewer.

If you want to use temperature as a swept variable, i.e. for temperature to appear on the x-axis, you run a DC sweep and then select temperature as the sweep variable, as shown in Fig. 15.3. The temperature will be swept from 0°C to 50°C in 1°C increments. The change in temperature is represented by $T - T_{\text{nom}}$.

Fig. 15.3 Performing a temperature sweep.

15.3. EXERCISES

Exercise 1

The specification for a $10\,k\Omega$ resistor is given as $200\,ppm/°C$. Therefore, $TC1 = 0.0002$.

1. Draw the circuit shown in Fig. 15.4 and double click on the resistor to open the Property Editor. Enter a linear coefficient value of 0.0002 in TC1 and **Display** both the property **Name and Value** of TC1 as shown in Fig. 15.5. Make sure you name the node as **VR, (Place > Net Alias)**.

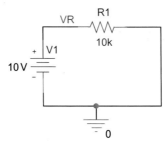

Fig. 15.4 Adding resistor temperature coefficients.

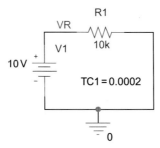

Fig. 15.5 Displaying TC1.

2. Set up a DC linear temperature sweep from $0°C$ to $+50°C$ in steps of $1°C$ as shown in Fig. 15.6.
3. Run the simulation.
4. In order to plot resistance, you need to plot the ratio of the voltage across R1 divided by the current flowing through R1. In Probe, select **Trace > Add** or .
5. At the bottom of the **Add Traces** window, there is an expression field in which you will define the resistance ratio. Select the V(VR) trace from the list of **Simulation Output Variables**; then, in the right-hand side **Functions or Macros**, select the divider symbol "/" and then select I(R1) from the **Simulation Output Variables**. You should see the expression as in Fig. 15.7.

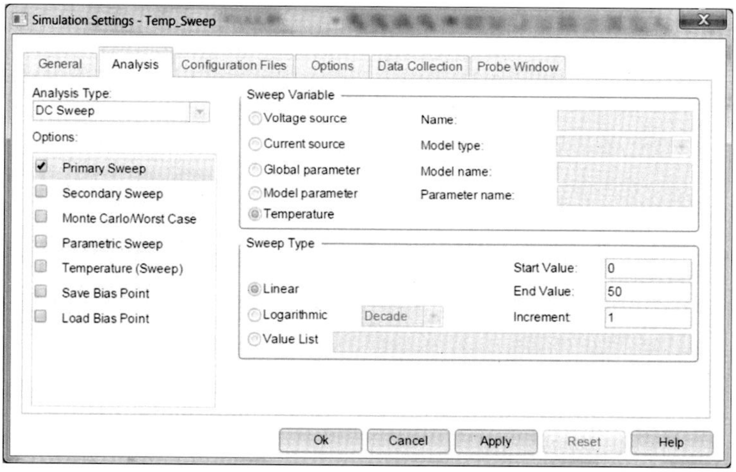

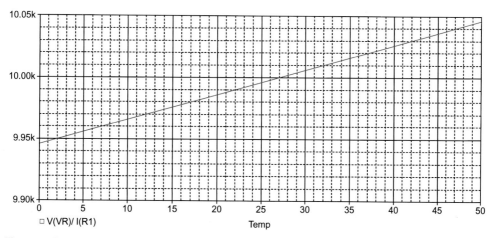

Fig. 15.6 DC temperature sweep.

Fig. 15.7 Trace expression for resistance of R1.

Fig. 15.8 Resistance change with a change in temperature.

Note that you can also type in the expression rather than select variables or functions. Probe will then display the expected increase in resistance with temperature (Fig. 15.8).

Turn the cursor on and verify that the value of the resistor is 10 kΩ at 27°C and 10,040 Ω at 47°C, which represents a rise in temperature of 20°C.

6. Reset TC1 to 0 and add and display the quadratic temperature coefficient, TC2 with a value of 0.001 (Fig. 15.9) and resimulate; but this time, rather than add in the trace expression, you can call up and restore the last display automatically.

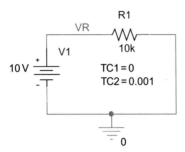

Fig. 15.9 Adding a quadratic temperature coefficient.

7. In Probe, select **Window > Display Control > Last Session > Restore**. You should see the quadratic response as shown in Fig. 15.10.

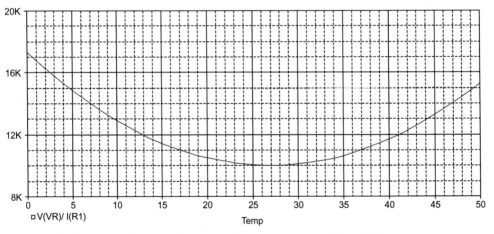

Fig. 15.10 Variation of resistance using quadratic temperature coefficient TC2.

Exercise 2

1. Draw the circuit shown in Fig. 15.11. The D1N914 diode can be found in either the diode or eval library.
2. Set up a nested DC sweep. For the primary sweep, sweep V1 from 0 to +10 V in steps of 0.01 V (Fig. 15.12). For the secondary sweep, sweep the temperature from −55°C to +75°C in steps of 10°C (Fig. 15.13).

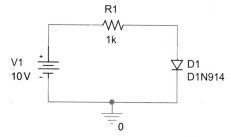

Fig. 15.11 Using a nested sweep to determine temperature effects on the diode.

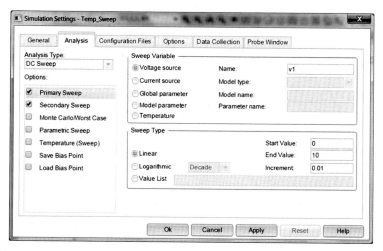

Fig. 15.12 Primary sweep.

Fig. 15.13 Secondary sweep.

3. We need to display the current through the diode versus the voltage across the diode. Select **Plot > Axis Settings > X Axis**. Select **Axis Variable…** and select V1(D1) (Fig. 15.14). Click on OK.

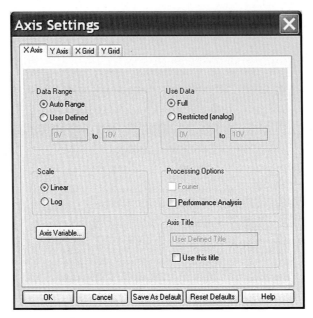

Fig. 15.14 Changing axis variable.

4. Add the trace for the current through the diode. **Trace > Add > I(D1)**. Fig. 15.15 shows the family of temperature current-voltage curves for D1.

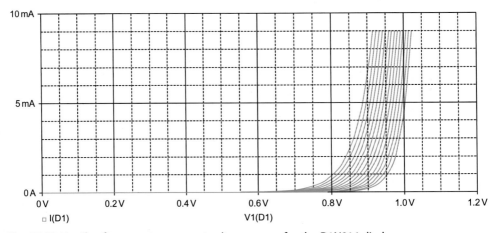

Fig. 15.15 Family of temperature current-voltage curves for the D1N914 diode.

CHAPTER 16

Adding and Creating PSpice Models

16.1. Capture Properties for a PSpice Part   217
16.2. PSpice Model Definition   219
16.3. Subcircuits   221
16.4. Model Editor   223
     16.4.1 Copying an Existing PSpice Model   224
     16.4.2 Model Import Wizard   224
     16.4.3 Downloading Models From a Vendor Website   228
     16.4.4 Encryption   229
     16.4.5 IBIS Translator   231
16.5. Exercises   231
     Exercise 1   231
     Exercise 2   234
     Exercise 3   238
     Exercise 4   239
     Exercise 5   240

PSpice models can be created and edited in the PSpice Model Editor, which can be started in standalone mode from the Start menu, PSpice > Simulation Accessories > Model Editor, or by highlighting a PSpice part in the schematic in Capture, rmb > Edit PSpice Model. When you edit a PSpice part from Capture, a copy of the PSpice model is created in a library file, which will have the same name as the project, i.e. <project name>.lib. This is so that the original PSpice model does not get modified. The copied library is written to the project file and can be seen as one of the Configured PSpice libraries in the Project Manager. Whenever you create a new Capture part for a PSpice model, you need to reference the library in which the model can be found by providing the path to the library file in the simulation profile under the **Configuration Files > Category > Library**.

16.1. CAPTURE PROPERTIES FOR A PSpice PART

For PSpice simulation, a Capture part needs to have four specific properties attached. These are the:
- Implementation: name of the model
- Implementation Path: left blank as model is searched for in the configured libraries in the simulation profile

- Implementation Type: PSpice Model
- PSpiceTemplate: provides the Capture part interface to the model or subcircuit

The above properties are automatically added when a Capture part is created either in Capture or in the Model Editor. Fig. 16.1 shows the attached properties for a Q2N3904 transistor.

	A
	⊞ **SCHEMATIC1 : PAGE1 : Q1**
Color	Default
COMPONENT	2N3904
Designator	
Graphic	Q2N3904.Normal
ID	
Implementation	Q2N3904
Implementation Path	
Implementation Type	PSpice Model
Location X-Coordinate	360
Location Y-Coordinate	110
Name	INS960
Part Reference	Q1
PCB Footprint	TO92
Power Pins Visible	⌐
Primitive	DEFAULT
PSpiceTemplate	Q^@REFDES %c %b %e @MODEL
Reference	Q1
Source Library	C:\CADENCE\SPB_16.3\TOOLS\C
Source Package	Q2N3904
Source Part	Q2N3904.Normal
Value	Q2N3904

Fig. 16.1 Q2N3904 attached properties.

The PSpiceTemplate is defined as:

Q^@REFDES %c %b %e @MODEL

where the first character Q defines a bipolar transistor. Other common device types are shown in Table 16.1. A complete table of device types can be found in the PSpice A/D Reference Guide.

The ^ is used by the netlister to define the hierarchical path to the device. The ^ is effectively replaced by the hierarchical path to the device. For example, the netlist in Fig. 16.2 is that of a hierarchical design consisting of a **Top** level design containing a **Bottom** hierarchical block. At the bottom level, there are two resistors R1 and R2, their hierarchical paths being defined by R_Bottom_R1 and R_Bottom_R2. There is also a resistor at the top level, R1.

Table 16.1 PSpice implementation definitions

Character	Device type	Pin order
R	Resistor	1,2
C	Capacitor	1,2
L	Inductor	1,2
D	Diode	Anode, cathode
Q	Transistor	Collector, base, emitter
M	MOSFET	Drain, gate, source, bulk
Z	IGBT	Collector, gate, emitter
I	Current source	+ve node, −ve node
V	Voltage source	+ve node, −ve node
X	Subcircuit	Node 1, node 2, ...node n

```
**** INCLUDING Top.net ****
* source HIERARCHY
R_Bottom_R1          N00522 N00469  1k TC=0,0
R_Bottom_R2          0 N00522  1k TC=0,0
V_V1            N00469 0 10V
R_R1            0 N00522  1k TC=0,0
```

Fig. 16.2 Hierarchical netlist.

@REFDES is the reference designator such as Q1, R2, etc. The % defines the pin names, the order of which is defined in Table 16.1 and the @MODEL references the PSpice model name.

16.2. PSpice MODEL DEFINITION

The basic definition for a PSpice model is given by:

.MODEL <model name> <model type>
+ ([<parameter name> = <value>

The model name must start with one of the PSpice device characters as shown in Table 16.1 and can be up to eight characters long. The model type is specific to the model; for example an NPN type is specific only to a bipolar transistor. Table 16.2 shows the associated model types for the PSpice models.

For example, the model definition for a Q2N3904 transistor is given by:

```
.model Q2N3904    NPN(Is=6.734f Xti=3 Eg=1.11 Vaf=74.03 Bf=416.4 Ne=1.259
+        Ise=6.734f Ikf=66.78m Xtb=1.5 Br=.7371 Nc=2 Isc=0 Ikr=0 Rc=1
+        Cjc=3.638p Mjc=.3085 Vjc=.75 Fc=.5 Cje=4.493p Mje=.2593 Vje=.75
+        Tr=239.5n Tf=301.2p Itf=.4 Vtf=4 Xtf=2 Rb=10)
*        National    pid=23         case=TO92
*        88-09-08 bam    creation
*$
```

The model name, Q2N3904, starts with a letter Q to signify that this is a bipolar transistor model. The model type is that of NPN and the parameter list is enclosed in { } brackets.

Table 16.2 PSpice model types

Model	Device type	Device
Qname	NPN	NPN bipolar
	PNP	PNP bipolar
	LPNP	Lateral PNP
Dname	D	Diode
Cname	CAP	Capacitor
Kname	CORE	Non-linear magnetic core
Lname	IND	Inductor
Mname	NMOS	N-channel MOSFET
	PMOS	
Jname	NJF	N-channel JFET
	PJF	P-channel JFET
Rname	RES	Resistor
Tname	TRN	Transmission line
Bname	GASFET	N-channel GAsFET
Zname	IGBT	N-channel IGBT
Nname	DINPUT	Digital input device
Oname	DOUTPUT	Digital output device
Wname	ISWITCH	Current-controlled switch
Uname	UADC	Multibit ADC
	UDAC	Multibit DAC
	UDLY	Digital delay line
	UEFF	Edge-triggered flip-flop
	UGATE	Standard gate
	UIO	Digital I/O model
	UTGATE	Tristate gate
Sname	VSWITCH	Voltage-controlled switch

The comment lines start with an asterisk, *, and gives information such as the semiconductor vendor, date of creation and the printed circuit board (PCB) footprint.

Most devices are defined using the basic model definition. However, the complete PSpice model definition is given by:

    .MODEL <model name> [AKO: <reference model name>]
    + <model type>
    + ([<parameter name> = <value> [tolerance specification]]*
    + [T_MEASURED=<value>] [[T_ABS=<value>] or
    + [T_REL_GLOBAL=<value>] or [T_REL_LOCAL=<value>]])

A Kind of (AKO) is used when you want to create a model based upon another model (referenced) but change some of the model parameters. For example, if you want to create a 2N3904 transistor model which has a minimum BF of 75 but want to keep the other model parameters the same, then the model definition would be:

    .model Q2N3904_minBF AKO:Q2N3904 NPN (BF=75)

In this way you can build up a family of transistors based on the original transistor but with different parameters.

In the model definition there are three parameters that relate to the temperature at which the model parameter values were calculated or measured:

T_MEASURED

T_ABS

T_REL_GLOBAL or T_REL_LOCAL

T_MEASURED is the temperature at which model parameters were measured or derived. Default is 27°C and, if set, this overrides TNOM.

T_ABS sets the absolute device temperature. No matter what the circuit temperature, the device temperature will be equal to T_ABS.

T_REL_GLOBAL is used to provide a temperature offset from the circuit temperature. The device temperature will be equal to the difference between the circuit temperature and T_REL_GLOBAL.

T_REL_GLOBAL is useful if you have a transistor that is going to be operating at a higher temperature than other transistors. For example, if you have a 2N3904 transistor that is going to be located near a heat source which is running at 5°C higher than ambient temperature, then you would add T_REL_GLOBAL=5°C to the end of the parameter list as shown below:

.model Q2N3904 NPN(Is=6.734f Xti=3 Eg=1.11 Vaf=74.03 Bf=416.4 Ne=1.259

+ Ise=6.734f Ikf=66.78m Xtb=1.5 Br=.7371 Nc=2 Isc=0 Ikr=0 Rc=1

+ Cjc=3.638p Mjc=.3085 Vjc=.75 Fc=.5 Cje=4.493p Mje=.2593 Vje=.75

+ Tr=239.5n Tf=301.2p Itf=.4 Vtf=4 Xtf=2 Rb=10 **T_REL_GLOBAL=5**)

* National pid=23 case=TO92

* 88-09-08 bam creation

*$

So when you run a temperature analysis at −55°C, 27°C and 125°C, the Q2N3904 will be simulated at −50°C, 32°C and 130°C.

16.3. SUBCIRCUITS

PSpice devices can also be represented as a network of components, which is the usual case for operational amplifiers (opamps) and voltage regulators. These devices are classed as subcircuits and the first letter in the PSpiceTemplate is an X. The PSpiceTemplate for the LF411 opamp is shown below:

X^@REFDES %+ %- %V+ %V- %OUT @MODEL

The order of the pins must match the PSpice subcircuit definition for the LF411 as shown below:

```
* connections:    non-inverting input
*                 | inverting input
*                 | | positive power supply
*                 | | | negative power supply
*                 | | | | output
*                 | | | | |
.subckt LF411    1 2 3 4 5
  c1    11 12 4.196E-12
  c2     6  7 10.00E-12
  css   10 99 1.333E-12
  dc     5 53 dy
  de    54  5 dy
  dlp   90 91 dx
  dln   92 90 dx
  dp     4  3 dx
  egnd  99  0 poly(2),(3,0),(4,0) 0 .5 .5
  fb     7 99 poly(5) vb vc ve vlp vln 0 31.83E6 -1E3 1E3 30E6 -30E6
  ga     6  0 11 12 251.4E-6
  gcm    0  6 10 99 2.514E-9
  iss   10  4 dc 170.0E-6
  hlim  90  0 vlim 1K
  j1    11  2 10 jx
  j2    12  1 10 jx
  r2     6  9 100.0E3
  rd1    3 11 3.978E3
  rd2    3 12 3.978E3
  ro1    8  5 50
  ro2    7 99 25
  rp     3  4 15.00E3
  rss   10 99 1.176E6
  vb     9  0 dc 0
  vc     3 53 dc 1.500
  ve    54  4 dc 1.500
  vlim   7  8 dc 0
  vlp   91  0 dc 25
  vln    0 92 dc 25
.model dx D(Is=800.0E-18 Rs=1m)
.model dy D(Is=800.00E-18 Rs=1m Cjo=10p)
.model jx NJF(Is=12.50E-12 Beta=743.3E-6 Vto=-1)
.ends
*$
```

Note

When you create a Capture part for a subcircuit, a rectangular box is drawn such that you can edit the box and draw your own graphics for that part using the Part Editor in Capture.

16.4. MODEL EDITOR

The Model Editor is used to view text model definitions and to display graphical model characteristics and model parameters. Fig. 16.3 shows the forward current versus voltage curve for a diode. When the Model Editor first loads the PSpice library, the model text description is shown. To see the graphical model characteristics, select **View > Extract Model**. Fig. 16.3 also shows a table in which data from a manufacturer's datasheet can be entered such that a new model can be created based upon the manufacturer's empirical data.

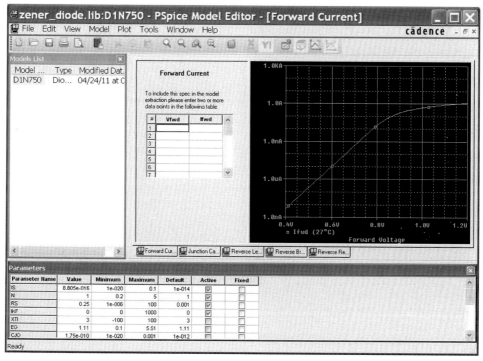

Fig. 16.3 Model Editor.

New models can be created by first selecting **File > New**, to create a library file, then **Model > New**, to create a model (Fig. 16.4).

There are 11 device models to choose from, and the option for **Use Device Characteristic Curves**, which represents the normal PSpice models, and **Use Template**, which relates to parameterized models. The Use Templates is for parameterized models, which are used in the PSpice Advanced Analysis software and are not covered in this text.

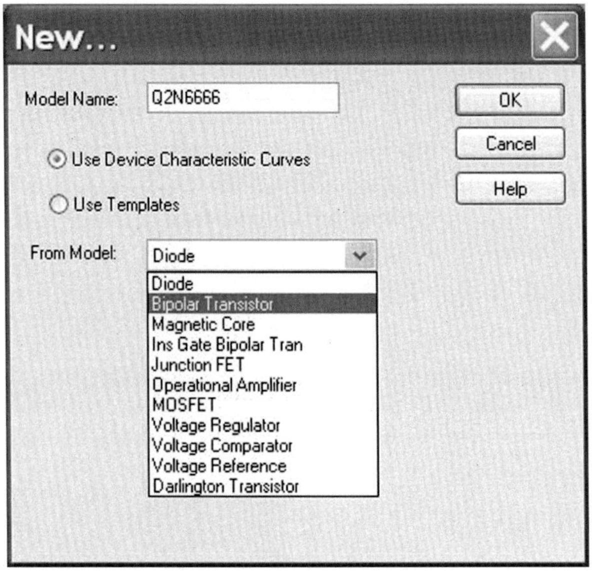

Fig. 16.4 Creating a new model.

Once a model has been created, a Capture part can automatically be generated. The first character of the model name and the model type defined in Table 16.2 determine which type of Capture part to generate. Before you generate a Capture part, you need to make sure that the correct schematic editor is selected. In **Tools > Options > Schematic Editor**, select Capture (Fig. 16.5).

As mentioned previously, the part generated will be determined by the first character in the model name and the model type. If the part to be generated is one of the standard PSpice devices, then you can use **File > Export to Capture Part Library**. As shown in Fig. 16.6, you select the libraries in which to save the PSpice model (.lib) and associated Capture part (.olb). A message window will appear, reporting whether the translation has been successful.

16.4.1 Copying an Existing PSpice Model

The Model Editor has the facility to make a copy of an existing PSpice model from an existing library. Selecting **Model > Copy From**, the **Copy Model** window shown in Fig. 16.7 is displayed. You browse to the **Source Library** and select the model from the displayed model list and enter the **New Model** name.

16.4.2 Model Import Wizard

The Model Import Wizard, **File > Import Wizard [Capture]**, allows you to view and select or replace Capture parts (symbols) for the models in a library one at a time.

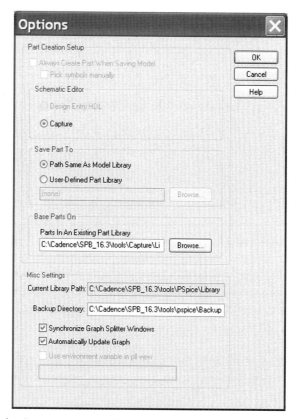

Fig. 16.5 Selecting the Capture schematic Editor in Options.

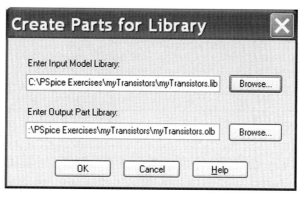

Fig. 16.6 Creating and saving the PSpice model and Capture part.

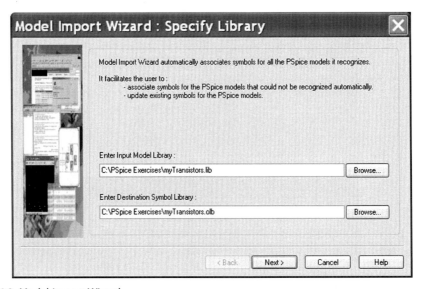

Fig. 16.7 Copying a PSpice model from an existing library.

You browse to the input model library and specify the destination symbol library as shown in Fig. 16.8. The good news is that the Model Import Wizard will display opamp symbols for opamps created in the Model Editor rather than rectangular boxes. The other good news is that you can browse for an existing Capture symbol to associate with the model.

Fig. 16.8 Model Import Wizard.

In Fig. 16.9 an international rectifier MOSFET has been copied from the IRF.lib library, modified and saved as myIRF540. You also have the option to view the model text, which is useful if you need to check the pin names.

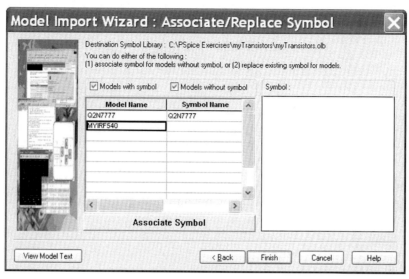

Fig. 16.9 Associate or replace a symbol to the model.

As no symbol exists for the MOS device yet, clicking on **Associate Symbol** will allow you to browse to a suitable library which contains a symbol for an NMOS transistor (Fig. 16.10).

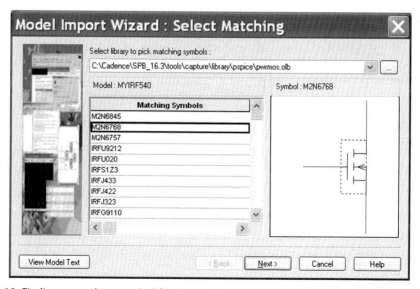

Fig. 16.10 Finding a matching symbol for the model.

Note

The MOS device is modeled as a subcircuit; if generated in the Model Editor without using the Model Import Wizard, it would be displayed as a rectangular box in Capture, which would then have to be edited.

In this example, the Capture symbol for an NMOS symbol was found in the pwrmos.olb library (Fig. 16.10). However, you now have to associate the model terminals (1, 2, 3) with the Capture symbol pins (d, g, s) (Fig. 16.11). All the symbol pin names will be listed.

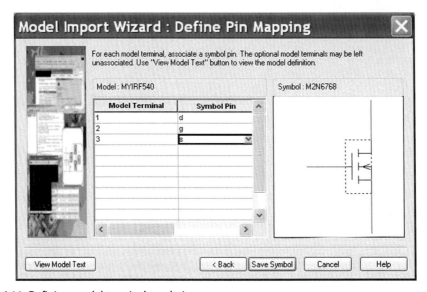

Fig. 16.11 Defining model terminals and pins.

When you select **Save Symbol** in Fig. 16.11, the models listed in the PSpice myTransistors.lib library file are shown (Fig. 16.12). Note that as all models have an associated capture symbol, the option to **Replace Symbol** is now shown. If there is no associated symbol, the option **Associate Symbol** is shown.

An example of using the Model Import Wizard can be found in Chapter 21, Exercise 2.

16.4.3 Downloading Models From a Vendor Website

The Model Editor is useful for displaying the characteristic curves for models, especially if the PSpice model has been downloaded from a vendor's website. By default, the Model Editor ignores the first line in a PSpice model file, so make sure the **.model** statement starts on the second line. Comment lines can be added to the model by starting the line

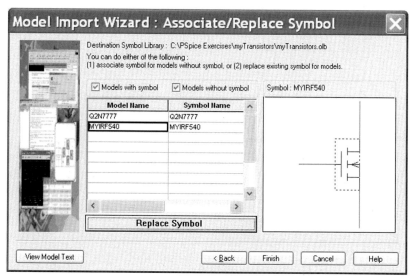

Fig. 16.12 Summary of models in myTransistors PSpice library.

with an asterisk, *, so it is always a good idea to add information regarding the model to the file, for example:

    * Q2N7777 transistor downloaded from semiconductor website 23.4.2011

A Capture part can also be generated in Capture from a PSpice model. In the Project Manager highlight the design file (.dsn) and select **Tools > Generate Part** to open the **Generate Part** window (Fig. 16.13). Select **Netlist/source file type** to **PSpice Model Library** and browse to the PSpice model in **Netlist/source file**. Then in **Implementation name** select the model name from the file. If you have downloaded a library of models, then all the individual model names will be available.

16.4.4 Encryption

There is an encryption facility in the Model Editor that allows you to encrypt PSpice models or libraries such that the models can be used for simulation but the model definitions cannot be viewed. To encrypt a model library, select **File > Encrypt Library** and in the **Library Encryption** window (Fig. 16.14) browse to the library to be encrypted and the folder where the library will be saved to.

    If you only want to encrypt part of the library, i.e. two of the PSpice library models, then add the following text to the beginning and end of the model definitions:

    $CDNENCSTART beginning of model text

    $CDNENCFINISH end of model text

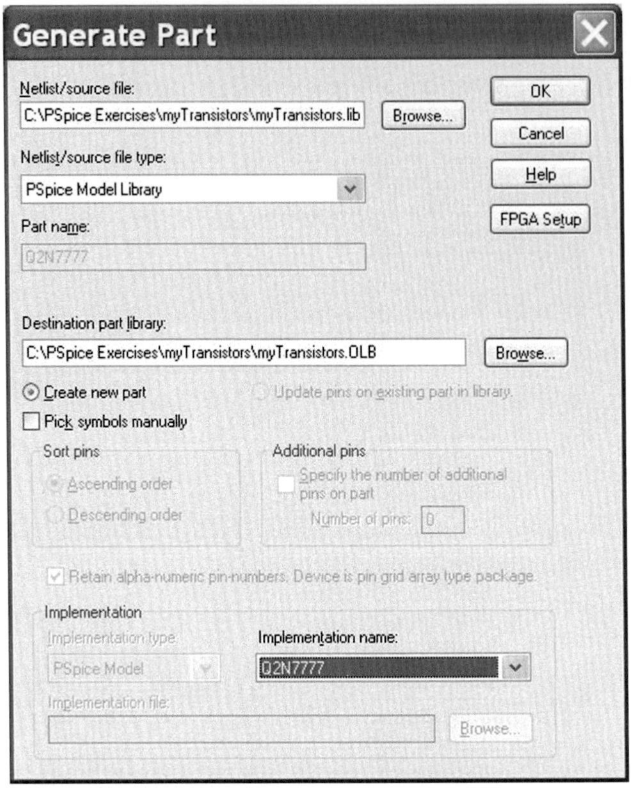

Fig. 16.13 Generating a Capture part from a PSpice library model.

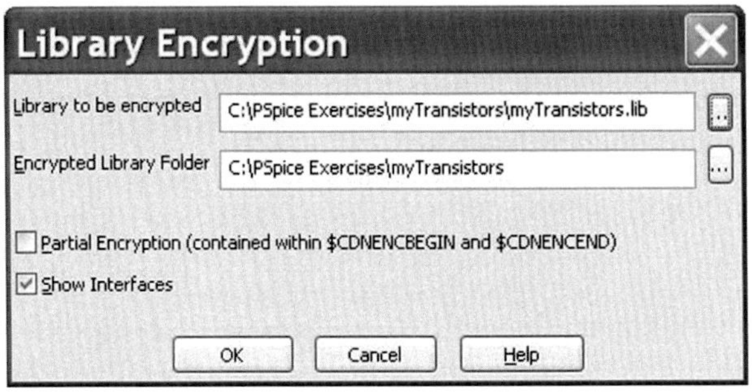

Fig. 16.14 Library encryption.

If **Partial Encryption** is not checked, then the **Show Interfaces** option allows you to encrypt the model text definition but still display the model interfaces, i.e. the model connecting pins such as emitter, base and collector for a bipolar transistor.

Note

Comment lines are not encrypted.

16.4.5 IBIS Translator

IBIS 1.1 models are supported in 16.5 whereas 16.6 supports IBIS models up to version 5.0.

In 16.6, IBIS and Cadence proprietary Device Model Language (DML) files can be translated into PSpice library files in the Model Editor. The translated IBIS I/O buffers are defined as macromodels.

IBIS (Input Output Buffer Information Specific) models are mainly used to analyse the transmission of digital high speed signals between ICs where the transmitted medium, usually copper traces, are modelled as transmission lines. The input and output buffers of ICs are characterised by tables of measured Voltage/Current and Voltage/Time data to provide a representation of I/O buffer behaviour without disclosing any proprietary information regarding the internal implementation of the I/O buffers. IBIS models based on these V/I and V/T tables can be used to analyse the Signal Integrity of the transmission of data by predicting the circuit response to mismatched impedance lines, crosstalk from adjacent transmission lines, overshoot, undershoot, ground bounce and simultaneous switching of digital high speed signals.

IBIS models provide a relatively accurate simulation model as they take into account structures for ESD protection and inherent parasitics associated with die bonding to the IC package pins. Compared to standard SPICE models, IBIS simulations run faster and do not suffer from non-convergence issues.

DML files are used by the existing suite of high speed simulation Cadence software tools for high speed analysis.

The IBIS translator is in the Model Editor and can be accessed from the top tool bar.

Model > IBIS Translator

The IBIS translator is not available in the demo Lite CD.

16.5. EXERCISES

Exercise 1

The breakdown voltage for a zener diode will be modified using the Model Editor.

1. Create a project called **zener_diode** and draw the circuit in Fig. 16.15. The 4V7 zener diode can be found in either the eval library or the diode library.

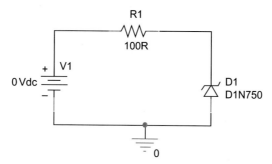

Fig. 16.15 Verifying the zener breakdown voltage.

2. Create a **PSpice > New Simulation Profile** and select the **Analysis type** to DC
    sweep. Create a simulation profile for a DC sweep for V1 from 1 to 10 V in steps of
    0.1 V (Fig. 16.16) to confirm the 4.7 V zener diode breakdown voltage. It does not
    matter if V1 has a voltage of 0 V as shown in the schematic, as a DC sweep is being
    performed.

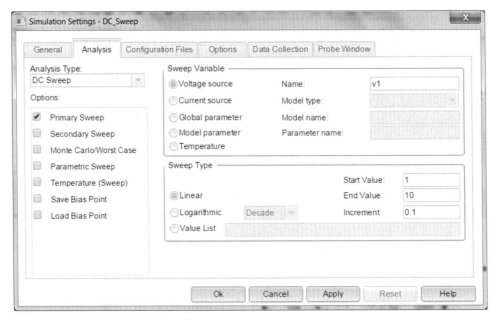

Fig. 16.16 DC sweep simulation profile.

3. Select the diode and **rmb > Edit PSpice Model** to open the Model Editor. Note
    the library name at the top of the Model Editor, **zener_diode.lib** (Fig. 16.17).
4. From the top toolbar, select **View > Extract Model.** In the first PSpice Model
    Editor window, click on **Yes**, then click on OK in the next window regarding
    parameters. The Model Editor will display the forward current characteristic for
    the diode as shown in Fig. 16.17.
5. Under the displayed curve, click on each of the five tabs to view the diode model
    characteristics. When you select a curve, the **Parameter** window (Fig. 16.18) dis-
    plays an **Active** check against each parameter that is associated with the displayed
    characteristic curve. Note the diode characteristic for the **Reverse Breakdown**
    curve and the active parameters.
6. In the Parameters section at the bottom of the Model Editor, scroll down and locate
    the diode breakdown parameter, BV. Change the value from 4.7 to 8.2 V and click

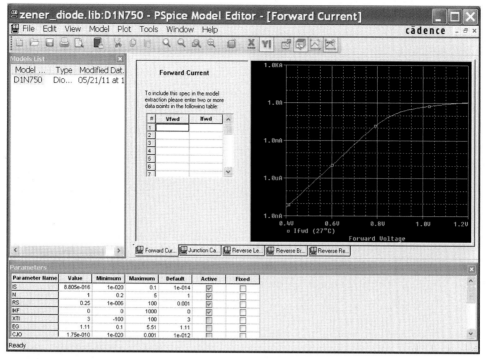

Fig. 16.17 Creating a zener.lib file.

Parameter Name	Value	Minimum	Maximum	Default	Active	Fixed
FC	0.5	0.001	10	0.5	☐	☐
ISR	1.859e-009	1e-020	0.1	1e-010	☐	☐
NR	2	0.5	5	2	☐	☐
BV	4.7	0.1	1000000	100	☑	☐
IBV	0.020245	1e-009	10	0.0001	☑	☐
TT	5e-009	1e-016	0.001	5e-009	☐	☐

Fig. 16.18 Model Editor showing extracted model parameter curves.

on the Fixed box (Fig. 16.19). Select the **Reverse Breakdown** tab under the curve, which should have changed to 8.2 V.

7. Save the library file, **File > Save** and close the Model Editor.
8. Edit the Simulation Profile. Select **Configuration Files > Library** and note that the zener_diode.lib has been added to the profile. The attached design icon indicates that the file is local only to the design (Fig. 16.20). The world icon , indicates that the file is global and can be seen by all designs. The nom.lib contains all the PSpice library files and hence is global to all designs.

Parameters

Parameter Name	Value	Minimum	Maximum	Default	Active	Fixed
FC	0.5	0.001	10	0.5	☐	☐
ISR	1.859e-009	1e-020	0.1	1e-010	☐	☐
NR	2	0.5	5	2	☐	☐
BV	8.2	0.1	1000000	100	☑	☑
IBV	0.020245	1e-009	10	0.0001	☑	☐
TT	5e-009	1e-016	0.001	5e-009	☐	☐

Fig. 16.19 Editing the breakdown voltage.

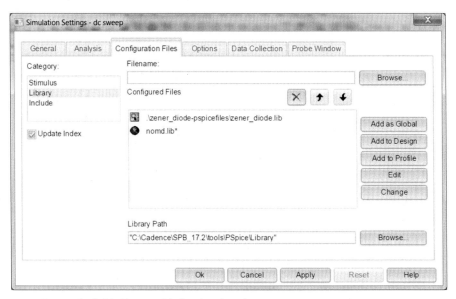

Fig. 16.20 Zener_diode.lib library added as local to design.

9. Select the zener_diode.lib file and click on **Edit**. The Model Editor will open up. Select the D1N750 in the **Models List** and the diode model parameter will appear as before. This enables model library files to be quickly viewed. Close the Model Editor and the simulation profile.
10. Rerun the simulation and confirm that the diode breakdown voltage is now 8.2 V.

Exercise 2

Whenever you create a new PSpice model and Capture part it is recommended that you create a new directory for your model. Do not install the new libraries in the Capture or PSpice folders. If a new OrCAD release is installed, then the PSpice and Capture libraries will be reinstalled and so any new models created will be lost.

For this exercise we will assume that a PSpice model for a transistor has been downloaded from a semiconductor's website. In order to recreate this scenario we will copy an existing transistor model from the bipolar.lib library to a new file, myTransistors.lib.

1. Using a text editor such as WordPad or Notepad, browse to the installed PSpice libraries, **<install path > Orcad 16.3\Tools\PSpice\Library**, and select the bipolar.lib or eval.lib PSpice library. Make sure you select **Files of type** to **All Files**.

2. In the library file, scroll down and select the Q2N3904 model definition as shown below (use Control F to find the Q2N3904):

```
.model Q2N3904        NPN(Is=6.734f Xti=3 Eg=1.11 Vaf=74.03 Bf=416.4 Ne=1.259
+                     Ise=6.734f Ikf=66.78m Xtb=1.5 Br=.7371 Nc=2 Isc=0 Ikr=0 Rc=1
+                     Cjc=3.638p Mjc=.3085 Vjc=.75 Fc=.5 Cje=4.493p Mje=.2593 Vje=.75
+                     Tr=239.5n Tf=301.2p Itf=.4 Vtf=4 Xtf=2 Rb=10)
*                     Nationalpid=23        case=TO92
*                     88-09-08 bam   creation
*$
```

3. Select the model text and copy and paste it into a new text file. Do not use rich text format (RTF) if using WordPad.

4. Edit the transistor model name to Q2N7777 and add a comment line to the first line 1 as shown below:

```
* example of a downloaded transistor model
.model Q2N3904        NPN(Is=6.734f Xti=3 Eg=1.11 Vaf=74.03 Bf=416.4 Ne=1.259
+                     Ise=6.734f Ikf=66.78m Xtb=1.5 Br=.7371 Nc=2 Isc=0 Ikr=0 Rc=1
+                     Cjc=3.638p Mjc=.3085 Vjc=.75 Fc=.5 Cje=4.493p Mje=.2593 Vje=.75
+                     Tr=239.5n Tf=301.2p Itf=.4 Vtf=4 Xtf=2 Rb=10)
*                     Nationalpid=23        case=TO92
*                     88-09-08 bam   creation
*$|
```

5. Save the file as **myTransistors.lib** in a folder called **myTransistors**. Make sure the file is saved as text and not RTF, otherwise control characters will be added to the model text.

6. Create a new PSpice project called myTransistors.

7. In the Project Manager make sure the myTransistors.dsn file is highlighted and select **Tools > Generate Part**. In the **Generate Part** window (Fig. 16.21) select:
   In **Netlist/source file:** type: select **PSpice Model Library**
   In **Netlist/source file:** browse to the myTransistors.lib file.
   In **Destination part library:** browse to the same folder where myTransistors.lib is.
   In **Implementation name:** there will only be one entry Q2N7777.
   Click on **OK**.

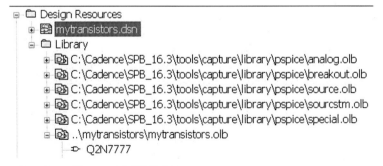

Fig. 16.21 Generating a Capture part.

8. The myTransistors.olb Capture library will be created and added to the library folder in the Project Manager (Fig. 16.22). Expand the library to see the Q2N7777 part.

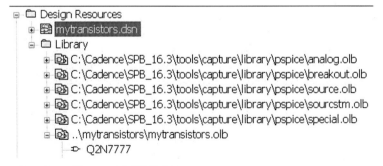

Fig. 16.22 myTransistors library with Q2N7777.

9. Open the schematic page and select **Place > Part** or press P on the keyboard (Fig. 16.23). The library **myTransistors** has automatically been added to the libraries and the **Part List** contains the transistor graphical symbol for an NPN transistor (Fig. 16.23).

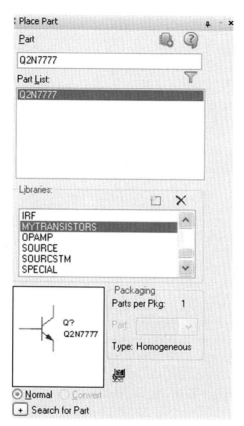

Fig. 16.23 myTransistors.olb library available in library list.

Also, the PSpice icon appears, indicating that the transistor has a PSpice model attached ⌦. You now need to make the myTransistors.lib file available for simulation in the Simulation Profile.

10. Create a new simulation profile, **PSpice > New Simulation Profile**. Under the **Configurations Files** tab, select **Category > Library** and browse to the folder where myTransistors.lib is. As mentioned in Exercise 1 Step 8, you can add the library files Global to the design, local to the design or to the Profile. In this exercise, add the file as Global (Fig. 16.24) and click on OK. The transistor is ready to be simulated and the myTransistors.lib library will be available for every new design.

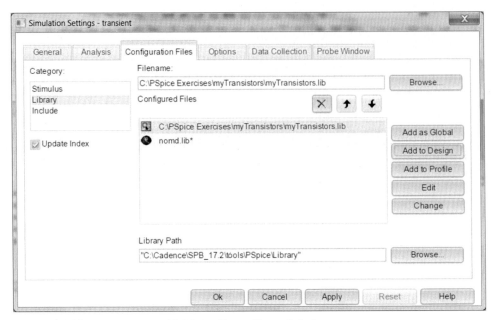

Fig. 16.24 Adding myTransistors.lib PSpice library file as Global.

Exercise 3

In Exercise 2, the model definition for the Q2N7777 has a Q for the first character in the model name, so this is recognized as a transistor and has an NPN as the model type. Hence, an NPN transistor symbol was generated. If the model name starts with an X for a subcircuit, a rectangular box is drawn which can then be edited in Capture using the Part Editor. You highlight the part in the library, **rmb > Edit Part**. Many of the power MOSFET models, for example, are defined as subcircuits.

However, in the Model Editor, the **Model Import Wizard** can be used to select an existing Capture graphic symbol for the model rather than having to edit the graphics in the Part Editor.

1. Open the Model Editor from the **Start** menu: **All Programs > Cadence (or OrCAD) > PSpice > Simulation Accessories > Model Editor**.
2. Select **File > Open** and browse to the myTransistors.lib library file. Click on Q2N7777 in the **Models List** and the model text will be displayed. Note that the first line which you added is not displayed.
3. Select **View > Extract model** and click on **Yes** in the Model Editor window. There will be eight characteristic curves for the transistor model.
4. Select **File > Export to Capture Part Library**. If the Capture option is not available, select **Tools > Options** and select Capture as the Schematic Editor (see Fig. 16.5).

In the **Create parts for Library** window, the output library folder will show the same location as the PSpice library file, as shown in Fig. 16.25. Click on OK and if a window appears asking to save the library, click on **Yes**. The translator window will appear and hopefully report no errors.

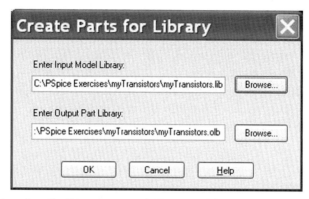

Fig. 16.25 Folder locations for Capture part and PSpice model library.

Exercise 4

You can also copy PSpice models from existing models in the Model Editor.
1. In the Model Editor, open the **bipolar.lib** library from the PSpice library folder.
2. Select **Model > Copy From** and in the **Copy Model** window enter Q2N3906X in **New Model**, select Q2N3906 from the library list and click on OK (Fig. 16.26).

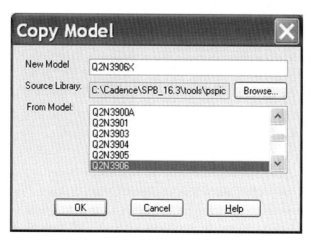

Fig. 16.26 Copying a PSpice model.

Exercise 5

New models can be created in the Model Editor. In the demo CD you can only create new diode models in the Model Editor. Models can be created using data from device characteristic curves from manufacturer's data sheets or alternatively, parameterized models can be created for use in PSpice Advanced Analysis (see Chapter 23). The **Use Templates** is for parameterized models that are used in the PSpice Advanced Analysis software, see Chapter 23.

The Use Templates is for parameterized models, which are used in the PSpice Advanced Analysis software and are not covered in this text. For this exercise, select **Use Device Characteristic Curves**.

1. In the Model Editor, select **File > New.**
2. Select **Model > New** and enter myDiode as the model name and select, **Use Device Characteristic Curves** (Fig. 16.27). Click on OK.

Fig. 16.27 Creating new PSpice models.

3. There are five characteristic curves for the diode model that can be displayed by clicking on the tabs at the bottom of the curves (see Fig. 16.28). You enter the data for each diode characteristic and then select, Tools > Extract Parameters. The extracted model parameters are displayed at the bottom of the Model Editor can be changed and the values fixed.

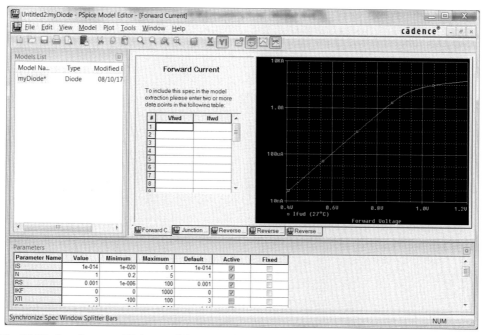

Fig. 16.28 Creating new PSpice models using characteristic curves.

4. Create a new model called myTransistor or myDiode2 if you are using the Lite version but this time, select the Use Templates option and select the Bipolar NPN type as shown in Fig. 16.29. Click on OK.

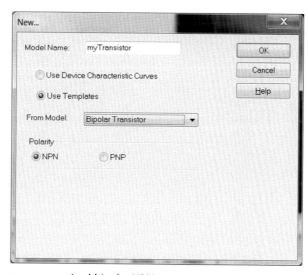

Fig. 16.29 Creating a parameterized bipolar NPN transistor.

5. Fig. 16.30 shows the parameterized properties for the transistor. You can enter tolerance and distribution spreads for each parameter as shown. These parameters are used in PSpice Advanced Analysis (Chapter 23).

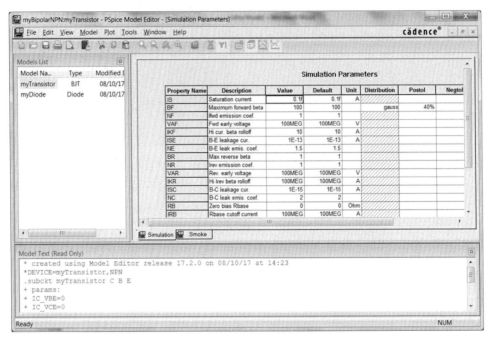

Fig. 16.30 Parameterized bipolar NPN transistor.

CHAPTER 17

Transmission Lines

Chapter Outline

17.1. Ideal Transmission Lines	243
17.2. Lossy Transmission Lines	244
17.3. Exercises	247
Exercise 1	247
Exercise 2	254

The signal integrity of high-speed signals via transmission lines is dependent on the frequency-related signal and dispersion losses of the transmission lines. Signal power loss is attributed to the increase in conductor resistance (skin effect) and the increase in dielectric conductance (dielectric loss) with an increase in frequency. Dispersion is the distortion of the signal wave shape resulting from delays introduced by the distributed frequency-dependent inductance and capacitance of the transmission line. Any reflected signals, due to impedance mismatch, will also exhibit loss and dispersion and subsequently degrade the performance of the transmission line.

Ideal and lossy transmission lines are modeled in PSpice using Tline distributed models and TLUMP lumped line segment models.

17.1. IDEAL TRANSMISSION LINES

The parameters required for an ideal transmission line are the characteristic impedance (Z0) and either the transmission line delay (TD) or the normalized line length (NL) which is the number of wavelengths along the line at a given frequency. You cannot enter TD and NL together. If you do not specify the frequency for NL, then the frequency defaults to 0.25 which represents the quarter wave frequency.

The time delay, TD, along a transmission line is given by:

$$\text{TD} = \frac{\text{LEN}}{v_{\text{p}}} \tag{17.1}$$

where TD is the transmission delay (s), LEN is the length of the transmission line (m), and v_{p} is the velocity of the propagating wave (propagating velocity) (m s^{-1}).

Analog Design and Simulation Using OrCAD Capture and PSpice
https://doi.org/10.1016/B978-0-08-102505-5.00017-3

For transmission lines, the propagation velocity is expressed as a percentage of the speed of light, such that:

$$v_p = c \times VF \qquad (17.2)$$

where VF is the velocity factor which has a value between 0 and 1, and c is the speed of light at $3 \times 10^8\,\mathrm{m\,s^{-1}}$.

The normalized transmission line length is given by:

$$NL = \frac{LEN}{\lambda} \qquad (17.3)$$

From $v = f\lambda$, the wavelength is given as:

$$\lambda = \frac{v_p}{f} \qquad (17.4)$$

Eq. (17.3) can be therefore rewritten as:

$$NL = LEN\frac{f}{v_p} \qquad (17.5)$$

where f is frequency (Hz) and λ is wavelength (m).

PSpice uses a T device from the **analog** library to model an ideal transmission line. Fig. 17.1A shows the Capture part for the T device and Fig. 17.1B the associated properties in the Property Editor.

So, for an ideal transmission line, if you do not know the delay time (TD) then you can enter values for NL and f and, as mentioned above, if you do not enter the frequency, then the default value of 0.25 is used, which represents the quarter wave frequency.

Initial conditions for voltage and current can be applied to the transmission line.

17.2. LOSSY TRANSMISSION LINES

Transmission lines can be considered to consist of a number of identical sections known as RLCG lumped line segments, as shown in Fig. 17.2. The R represents the line resistance, L the line inductance, C the dielectric capacitance and G the dielectric conductance. For long transmission lines, one solution would be to use a number of lumped RLCG segments connected together. PSpice provides lumped line segments of up to 128 segments in the TLine library. However, lumping together large line segments can lead to long simulation times.

Simpler RC transmission line models are also available in the TLine library, as are over 40 coaxial cable models and twisted wire pair models.

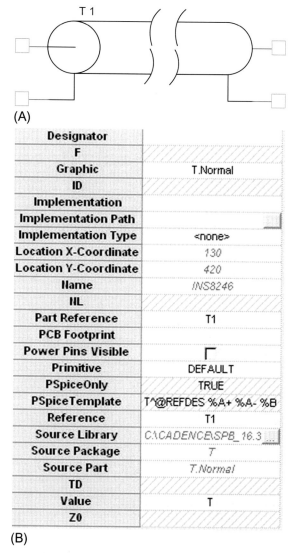

Designator	
F	
Graphic	T.Normal
ID	
Implementation	
Implementation Path	
Implementation Type	<none>
Location X-Coordinate	130
Location Y-Coordinate	420
Name	INS8246
NL	
Part Reference	T1
PCB Footprint	
Power Pins Visible	
Primitive	DEFAULT
PSpiceOnly	TRUE
PSpiceTemplate	T^@REFDES %A+ %A- %B
Reference	T1
Source Library	C:\CADENCE\SPB_16.3
Source Package	T
Source Part	T.Normal
TD	
Value	T
Z0	

(B)

Fig. 17.1 Ideal transmission line Tline: (A) Capture part: T device and (B) associated Tline properties.

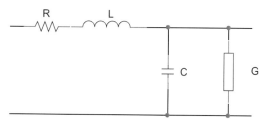

Fig. 17.2 RLCG lumped line segment of a transmission line.

An alternative approach for lossy transmission lines is to use a distributed model which relies on an impulse response convolution method to determine the transmission line response. Fig. 17.3 shows the TLOSSY PSpice device and the associated properties in the Property Editor.

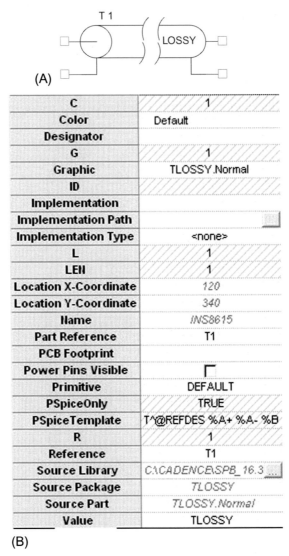

C	1
Color	Default
Designator	
G	1
Graphic	TLOSSY.Normal
ID	
Implementation	
Implementation Path	
Implementation Type	<none>
L	1
LEN	1
Location X-Coordinate	120
Location Y-Coordinate	340
Name	INS8615
Part Reference	T1
PCB Footprint	
Power Pins Visible	
Primitive	DEFAULT
PSpiceOnly	TRUE
PSpiceTemplate	T^@REFDES %A+ %A- %B
R	1
Reference	T1
Source Library	C:\CADENCE\SPB_16.3 ...
Source Package	TLOSSY
Source Part	TLOSSY.Normal
Value	TLOSSY

(B)

Fig. 17.3 Lossy transmission line TLOSSY: (A) Capture part: TLOSSY device and (B) associated TLOSSY properties.

The length of the transmission line is represented by the LEN property and the R, L, C and G properties are specified as per unit length.

Note

The maximum internal time step generated for distributed transmission line models is limited to one half of the transmission line delay, TD. Therefore, for short transmission lines, the simulation time may be considerably longer for distributed line models compared to using a lumped line model for a short transmission line.

17.3. EXERCISES

The following exercises demonstrate the basic transmission line characteristics for different load terminations.

Exercise 1

Matched Load for RL

1. Draw the circuit in Fig. 17.4. The transmission device T is from the **analog** library and the voltage pulse is from the **source** library. When you place the load resistor, RL, on the schematic, by default, pin 1 is on the left-hand side. Rotate resistor RL three times such that pin 1 is at the top, which connects to T1. By convention, current flowing into pin 1 is defined as positive, such that a measured negative current at pin 1 represents a current flowing out of pin 1.

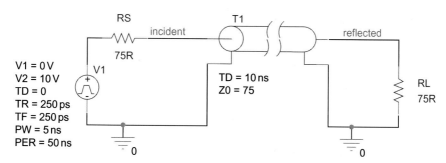

Fig. 17.4 Matched source and load impedance transmission line.

2. Double click on T1 to open the Property Editor and add and display the property values as shown in Fig. 17.5.

TD	10 ns
Value	T
Z0	75

Fig. 17.5 Adding property values for TD and Z0.

Highlight both TD and Z0 by holding down the control key, select **Display** and in the **Display** Properties window, select **Display > Name and Value** as shown in Fig. 17.6.

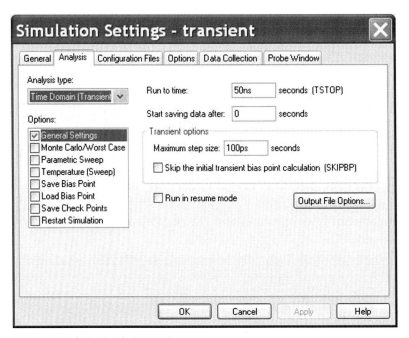

Fig. 17.6 Display the TD and Z0 property values.

3. Create a transient analysis simulation profile with a Run to time of 50 ns and a Maximum step size of 100 ps (Fig. 17.7).

Fig. 17.7 Transient analysis simulation settings.

4. Add voltage markers on the incident and reflected nodes (Fig. 17.8) and run the simulation.

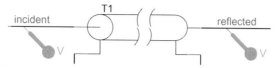

Fig. 17.8 Placing voltage markers.

5. Fig. 17.9 shows the source pulse and the delayed load pulse after 10 ns. Initially, the source resistor and transmission line act as a potential divider, so the voltage amplitude divides down to 5 V.

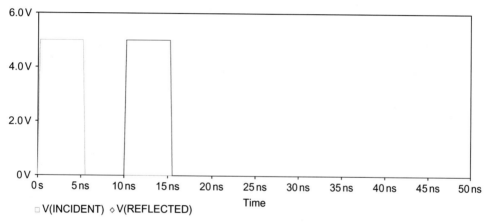

Fig. 17.9 Matched transmission line voltage waveforms.

6. Delete the voltage markers in Capture and place current markers on the input pin (incident) to T1 and the top pin of the load resistor, RL (Fig. 17.10).

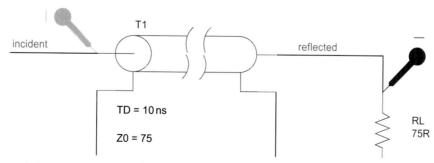

Fig. 17.10 Placing current markers.

There is no need to rerun the simulation as the waveform display in Probe will automatically be updated. You should see that both currents have the same magnitude and are separated by the specified 10 ns delay (Fig. 17.11).

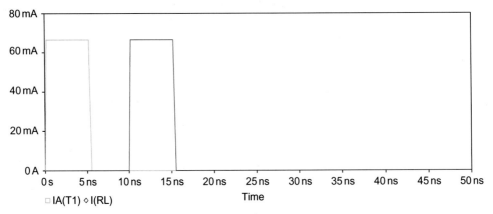

Fig. 17.11 Matched transmission line current waveforms.

As the load impedance is equal to the impedance of the transmission line, the line is said to be matched. There are no voltage or reflections.

RL Replaced With a Short Circuit

7. Delete the current markers and replace the load resistance with a small value resistance of 1 μΩ to represent a short circuit. Place voltage markers on the incident node and the top of the short circuit resistor as shown in Fig. 17.12 and rerun the simulation.

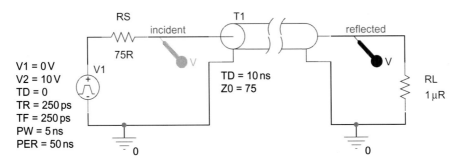

Fig. 17.12 Short circuit load.

You should see the transmission line response in Fig. 17.13. With a short circuit the load voltage will be 0 V and the incident voltage wave will be reflected but 180° out of phase with the incident wave.

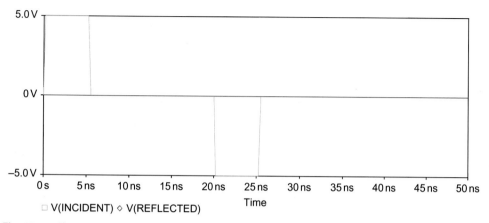

□ V(INCIDENT) ◇ V(REFLECTED)   Time

Fig. 17.13 Short circuit transmission line voltage waveforms.

8. In Capture, delete the voltage markers and place current markers on the input pin (incident) to T1 and the top pin of the RL resistor (Fig. 17.14).

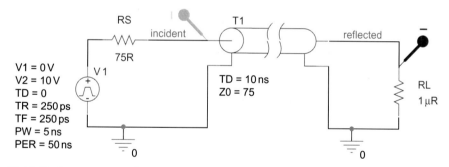

Fig. 17.14 Placing current markers.

There is no need to rerun the simulation as the waveform display in Probe will automatically be updated. You should see that the incident current wave is reflected with the same amplitude such that the reflected wave is double the magnitude of the incident current wave (Fig. 17.15).

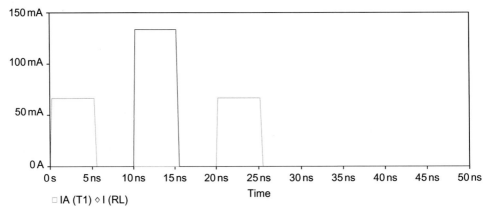

□ IA (T1) ◇ I (RL)

Fig. 17.15 Short circuit transmission line current waveforms.

RL Replaced With an Open Circuit

9. Delete the current markers and change the load resistance to a 1 TΩ resistor to represent an open circuit. Place voltage markers on the incident node and the top of the short circuit resistor (Fig. 17.16) and rerun the simulation.

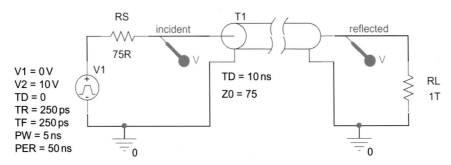

Fig. 17.16 Placing voltage markers.

You should see the response shown in Fig. 17.17. The reflected voltage is equal to the source voltage, so the reflected voltage is double in magnitude to that transmitted.

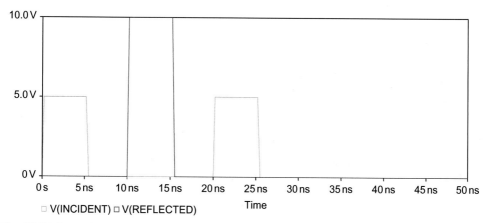

Fig. 17.17 Open circuit transmission line voltage waveforms.

10. Delete the voltage markers in Capture and place current markers on the input pin (incident) to T1 and the top pin of resistor RL. There is no need to rerun the simulation as the waveform display in Probe will automatically be updated. As the output is an open circuit, no current will flow. The current at the open circuit is reflected back with the same magnitude but is 180° out of phase, as seen in Fig. 17.18.

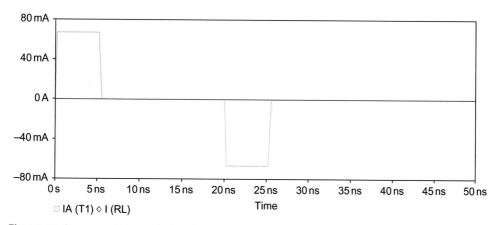

Fig. 17.18 Open circuit transmission line current waveforms.

Exercise 2

Standing Wave Ratio (SWR)

Fig. 17.19 shows a lossless transmission line with a short circuit. As shown in Fig. 17.13, the incident voltage is reflected with the same amplitude but 180° out of phase. The incident and reflected waves will sum together to produce a standing wave, otherwise known as a stationary wave, as will be demonstrated below.

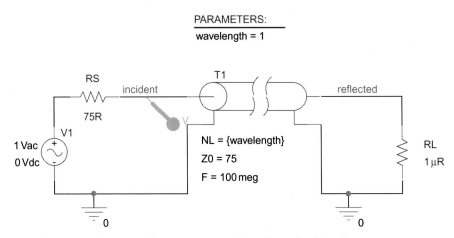

Fig. 17.19 Demonstrating a standing wave pattern for a short circuit load.

SWR for Short Circuit

1. Delete the current markers and change the value of RL to 1 μR for a short circuit. Delete the voltage pulse, V1, and replace with a VAC source from the source library.
    As mentioned previously, you cannot use TD and NL together, so you can either delete the TD property in the Property Editor or replace the transmission line with a new part.
2. Delete the transmission line T1.
3. Place a T part from the analog library.
4. We are going to vary the value of the transmission line property NL, so we need to parameterize the property value of NL in the Property Editor. Double click on the T part to open the Property Editor. Highlight the NL property value box which has shaded lines and enter {wavelength} (Fig. 17.20). As soon as you start typing, the shaded lines in the NL value box will disappear. The {} brackets represent a placeholder for a variable parameter. Do not close the Property Editor.

NL	{wavelength}

Fig. 17.20 Adding a parameterized value to the wavelength property.

5. It is a good idea to display new properties. In the **Property Editor**, highlight the wavelength property and select Display (or **rmb > Display**) and select **Name and Value** as shown in Fig. 17.21. Do not close the Property Editor.

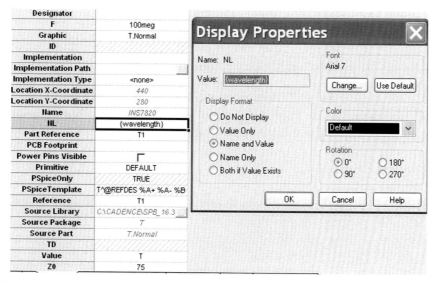

Fig. 17.21 Displaying name and value of NL property.

6. Set and display, as in Step 5, the name and property values for a frequency, *F*, of 100 megHz and a characteristic impedance, Z0, of 75R, as shown in the circuit diagram Fig. 17.19. Close the Property Editor.

7. A default value for the wavelength parameter needs to be defined. Add a **Param** part from the **special** library and double click on the part to open up the Property Editor. Select **New Row** (or New Column) and enter the property **Name** as wavelength and property **Value** as 1 shown in Fig. 17.22.

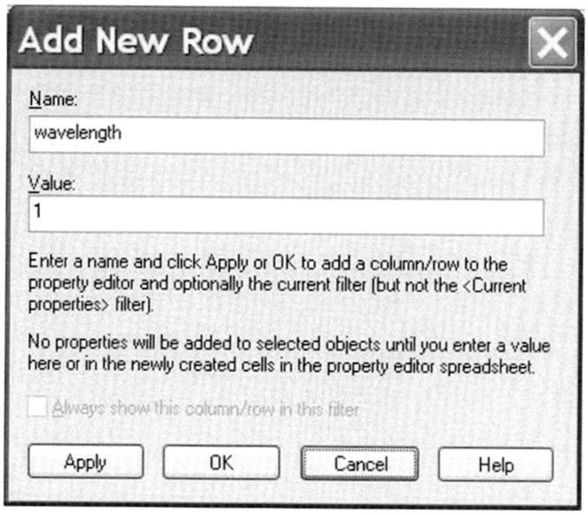

Fig. 17.22 Adding a new wavelength property with a default value of 1.

8. Display the name and value of the wavelength property and close the Property Editor.
9. Your schematic should now appear as shown in Fig. 17.19.

Note

It is easier to edit property values that are displayed on the schematic rather than having to keep opening the Property Editor.

---

10. You will need to set up a **Parametric** simulation sweep together with an **AC analysis**. Create a new PSpice simulation profile, **PSpice > New Simulation Profile**, and select the analysis for **AC Sweep/Noise** from 100 to 200 megHz using a **Linear** sweep with the **Total Points** equal to 1 (Fig. 17.23). Click on **Apply** but do not exit the simulation profile.

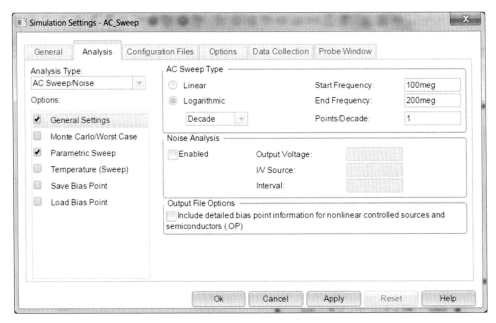

Fig. 17.23 AC sweep settings.

In the **Options** box, select **Parametric Sweep** and set up a global parametric sweep of the **wavelength** property from 0 to 1 in steps of 0.01, as shown in Fig. 17.24. Click on OK.

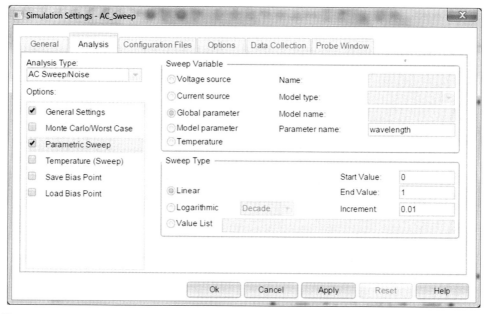

Fig. 17.24 Parametric sweep settings.

11. Place a voltage marker on the incident node and run the simulation. In the **Available Sections** window, make sure all sections are highlighted and click on OK. You should see the standing wave pattern in Fig. 17.25.

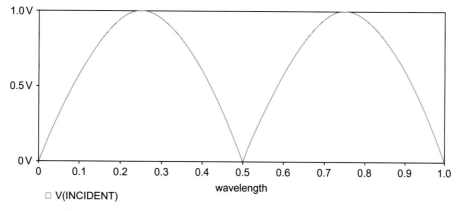

Fig. 17.25 Standing wave pattern for a short circuit load.

SWR for Open Circuit

For an open circuit, the voltage is reflected back with the same amplitude but is 180° out of phase.

12. Modify the value of the load resistor in Fig. 17.26 to 1T to represent an open circuit and simulate the circuit using the same simulation profile.

   You should see the standing wave as shown in Fig. 17.27.

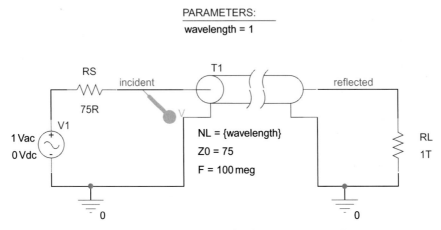

Fig. 17.26 Demonstrating a standing wave pattern for an open circuit load.

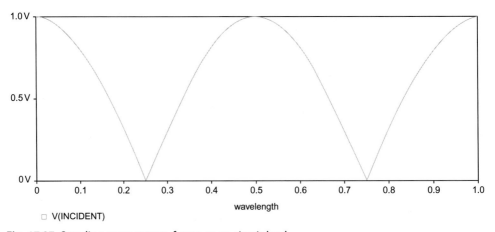

Fig. 17.27 Standing wave pattern for an open circuit load.

CHAPTER 18

Digital Simulation

Chapter Outline

18.1. Digital Device Models   259
18.2. Digital Circuits   260
18.3. Digital Simulation Profile   262
18.4. Displaying Digital Signals   263
18.5. Exercises   264
     Exercise 1   264
     Exercise 2   268
     Exercise 3   272

PSpice uses the same simulation engine for both analog and digital parts. Digital TTL and CMOS parts are modeled as subcircuits and include the common digital functions such as gates, registers, flip-flops, inverters, etc. Within each subcircuit, a digital primitive makes up the gate function (AND, OR, etc.) and defines the timing and interface specification for the gate function. Other digital devices include: delay lines, AtoD, DtoA, RAM, ROM, and Programmable Logic Arrays.

18.1. DIGITAL DEVICE MODELS

A model definition for a 2-input CMOS NAND gate is shown here.

```
* CD4011B  CMOS NAND GATE QUAD 2 INPUTS
*
* The CMOS Integrated Circuits Data Book, 1983, RCA Solid State
* tvh  09/29/89  Update interface and model names
*
.subckt CD4011B A B J
+    optional: VDD=$G_CD4000_VDD VSS=$G_CD4000_VSS
+    params: MNTYMXDLY=0 IO_LEVEL=0
U1 nand(2) VDD VSS
+    A B   J
+ D_CD4011B IO_4000B MNTYMXDLY={MNTYMXDLY} IO_LEVEL={IO_LEVEL}
.ends
```

Analog Design and Simulation Using OrCAD Capture and PSpice
https://doi.org/10.1016/B978-0-08-102505-5.00018-5

259

The first five lines are comments giving a description of the part and a reference to the data source. On line six, is the subcircuit definition of the CD4011B with three pins A, B, and J. The global power supply is defined by VDD=$G_CD4000_VDD and VSS=$G_CD4000_VSS. The optional parameters, MNTYMXDLY=0 defines the minimum, typical, and maximum delay and the IO_LEVEL which defines one of four AtoD or DtoA interface subcircuits if the digital device is connected to an analog device.

U1 defines a two input nand(2) primitive which has input terminals; VDD, VSS, A, B, and J. The "+" signifies a continuation to the next line. The next line (line 11) declares two models, the timing model, D_CD4011B, which defines the timing characteristics such as propagation delay, setup, and hold times and the I/O model, IO_4000B, which defines the loading and driving characteristics for the gate. Subcircuits always end with a ".ends" statement as in line 12. The model D_CD4011B can be found in the CD4000.lib and the model IO_4000B can be found in the dig_io.lib. More detailed information can be found in the PSpice reference manual.

18.2. DIGITAL CIRCUITS

Digital gates by default do not show their power supply pins because this would require a relatively large number of wires to connect all the gates to the power supply which would overcomplicate the circuit. Instead, TTL and CMOS devices are connected to global power supply nodes which are not displayed and by default are set to 5 V. Different power supplies can be set to accommodate the 3–18 V voltage supply range for CMOS devices. This will not affect the input thresholds and output drives for CMOS devices but the propagation delays will still be defined for a 5 V power supply. For accurate propagation delays, the timing models will have to be modified.

To set digital logic levels on IC pins, it is recommended to use digital HI and LO symbols from the **Place > Power** menu and to use digital PULLUP resistors from the **dig_misc** library to tie a pin high or low via a resistor. **No Connect** symbols, from the **Place** menu can be used to identify unconnected pins. Fig. 18.1 shows the respective Capture symbols and parts.

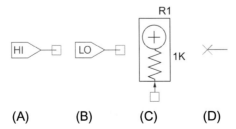

(A)   (B)   (C)   (D)

Fig. 18.1 (A) digital HI, (B) digital LO, (C) digital pullup, and (D) no connect.

In Fig. 18.2, a digital clock signal is applied to the input of an 8-bit binary counter (U1A and U1B). In order to enable the counter, the CLR input is tied low by using a digital LO symbol. Each counter output is connected to an 8-bit bus using bus entry points, **Place > Bus Entry**, select the icon ⌐ ⌐ or press E on the keyboard.

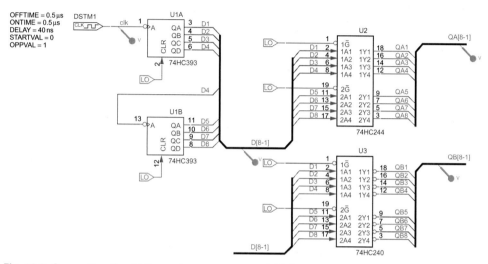

Fig. 18.2 Connecting the 4-bit counter outputs to an 8-bit bus.

Note

From version 16.3, connecting pins can automatically be drawn to a bus. Draw a bus and then select, **Place > Auto Wire > Connect to Bus**. Click on a connecting pin and then click on the bus (you will be prompted to enter the name of the net) then the bus. The bus entry point and wire will automatically be drawn.

Each wire connected to a bus entry point has been labeled, D1, D2, etc. and the bus itself has a net name of D[8–1], the order of which is msb-lsb. The bus on the data inputs to U3 is also named as D[8–1] and will therefore be connected to the 8-bit bus. The bus can also be labeled as; D[7–0] or D[7..0], it is your preference. Only signals of the same type can be grouped together on a bus, mixed busses cannot be defined in Capture. However, in Probe, signals of different types can be collected together and displayed as a bus waveform. Markers can be placed on a bus as well as wires.

18.3. DIGITAL SIMULATION PROFILE

From version 17.2 onward, digital simulation options are presented differently. For pre 17.2 software releases, digital simulation options can be found in the simulation profile under the **Options** tab and selecting **Category: > Gate-level Simulation** as shown in Fig. 18.3. The **Timing Mode** option lets you select the Minimum, Typical, Maximum, or Worst–case timing characteristics for the digital devices. There are four I/O AtoD and DtoA interfaces you can select and most importantly of all, you can initialize all flip-flops to either X for do not care, logic 0 or logic 1. There is also the option to suppress simulation error messages as PSpice will report any digital timing hazards or timing violations.

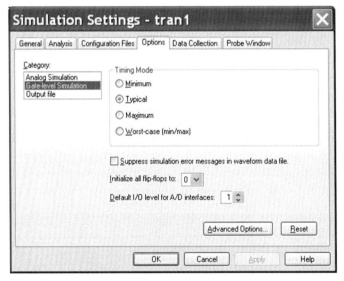

Fig. 18.3 Digital simulation options (pre 17.2).

For post 17.2, the digital simulation options are found under **Options > Gate Level Simulation > General**, see Fig. 18.4. A description of each option is displayed in the bottom window when you click on the parameter name.

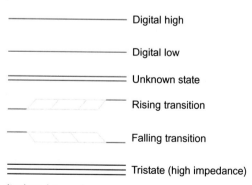

Fig. 18.4 Digital simulation options (post 17.2).

18.4. DISPLAYING DIGITAL SIGNALS

Digital signals are shown as either high or low logic levels. However, for regions of ambiguity where the transition time is not precisely known, the rising and falling transitions are shown in yellow as shown in Fig. 18.5. Unknown states are shown as two red lines and high impedance states are shown as three blue lines.

Digital high

Digital low

Unknown state

Rising transition

Falling transition

Tristate (high impedance)

Fig. 18.5 Digital signals displayed in probe.

Note

One common mistake is not to initialize registers (flip-flops) in a circuit which will result in the two red lines appearing representing an unknown state. Make sure you initialize the flip-flops as shown in Figs. 18.3 and Fig. 18.4.

You can group digital signals together and displayed as a bus in the Probe window. The bus name can be created in the **Trace Expression** field in the **Trace > Add Trace** window. Up to 32 digital signals can be listed with the order msb to lsb with a radix of either; hexadecimal (default), decimal, octal, or binary. For example,

{D4 D3 D2 D1};myBus;d   will display D4 to D1 (msb-lsb) labeled as myBus with decimal numbers

{WR RD CE};control;b   will display bus control in binary

Fig. 18.6 shows a QB[8:1] bus shown by default in hexadecimal. The Dbus has been created from a collection of digital signals and is shown in hexadecimal, decimal, and binary.

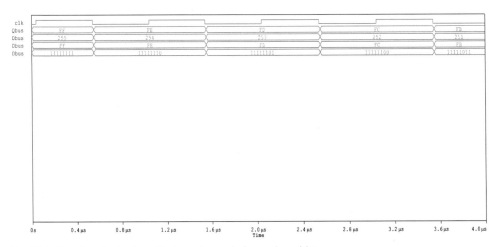

Fig. 18.6 Bus signals displayed in hexadecimal, decimal, and binary.

18.5. EXERCISES

Exercise 1

You will verify the output sequence of a modulus 3 synchronous counter.

1. Create a project called **Mod 3 Counter**. Rename SCHEMATIC1 to **counter** and draw the modulus 3 synchronous counter in Fig. 18.7. The digital flip-flops and the OR gate are from the CD4000 library. The digital HI and LO symbols are from

the **Place** menu or press "F" on the keyboard. The digital stimulus is **digClock** from the source library.

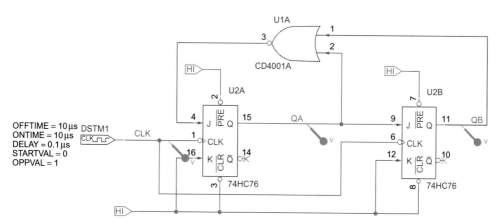

Fig. 18.7 Modulus 3 synchronous counter.

2. Name the nodes as shown. **Place > Net Alias** or press "N" on the keyboard.
3. You need to initialize the flip-flops to 0. If you have pre 17.2 software version, then follow step 3.1 otherwise for post 17.2 versions, follow step 3.2.

    3.1 Set up a PSpice simulation profile for 100 µs and select **Options > Category: Gate-level simulation** and set **Initialize all flip-flops** to 0, see Fig. 18.8.

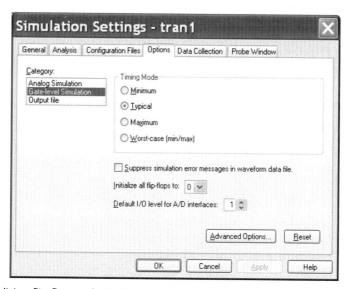

Fig. 18.8 Initializing flip-flops to logic 0 (pre 17.2).

3.2 Set up a PSpice simulation profile for 100 μs and select **Options > Gate Level Simulation > General**. Set the **Value** for DIGINITSTATE to 0 from the pull down menu, see Fig. 18.9.

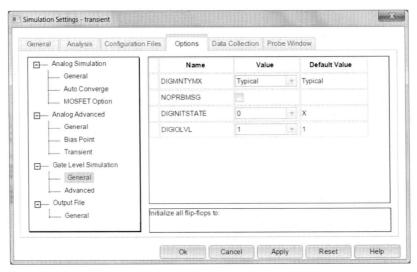

Fig. 18.9 Initializing flip-flops to logic 0 (post 17.2).

4. Place voltage markers on the CLK, QA, and QB nodes.
5. Run the simulation. The trace names will appear in the order that you placed the voltage probes. In Probe, rearrange the trace names such that the CLK is at the top, then QA and QB. This can be done by selectively cutting and pasting. Select CLK and press control-X which will delete the trace, then press control V to paste the trace name. Note that both traces QA and QB are initialized to logic 0.
6. Turn on the cursor and move the cursor along the waveforms. The corresponding logic levels should appear on the *y*-axis as in Fig. 18.10. Note that the output of the flip-flop only changes on the falling edge of the clock signal. The CD4027 flip-flops are negative edge triggered.

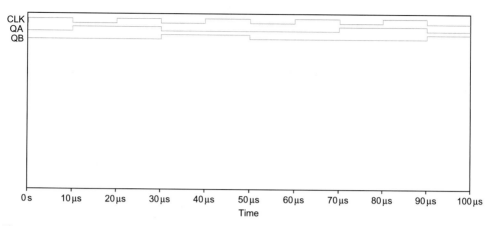

Fig. 18.10 Digital counter waveforms.

7. You will add a bus to show the binary count. Select **Trace > Add** and in the Trace Expression field enter
   {QB,QA};count_b;b
   and click on OK to display the binary count.
8. Select **Trace > Add** and in the Trace Expression field enter
   {QB,QA};count_d;d
   and click on OK to display the decimal count.
9. Select **Trace > Add** and in the Trace Expression field enter
   {QB,QA};count_h;h
   and click on OK to display the hexadecimal count.
10. Your Probe waveforms should resemble those shown in Fig. 18.11 which is that of a modulus 3 counter with a count sequence of 0,1,2,0,1,2,0, etc.

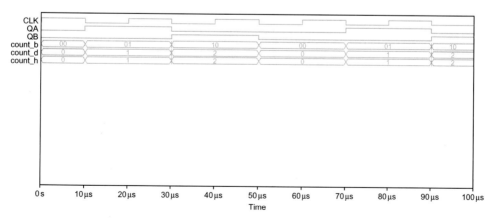

Fig. 18.11 Counter output digital waveforms.

Exercise 2

The circuit in Fig. 18.12 is an example of how to connect signals to a bus and how to select the different A and B sections of an IC. A clock signal is divided down by two 4-bit binary counters, U1A and U1B, and then passed to two octal buffers U2 and U3 via an 8 bit bus. U3 is an inverting octal buffer.

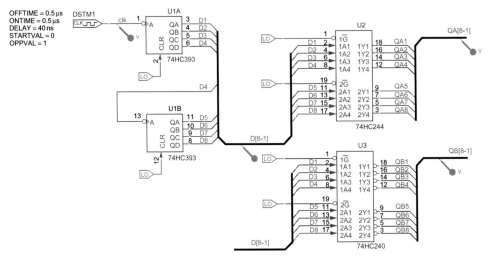

Fig. 18.12 Connecting signals to a bus.

1. The 74HC393 has two identical sections designated A and B shown in the **Packaging** window in the **Place Part** menu in Fig. 18.13. When you select Part B, the pin numbers will change accordingly.

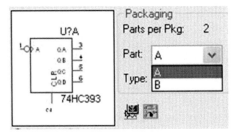

Fig. 18.13 Selecting different parts (sections) from a package.

2. Place the parts for the circuit in Fig. 18.12 but do not connect the wires or bus yet. DSTM1 is a DigClock source from the **source** library and the HC devices can be found in the 74HC library. The HI and LO symbols are from the **Place > Power** menu.

3. Draw the busses in the circuit. To draw an angled bus, hold down the shift key and left mouse click to define the angle and then draw the bus.

4. Starting with U1A pin 3, place a bus entry on the bus as shown in Fig. 18.12. Draw a wire from the bus entry to pin 3 of U1A.

5. Select the wire and place a net alias (press N) labeled D1 on the wire. Press escape or **rmb > End Mode**.

6. Draw a selection box around the wire, net name, and bus entry point as shown in Fig. 18.14.

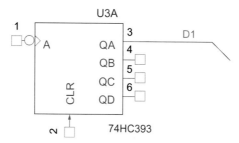

Fig. 18.14 Selecting the wire, bus entry, and net name.

7. Hold the control key down, place the cursor on the wire, and drag it down so it connects to pin 4. The net name automatically increments to D2. With the wire still highlighted, press F4 twice and two more nets, D3 and D4 will appear.

Note

From release 16.3, there is now the option to Auto Wire two points **Select Place > Auto Wire** , select two points and the wire is drawn automatically. Another new feature is the **Place > Auto Wire > Connect to bus**. You click on a connecting pin and then the bus such that the wire and bus entry will be connected automatically. You will also be prompted for the net name.

If you have release 16.3 or later go to (8) else continue to (9).

8. Select **Place > Auto Wire > Connect to bus** and wire up the rest of the IC pins as in Fig. 18.12.

9. The busses are labeled the same as for wires using, **Place Net >** Alias. Make sure all three busses are labeled correctly, [msb–lsb].

10. As explained in Exercise 1, the presentation of the digital simulation options has changed between pre 17.2 and post 17.2 releases. For pre 17.2 follow step 10.1 otherwise for post 17.2 releases, follow step 10.2.

10.1 Set up a simulation profile for a transient analysis with a run to time of 10 μs. Select **Options > Category: Gate-level simulation** and set **Initialize all flip-flops** to 0, see Fig. 18.15.

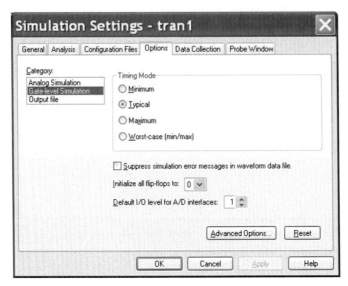

Fig. 18.15 Initializing flip-flops to logic 0 (pre 17.2).

10.2 Set up a simulation profile for a transient analysis with a run to time of 10 μs. Select **Options > Gate Level Simulation > General**. Set the **Value** for DIGINITSTATE to 0 from the pull down menu, see Fig. 18.16.

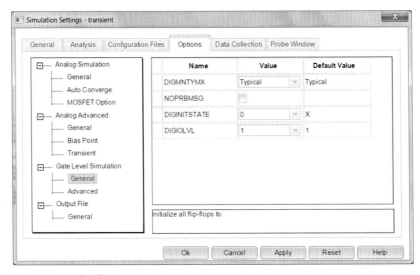

Fig. 18.16 Initializing flip-flops to logic 0 (post 17.2).

11. Place voltage markers on each bus as shown and run the simulation.

12. You should see the count increase for bus D[8–1] and QA[8–1]. As U3 is an inverting buffer, the count for QB[8–1] will start at FF and decrease accordingly. Compare your waveforms with those shown in Fig. 18.17.

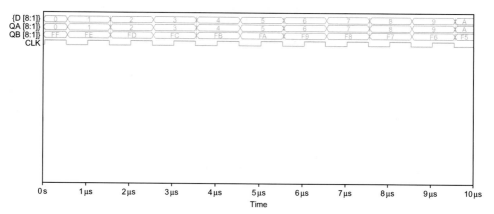

Fig. 18.17 Digital bus waveforms.

13. Delete all the traces, **Trace > Delete All Traces**.

14. Select **Trace > Add Trace**. On the right hand side in the Functions or Macros, select the curly brackets {} the **Trace Expression** box will contain {} with the cursor sitting in the middle waiting for you to select a trace variable name.

15. Select D4 followed by D3, D2, and D1. Add the following text to the expression and click on OK.

    {D4,D3,D2,D1};myBus;d

16. Create a bus called **nibble** consisting of QA4, QA3, QA2, and QA1 and display the bus in binary.

    {QA4 QA3 QA2 QA1};nibble;b

    You should see the waveforms as in Fig. 18.18.

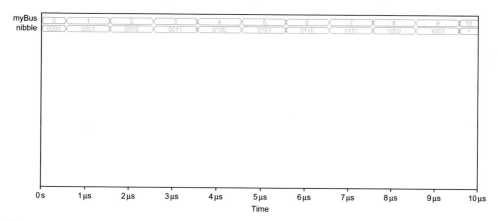

Fig. 18.18 Custom made busses.

17. In Capture disable U2 by setting the enable pins $(\overline{G})$, 1 and 19 high by using a $D_HI symbol from **Place > Power**. Run the simulation.

> **Tip**
>
> Select the $D_HI symbol from the most recently placed part list, see Fig. 18.19. If the symbol is not in the pull down menu, type $D_HI in the box and press return.

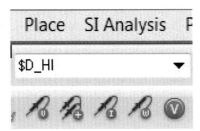

Fig. 18.19 Most recently placed part.

18. Run the simulation.
19. The trace for QA[8–1] is displayed with a Z inside indicating a high impedance tri-state output, see Fig. 18.20.

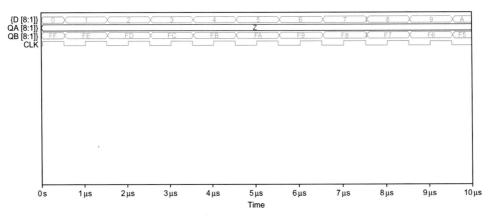

Fig. 18.20 High impedance shown on U2's outputs.

Exercise 3

PSpice reports and plots timing violations relating to setup times, hold times, and minimum pulse width. By decreasing the clock pulse width we can investigate the reporting of these errors.

1. Change the clock OFFTIME to 0.01 μs and ONTIME to 0.01 μs.
2. Reduce the simulation time from 10 to 1 μs.
3. Run the simulation.

4. The Simulation Message window will appear as shown in Fig. 18.21.

Fig. 18.21 Number of simulation messages.

5. Click on **Yes** and you will see a list of Warnings (Fig. 18.22).

Fig. 18.22 List of warning messages.

6. The **Minimum Severity Level** pull down menu lists severity levels for Fatal, Serious, Warning, and Info. For Fatal severity, the simulation stops. Leave the level on Warning and click on **Plot**.

7. The violation information displayed gives the time at which the violation occurred and with which device. The message not only gives you the measured violation timing value but also the specified timing value. PSpice also plots the occurrence of the violated timing waveforms.

Selecting each Time-Message will open up a new Probe plot window. All the reported violations can be viewed in the output file, **VIEW > Output File**.

DIGITAL Message ID#1 (WARNING):

WIDTH/MIN-HIGH Violation at time 50 ns

Device: X_U1A.UHC393DLY

Minimum high WIDTH = 20 ns

NODE: X_U1A.A, measured WIDTH = 10 ns

CHAPTER 19

Mixed Simulation

Chapter Outline

19.1. Exercises   276
    Exercise 1   276
    Exercise 2   279

PSpice uses the same simulation engine for analog and digital circuits. The simulation results in Probe share the same time axis but are split into separate analog and digital plot windows. Analog and digital components in a circuit are connected together at nodes. In PSpice there are three types of connecting nodes: analog, where all connected parts are analog; digital, where all connected parts are digital; and interface, where there is a mixture of analog and digital parts. Interface nodes are automatically separated into one analog node and one or more digital nodes by inserting analog and digital interface subcircuits, which are either analog to digital (A to D) or digital to analog (D to A) interface subcircuits. These subcircuits will also have their own power supply. As this process is automatic and runs behinds the scenes, we do not normally have to worry about the interface subcircuits, although they are available as traces in Probe.

Fig. 19.1 shows an analog comparator with an open collector transistor connected to a digital gate. The pull-up resistor is connected to the digital power supply and the output ground for the comparator is connected to digital ground. Fig. 19.2 shows the digital waveforms being plotted in the upper area of Probe and the analog waveforms plotted in the lower area.

Analog Design and Simulation Using OrCAD Capture and PSpice
https://doi.org/10.1016/B978-0-08-102505-5.00019-7

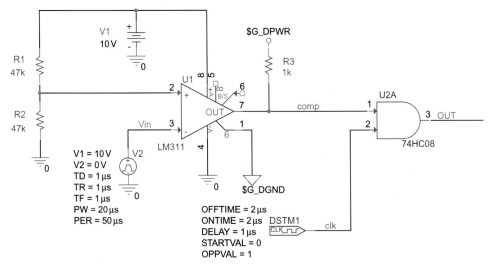

Fig. 19.1 Analog comparator switching a digital gate.

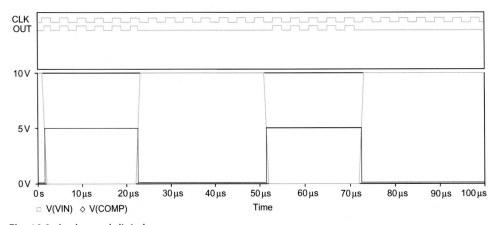

Fig. 19.2 Analog and digital traces.

Mixed analog and digital circuits follow the same procedure for placing parts, creating a simulation profile and simulation.

19.1. EXERCISES

Exercise 1

Fig. 19.3 shows an AD7224 digital to analog converter (DAC) with an input digital data word of 0111 1111. From the manufacturer's data sheet, the output voltage is given by:

$$V_o = \frac{V_{REF} \times 127}{256} = 4.96\,V \qquad (19.1)$$

The timing cycles for the DAC have been set up according to the manufacturer's datasheet.

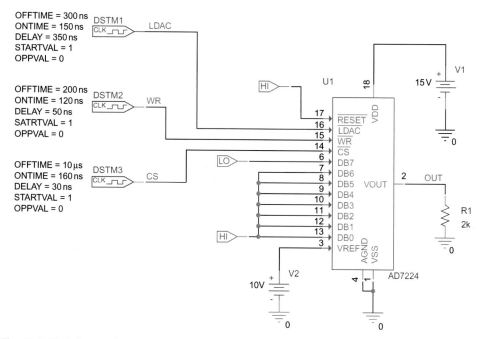

Fig. 19.3 Digital to analog conversion using the AD7224.

1. Draw the circuit in Fig. 19.3. The AD7224 can be found in the DATACONV library and the DigClock stimuli can be found in the source library.

Note

As explained in the Digital Simulation Chapter, the presentation of the digital simulation options in the PSpice Simulation Settings have changed between pre 17.2 and post 17.2 releases.

If you have a pre 17.2 release (16.6 or before) then set up a simulation profile for a transient analysis with a **Run to Time** of 5 μs then select **Options > Category: Gate–level simulation** and set **Initialize all flip-flops** to 0, see Fig. 19.4. Close the Simulation Settings.

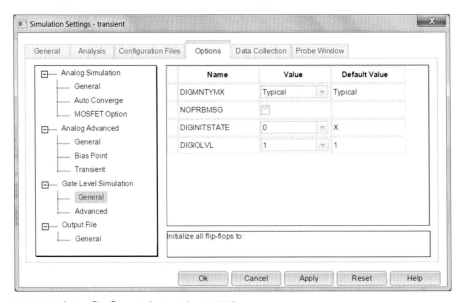

Fig. 19.4 Initializing flip-flops to logic 0 (pre 17.2).

If you have a post 17.2 release (17.2 and onwards) then set up a simulation profile for a transient analysis with a **Run to Time** of 5 μs then select **Options > Gate Level Simulation > General**. Set the **Value** for DIGINITSTATE to 0 from the pull down menu, see Fig. 19.5. Close the Simulation Settings.

Fig. 19.5 Initializing flip-flops to logic 0 (post 17.2).

2. Place voltage markers on the nets, LDAC, WR, CS and OUT.
3. Run the simulation.
4. In Probe, you will see that the upper plot is for the digital signals and the lower plot is for the analog OUT signal (Fig. 19.6).

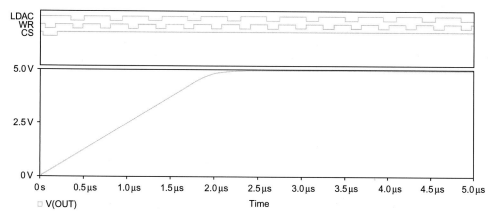

Fig. 19.6 Analog and digital waveforms.

5. Turn the cursor on and check that the output voltage is 4.96 V as calculated.

Exercise 2

Fig. 19.7 shows the ubiquitous NE555 timer, used in countless applications. The timing equations for the NE555 are given as:

$$f = \frac{1.44}{(RA + 2RB)C} \tag{19.2}$$

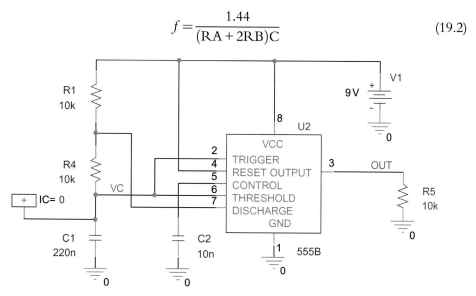

Fig. 19.7 NE555 clock oscillator.

$$\text{Duty cycle} = \frac{RA + RB}{RA + 2RB} \qquad (19.3)$$

Using the components as shown in Fig. 19.7 will give a calculated clock frequency of 218 Hz and a duty cycle of 0.67.

1. Create a new project called **Clock Oscillator**. Rename SCHEMATIC1 to **clock** and draw the circuit in Fig. 19.7; the 555 can be found in the anl_misc library. There are three versions, 555alt, 555B and 555C, which have the pins arranged differently. Do not forget to place an initial condition, IC1, from the special library on C1.
2. Create a transient simulation profile for 20 ms. Place markers on VC and on OUT.
3. Run the simulation.
4. Display the cursors and determine the period of oscillation and hence the clock frequency.
5. Determine the duty cycle, which is the on time divided by the off time.
6. Confirm your measurements by selecting **Trace > Evaluate Measurement** and Period(1) then selecting V(OUT).
   ```
   Period(V(OUT))
   ```
7. Select **Trace > Evaluate Measurement** and Period_XRange, (1,begin_x,end_x) then select V(OUT), and then enter 5 and 20 m.
   ```
   Period_XRange(V(out),5m,20m)
   ```
8. Confirm your measurements by selecting **Trace > Evaluate Measurement** and DutyCycle(1), then selecting V(OUT).
   ```
   DutyCycle(V(OUT))
   ```
9. If the results of the measurements are not shown, select **View > Measurement Results**. Your results should be similar to those shown in Fig. 19.8.

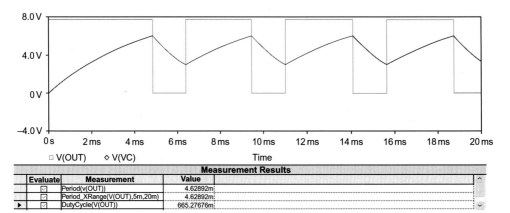

Evaluate	Measurement	Value
☑	Period(v(OUT))	4.62892m
☑	Period_XRange(V(OUT),5m,20m)	4.62892m
☑	DutyCycle(V(OUT))	665.27676m

Fig. 19.8 Resultant waveforms and measurement values.

The Cadence\OrCAD software installation includes a good selection of analog, digital and mixed example circuits in the anasim, digsim and mixsim directories. These can be found in the installed directory, for example:

```
<install path>\Cadence\SPB_16.3\tools\pspice\capture_samples
<install path>\Cadence\OrCAD_16.3\tools\pspice\capture_samples\
```

CHAPTER 20

Creating Hierarchical Designs

Chapter Outline

20.1. Hierarchical Ports and Off-Page Connectors   285
20.2. Hierarchical Blocks and Symbols   287
    20.2.1 Hierarchical Blocks   287
    20.2.2 Hierarchical Symbols   288
20.3. Passing Parameters   289
20.4. Hierarchical Netlist   290
20.5. Exercises   291
    Exercise 1   291
    Exercise 2   294
    Exercise 3   300
    Exercise 4   302
    Exercise 5   306

Capture designs can either be flat, in which signals are connected across pages in the design, or hierarchical, in which the design is partitioned into blocks and signals transverse up and down the hierarchy. Flat designs are represented in the Project Manager as having a single schematic folder with a number of associated pages, whereas hierarchical designs will have more than one schematic folder (Fig. 20.1). Each schematic folder in the hierarchy will be represented by a hierarchical block in a schematic. By selecting a hierarchical block, you select the underlying schematic and effectively descend the hierarchy.

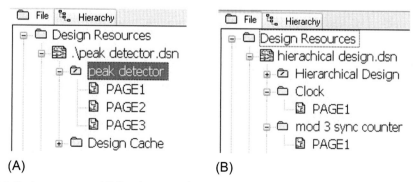

Fig. 20.1 Project structure: (A) flat design and (B) hierarchical design.

Analog Design and Simulation Using OrCAD Capture and PSpice
https://doi.org/10.1016/B978-0-08-102505-5.00020-3

For the flat design in Fig. 20.1, there is one schematic folder and three pages. For the hierarchical design, there are three schematic folders in the hierarchy each with their own schematic page or pages.

The Project Manager in Fig. 20.2A shows two schematic folders, **Top** and **Bottom**. The associated schematics are shown in Fig. 20.2B and C, respectively. The **Top** schematic (Fig. 20.2B) contains a hierarchical block called **Bottom** which has two hierarchical pins, IN and OUT. To descend the hierarchy, you highlight the block and **rmb > Descend Hierarchy**, or you can double click on the block and the **Bottom** schematic will be displayed. The connection between the block and the schematic is provided by the hierarchical pins on the Bottom block having the same name as the hierarchical ports in the Bottom schematic, IN and OUT.

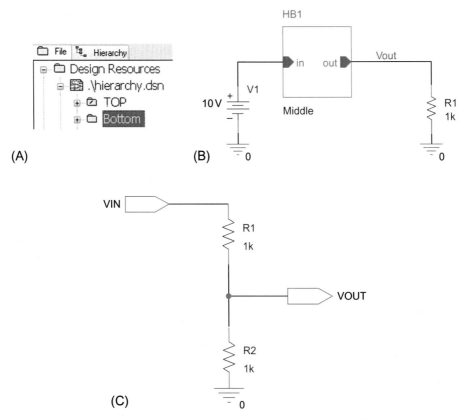

Fig. 20.2 Hierarchical project: (A) Project Manager; (B) top schematic; and (C) bottom schematic.

In the Project Manager, there is also a Hierarchy tab next to the default **File** tab. By selecting the **Hierarchy** tab, the location of individual parts in the design can be displayed. Fig. 20.3 shows that there is a resistor R1, a voltage source V1 and a Hierarchical block HB1 in the Top level schematic and two resistors in the Bottom schematic.

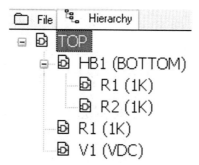

Fig. 20.3 Design hierarchy showing location of individual components.

20.1. HIERARCHICAL PORTS AND OFF-PAGE CONNECTORS

As in the case for flat designs, there is normally one folder and one or more pages. In order to connect signals across the pages, off-page connectors are used: **Place > Off-Page Connectors** (Fig. 20.4).

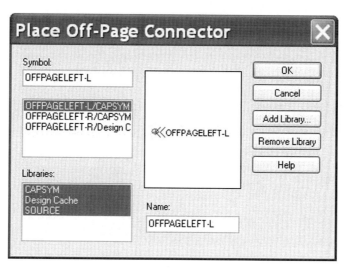

Fig. 20.4 Off-page connectors.

Two types are used to indicate the direction of the data flow, i.e. input to output. When a wire connects to an off-page connector, the net name of the wire inherits the name of the connector.

Hierarchical ports connect signals between levels of hierarchy: **Place > Hierarchical Ports** (Fig. 20.5). As with off-page connectors, a wire connected to a hierarchical port inherits the name of the port.

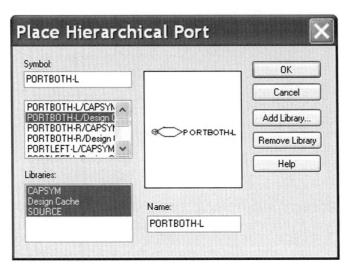

Fig. 20.5 Place hierarchical ports.

Different hierarchical ports are available which represent the type of port and direction of data flow. Fig. 20.6 shows the types of port available. For example, PORTRIGHT-R is a port that points to the right and has a connection on the right-hand side. Which port you use is entirely your choice.

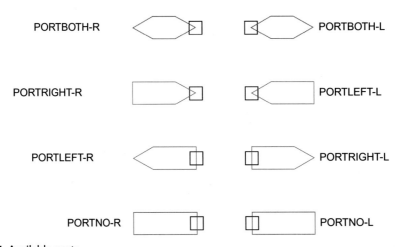

Fig. 20.6 Available ports.

20.2. HIERARCHICAL BLOCKS AND SYMBOLS

Hierarchical blocks are normally used for top-down designs where the block is drawn on the top-level schematic and associated signal pins are added. Pushing down into the block (descending the hierarchy), the referenced schematic contains the same number of ports as pins with the same associated signal names. The hierarchical blocks cannot be saved to a library as they are drawn "on the fly" and saved within the schematic file.

Hierarchical symbols are normally used for bottom-up designs where the schematic is drawn first and ports are added to the input and output signals. A symbol is then created with the same number of signal pins and associated names. These hierarchical symbols can be saved to a library for use in other designs.

20.2.1 Hierarchical Blocks

Hierarchical blocks are created "on the fly" in the schematic: **Place >Hierarchical Block** (Fig. 20.7).

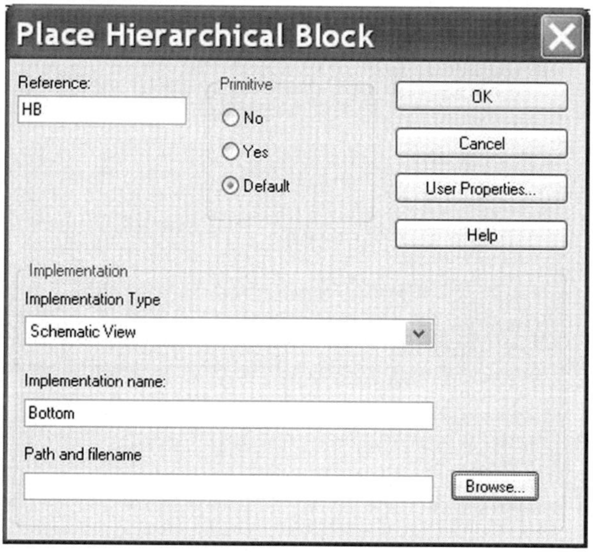

Fig. 20.7 Creating a hierarchical block.

You define the **Reference** designator, which is your choice, and you then have the option to select the **Implementation Type** and **Implementation Name**. The implementation type defines what the block is referencing and can be any of the types shown in Fig. 20.8. Normally for PSpice projects, the **Schematic View** is selected and the **Implementation Name** is the name of the schematic.

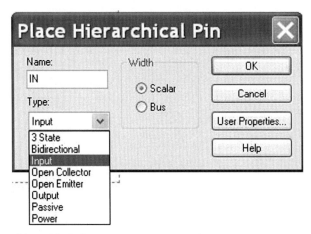

Fig. 20.8 Implementation types.

You then draw a rectangle for the block and with the block still highlighted select **Place > Hierarchical Pin**. In the **Place Hierarchical Pin** box, you define a name for the pin and its type from the pull-down menu as shown in Fig. 20.9. The **Width** option allows you to place a pin representing a bus or a scalar for a single pin. You then place the pin anywhere on the perimeter of the block.

Fig. 20.9 Defining a hierarchical pin.

20.2.2 Hierarchical Symbols

This is where a circuit is effectively symbolized and a Capture part is generated to represent that circuit. Hierarchical ports are added to the circuit which will appear as pins on the hierarchical symbol. In Capture, a part is generated by selecting Tools > Generate Part (Fig. 20.10). You select the Netlist/source file type as a Capture Schematic/Design file (.dsn), select the name and location of the Part library and select the name of the schematic folder (Source Schematic name), and a hierarchical symbol will be generated.

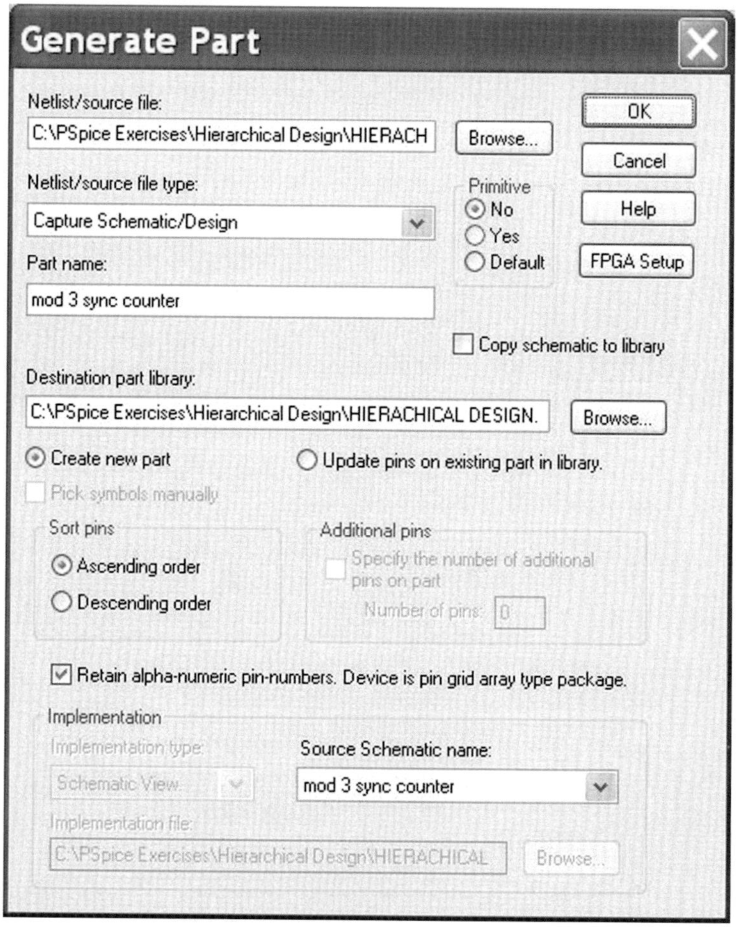

Fig. 20.10 Generating a capture hierarchical symbol.

20.3. PASSING PARAMETERS

Parameters can be passed between levels of hierarchy using the **Subparam** part from the **Special** library. This allows different parameters to be passed to hierarchical blocks or symbols. For example, you may have a filter block where the gain of the filter is programmed by a single resistor. Using the **Subparam** part, different resistor values can be passed down to each filter to set up different filter gains. Fig. 20.11 shows one such implementation where a different value for RVAL sets different filter gains for HB1 and HB2.

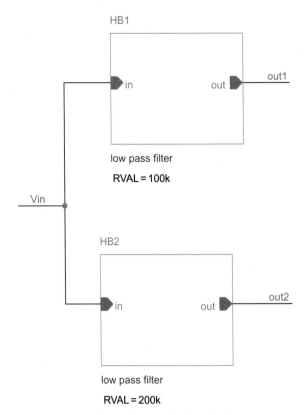

Fig. 20.11 Passing subparam RVAL values to hierarchical symbols.

20.4. HIERARCHICAL NETLIST

PSpice can generate a hierarchical netlist such that instantiated subcircuit definitions will only appear once in the netlist. In Fig. 20.12, there are two subcircuits declared at the top level, X_U1 and X_U2, which reference the subcircuit, osc125Hz. Rather than include the text for both subcircuit definitions in the netlist, only reference calls are made to the subcircuit.

```
* source HIERARCHY
V_V1            N00673 0 12V
V_V2            N02404 0 -12V
X_U1 OUT1 N00673 N02404 osc125Hz PARAMS: RVAL=160k
X_U2 OUT2 N00673 N02404 osc125Hz PARAMS: RVAL=160k

.SUBCKT osc125Hz OUT VCC VSS PARAMS: RVAL=160K
C_C1            N24151 0  0.01u IC=0 TC=0,0
R_R1            N24187 OUT  160k TC=0,0
R_R2            N24151 OUT  160k TC=0,0
X_U1A           N24187 N24151 VCC VSS OUT AD648A
R_R3            0 N24187  910k TC=0,0
.IC             V(N24151 )=0
.ENDS
```

Fig. 20.12 Hierarchical netlist.

20.5. EXERCISES

Whenever you create a hierarchical design, it is recommended that no power supplies be included in underlying schematics. Power supply ports should be included on the hierarchical symbol and blocks such that the overall power supply connections are made at the top level where they are visible. The ground symbol, though, is global to all designs and so this does not need a port unless you are using separate digital and analog grounds.

Exercise 1

You will create a hierarchical top-down design shown in Fig. 20.13, where the top block references an underlying bottom schematic. Hierarchical blocks cannot be saved to a library.

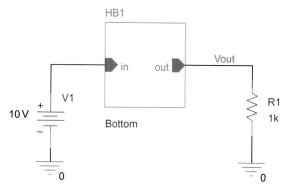

Fig. 20.13 Block diagram project.

1. Create a new project called Top Down Design.
2. In the Project Manager, rename SCHEMATIC1 to Top.
3. Create a new schematic called bottom, highlight the Top Down Design.dsn file and **rmb > New Schematic** and name it Bottom.
4. In the Project Manager, select the schematic Bottom, **rmb > New Page** and accept the default Page1 name.
5. Your Project Manager should look like that shown in Fig. 20.14.

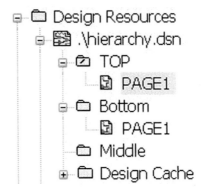

Fig. 20.14 Hierarchical design folders.

6. Open the schematic for Bottom and draw the circuit in Fig. 20.15. For the hierarchical ports, **Place > Hierarchical Port** and place a Hierarchical port (PORTRIGHT-L) on the output node **out** and a PORTRIGHT-R on nodes **in** and **out**. Save and close the schematic page.

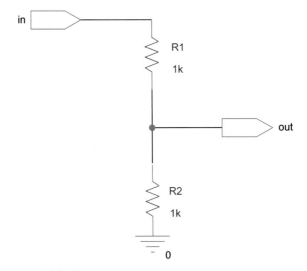

Fig. 20.15 Resistor potential divider.

7. Open the schematic for the Top folder.
8. In the Project Manager, select **Place > Hierarchical Block** and enter a value for the reference designator. In **Implementation Type**, select **Schematic View** and enter **Bottom** as the **Implementation name** as shown in Fig. 20.16. Leave the **Path and filename** blank as the schematic is part of the project. Click on OK.

Place Hierarchical Block

Reference:
HB1

Primitive
○ No
○ Yes
⦿ Default

OK
Cancel
User Properties...
Help

Implementation
Implementation Type
Schematic View

Implementation name:
Bottom

Path and filename

Browse...

Fig. 20.16 Defining the hierarchical block and the referenced schematic.

9. When the cursor changes to a cross-hair, mouse click once and then draw a rectangular block. The hierarchical port names on the bottom schematic will appear as hierarchical pins on the block. Move the pins as shown in Fig. 20.17.

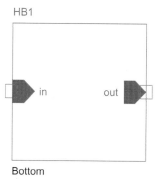

HB1

in   out

Bottom

Fig. 20.17 Hierarchical block with pins added.

10. Select the block and **rmb > Descend Hierarchy** (or double click on the block) and the underlying Bottom schematic will appear.
11. In the schematic, **rmb > Ascend Hierarchy** will take you back up to the top level.
12. Place a V_{DC} voltage source on the input and a resistor on the output as shown in Fig. 20.18. Place a voltage marker on the output and run a bias simulation and confirm the output voltage of 5 V.

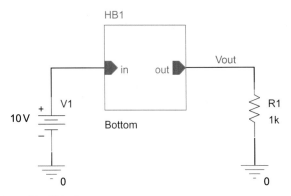

Fig. 20.18 Testing hierarchical block.

Exercise 2

Creating a Hierarchical Symbol Which Can Be Saved to a Library

1. Create a new project called Hierarchy. In the Project Manager, rename SCHE-MATIC1 to osc125Hz (Fig. 20.19).

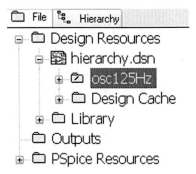

Fig. 20.19 Rename SCHEMATIC1 folder to osc125Hz.

2. Draw the circuit in Fig. 20.20. The AD648 opamp is from the opamp library. If you are using the eval version, use the uA741 opamp from the eval library. For the hierarchical ports, **Place > Hierarchical Port** (Fig. 20.21) and place a hierarchical port (PORTRIGHT-L) on the output node **out** and a PORTRIGHT-R for the VCC and VSS power supply connections. Name the power ports VCC and VSS, respectively, and draw a short wire to each port as shown in Fig. 20.20.

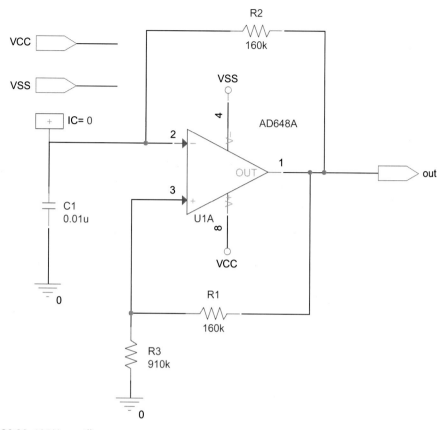

Fig. 20.20 125 Hz oscillator.

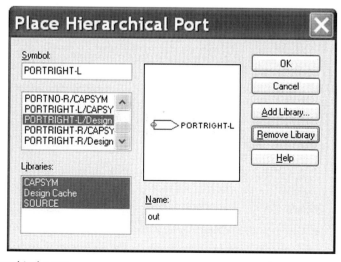

Fig. 20.21 Hierarchical ports.

The wires connected to the hierarchical ports for VCC and VSS will automatically be named VCC and VSS, respectively, and hence will be connected to the opamp power pins.

In order for the circuit to oscillate, an initial condition of 0 V needs to be placed on the capacitor. Place an IC1 from the special library on the capacitor C1, as shown in Fig. 20.20.

> **Tip**
> When you first place the hierarchical ports and the net name is some distance away from the port, rotate the port four times and the net name will be placed closer to the port.

3. Save the Project.
4. Create a hierarchical symbol by highlighting the dsn file (Hierarchy.dsn) in the Project Manager and select **Tools > Generate Part**.
5. In the **Generate Part** window, select **Netlist/source file type**: to **Capture Schematic/Design**. Then in the **Netlist/source file**, browse to the **Hierarchy** project folder and select the Hierarchy.dsn file. Do not exit yet.
6. You need to create a library for the new hierarchical part, so in the **Destination Part Library**, call the library Hierarchy and save it in the PSpiceExercises folder or a folder of your choice.
7. Note that the **Source Schematic name** now shows the schematic name osc125Hz rather than SCHEMATIC1. Your **Generate Part** window should look similar to that shown in Fig. 20.22.

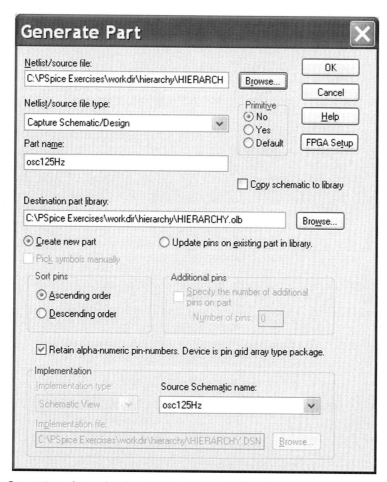

Fig. 20.22 Generating a hierarchical symbol for the osc125Hz schematic.

8. If you are using version 16.3, the **Split Part Section Input spreadsheet** will open. Click on **Save** and OK.

9. In the Project Manager, you should see the hierarchy.olb library file added to the **Outputs** folder. Move the hierarchy.olb from the **Outputs** folder (or you can cut and paste) to the **Library** folder as shown in Fig. 20.23 and expand the library to see the osc125Hz library part.

Fig. 20.23 The Hierarchy.olb Capture library has been added to the Library folder.

10. Double click on the osc125Hz part in the Hierarchy library, which will open the Part Editor and display the part as shown in Fig. 20.24.

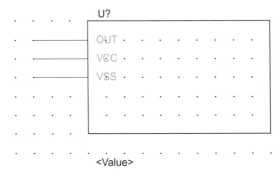

Fig. 20.24 The generated osc125Hz symbol.

11. Double click on **<Value>** and enter a Value of osc125Hz as shown in Fig. 20.25.

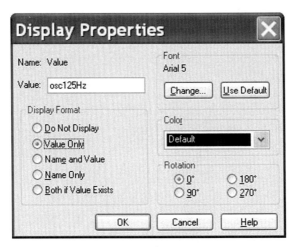

Fig. 20.25 Entering the displayed value property of osc125Hz.

12. Double click on the pin connected to OUT, and in **Pin Properties**, select the **Shape** as Short, **Type** as Output and pin **Number** as 1 (Fig. 20.26) and click on OK.

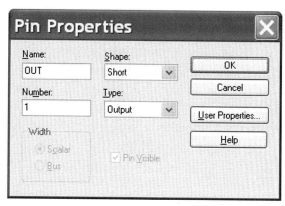

Fig. 20.26 Pin properties for the osc125Hz.

13. Change the VCC pin shape to **Short** of type **Input** and pin number 2.
14. Change the VSS pin shape to **Short** of type **Input** and pin number 3.
15. Double click anywhere in the Part Editor to display the **User Properties** box (Fig. 20.27). Highlight **Pin Numbers Visible** and select True from the pull-down menu. Click on OK.

Fig. 20.27 User properties.

16. Move the output pin from the left- to the right-hand side and resize the part. Your osc125Hz symbol should look similar to the symbol in Fig. 20.28.

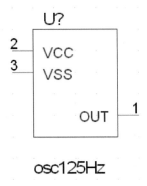

Fig. 20.28 Modified osc125Hz symbol.

17. Double click anywhere in the Part Editor to display the User Properties window, which displays the generated properties. Click on OK. Close the Part Editor by closing the Hierarcy.olb window (click on the upper right-hand cross in Windows).

Exercise 3

Testing the Oscillator in a Hierarchy

1. We need to test the osc125Hz hierarchical symbols. Create a new schematic by selecting **Hierarchy.dsn** and **rmb > New Schematic** and name it **Test Osc125Hz**. By default, the new schematic is named SCHEMATIC1. Highlight SCHEMATIC1 and **rmb > Rename** to **Test Osc125Hz**.

2. Highlight **Test Osc125Hz** and **rmb > New Page**, accept the name **PAGE1** and click on OK. Your Project Manager window should look like that in Fig. 20.29.

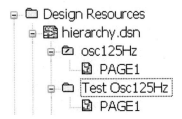

Fig. 20.29 Project Manager schematic folders.

3. Double click on PAGE1 in Test Osc125Hz and draw the circuit diagram in Fig. 20.30. The osc125Hz can be found in the **Hierarchy** library. Make sure V1 is 12V and V2 is −12V.

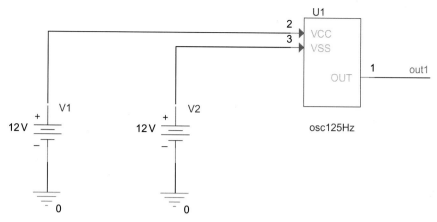

Fig. 20.30 Test circuit for osc125Hz.

4. Create a simulation profile called transient and run the simulation for 200 ms.
5. Try to place a voltage marker on the net **out1**. You will get a message as shown in Fig. 20.31.

Fig. 20.31 Hierarchy message.

Note

The Test Osc125Hz should be at the top level known as the root schematic. You can run a simulation on the Osc125Hz oscillator but not on the Test Osc125Hz as it is not in the hierarchy. This is an ideal way to progressively simulate a circuit from the bottom upwards to test each level of the hierarchy.

---

6. In the Project Manager, highlight the **Test Osc125Hz** folder and **rmb > Make Root**. You will get another message saying that the design must be saved first.
7. Save the design and highlight the **Test Osc125Hz** and **rmb > Make Root**. This time the **Test Osc125Hz** will be placed at the top of the hierarchy and a slash symbol will appear in the yellow folder (Fig. 20.32). If you do not see the response as shown, check the power supplies and make sure that you placed an initial condition on C1.

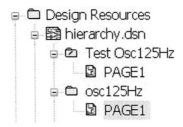

Fig. 20.32 Making Test Osc125Hz the root schematic.

8. You will have to create another simulation profile before you place the voltage marker.
9. Run the simulation. You should see the oscillator response as in Fig. 20.33.

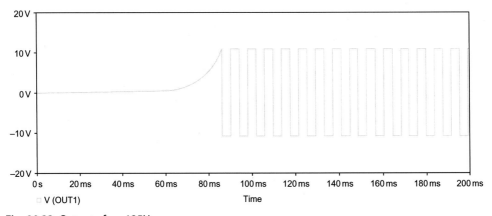

Fig. 20.33 Output of osc125Hz.

Exercise 4

You will create a subparameter in the osc125Hz circuit which will appear on the hierarchical symbol such that a value can be entered on the symbol and passed down to the schematic.

1. Modify the osc125Hz circuit as shown in Fig. 20.34. The **subparam** part can be found in the **special** library.

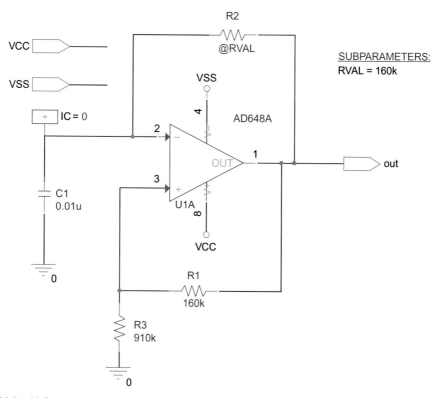

Fig. 20.34 Making R2 a hierarchical parameter.

2. Double click on the subparam parameter and in the Property Editor, add a new row (or column) adding RVAL with a default value of 160k. Select RVAL and **rmb > Display** (Fig. 20.35). In the **Display Properties** window, select **Name and Value** (Fig. 20.36).

RVAL	160k	
Source Library	*C:\CADENCE\SPB_ 16.*	Pivot
Source Package	*SUBPARAM*	
Source Part	*SUBPARAM.Norma*	Edit...
Value	SUBPARAM	Delete Property
		Display...

◄ ► \ **Parts** ⌐ Schematic Nets ⌐ ⌐

Fig. 20.35 Creating an RVAL property with a value of 160k.

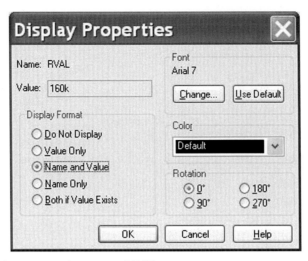

Fig. 20.36 Displaying name and property of RVAL.

3. In the schematic, replace the value of R2 with @RVAL as shown in Fig. 20.34.
4. Save the schematic.
5. Generate a Part for the new osc125Hz as in Exercise 2 and modify the pins in the Part Editor as described in Exercise 2.

Note

The schematic will still contain the previous osc125Hz part, so you will need to delete the old part and replace it with the new part. Alternatively, you can update the part in the Design Cache. The Design Cache is effectively a library which contains all the parts in the schematic. When you delete a part from the schematic, the Design Cache still holds that part until you select **Cleanup Cache**. The Design Cache can be used to update parts and replace parts. From version 16.6, the **Replace Cache** can now be applied to more than one part. Multiple parts can be selected by holding down the Control or Shift key.

---

6. In the Project Manager, expand the Design Cache and select the osc125Hz (Fig. 20.37).

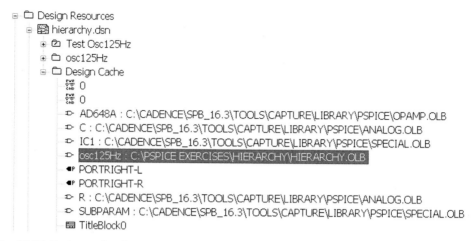

Fig. 20.37 Design cache showing parts in the schematic.

7. Select **rmb > Update Cache**. Click on YES to Update Cache and click on OK when asked if you want to save the design.
8. Your Test Osc125Hz schematic should contain the new osc125Hz part as shown in Fig. 20.38.

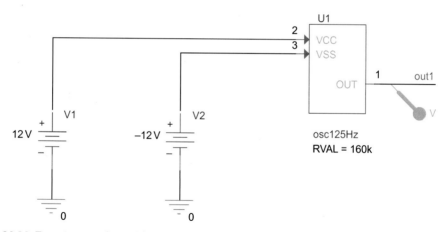

Fig. 20.38 Test circuit with modified osc125Hz part.

9. Modify the RVAL to 100k and run the simulation. You should see a different oscillator period displayed in Probe, as shown in Fig. 20.39.

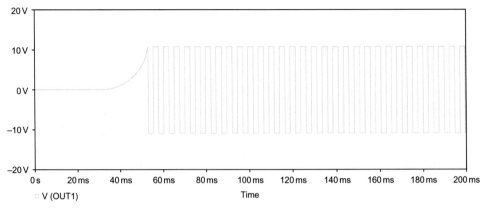

Fig. 20.39 Using a value of 100k for the oscillator.

Exercise 5

The **Digital Counter** hierarchical design presented here will be used in the Test Bench in Chapter 22.

Fig. 20.40 shows a hierarchical design put together based on the 555 clock oscillator in Chapter 19 connected to the mod 3 sync counter in Chapter 18. The modified clock and counter circuits that will be used are shown in Figs. 20.41 and 20.42, respectively.

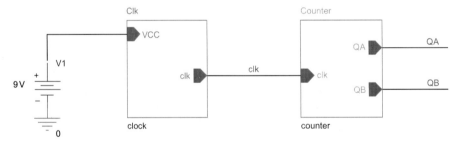

Fig. 20.40 Hierarchical design of digital counter.

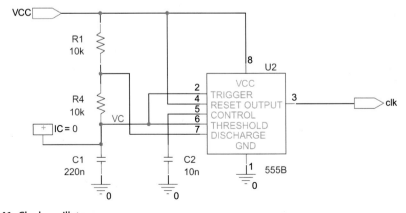

Fig. 20.41 Clock oscillator.

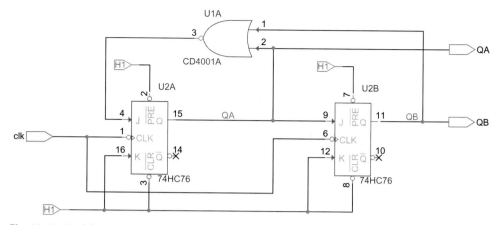

Fig. 20.42 Modulus 3 counter.

1. Create a new project **Digital Counter** and rename SCHEMATIC1 to **Digital Counter**.
2. Open the project **Clock Oscillator** from Chapter 19, Exercise 2.
3. Place the two Project Managers side by side. You will copy and paste the **clock** schematic folder from the **Clock Oscillator** Project Manager to the **Digital Counter** Project Manager.

   Highlight the **clock** schematic folder and press control-C. Highlight **Digital Counter.dsn** and press control-V. Close the Clock Oscillator project.

   If you did not rename SCHEMATIC1 in the **Clock Oscillator** to clock, select **rmb > Rename**. Close the **Clock Oscillator** project.
4. Open the **Mod 3 Counter** project in Chapter 18, Exercise 1, and as in Step 3, copy the **counter** schematic to the **Digital Counter** Project Manager. Close the **Mod 3 Counter** project. Your Project Manager should appear as in Fig. 20.43.

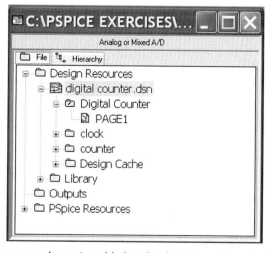

Fig. 20.43 Clock and counter schematics added to the digital counter project.

You will create a hierarchical design with two hierarchical blocks at the top level referencing the clock and counter circuits as seen in Figs. 20.40 and 20.41. Hierarchical ports will be added to the clock and counter circuits such that hierarchical pins will automatically be added when the hierarchical blocks are created.

5. Open the **clock** schematic and delete the 9 V voltage source, V1, and its associated 0 V ground symbol. Delete the load resistor connected to pin 3 of the 555 and delete the net name **out** (if added).

6. Select **Place > Hierarchical Port**. Select the PORTRIGHT–L and name it **clk** (Fig. 20.44). Click on OK and add the port to the clock output node. Place a hierarchical port for VCC on the VCC node. It really is your choice as to which type of hierarchical port symbol you use for input and output ports. Your circuit should be similar to that in Fig. 20.41.

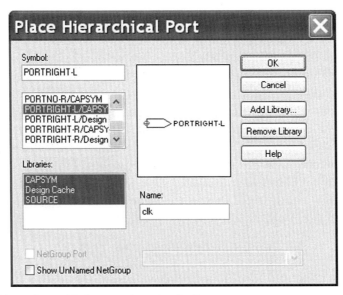

Fig. 20.44 Adding a hierarchical port and naming it clk.

7. Double click on the **clk** port to open the **Property Editor**. Click on the pull-down menu for the **Type** property value and select **Output** (Fig. 20.45). Close the **Property Editor**. The VCC port will be of type **input** by default. Close and save the schematic.

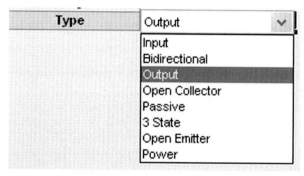

Fig. 20.45 Hierarchical port types.

Note

By default, hierarchical input ports are placed on the left-hand side and output ports on the right-hand side of hierarchical blocks. You can always change the port types by double clicking on the hierarchical block pins.

8. Open the counter schematic and remove the digital clock source. Place hierarchical ports on the clk, QA and QB wires as shown in Fig. 20.42. As in Step 7, change the QA and QB hierarchical ports to type **output**. Close and save the schematic.

9. Open the top-level **Digital Counter** schematic, select **Place > Hierarchical Block** and enter the details as shown in Fig. 20.46. Leave the **Path and filename** empty. Click on OK and draw a rectangular block. The VCC and clk pins will appear as seen in Fig. 20.40.

Fig. 20.46 Creating a hierarchical block.

If no pins appear on the block, double click on the hierarchical block to open the **Property Editor** and check to see that the **Implementation** value is named correctly as **clock**.

Note
If you have forgotten, for example, to add a VCC port, add the port in the clock schematic and **rmb > Ascend Hierarchy**. Highlight the clock block and **rmb > Synchronize Up**. The VCC port will be added as a hierarchical pin to the block.

---

10. Draw another hierarchical block as in Step 9 and name this **counter** (Fig. 20.47). Click on OK and position the hierarchical pins as shown in Fig. 20.40.
11. Complete the circuit as in Fig. 20.40.

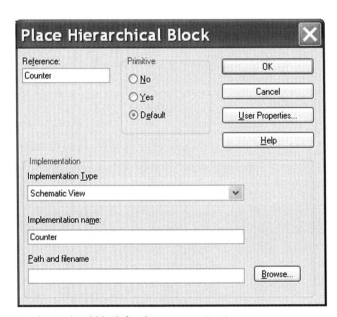

Fig. 20.47 Creating a hierarchical block for the counter circuit.

CHAPTER 21

Magnetic Parts Editor

Chapter Outline

21.1. Design Cycle   311
21.2. Exercises   312
     Exercise 1   312
     Exercise 2   325
     Exercise 3   328

The Magnetic Parts Editor (MPE) is used for the design of transformers and inductors in switched mode power supply topologies. In particular, the MPE provides a complete transformer design cycle for forward converters, both single and double switch, and a flyback converter operating in discontinuous conduction mode. At the end of the design cycle, the MPE generates a comprehensive data sheet which manufacturers can use for the fabrication of transformers and inductors. The MPE also generates the inductor and transformer PSpice simulation models.

Included with the MPE is a database of commercially available magnetic parts such as wire types, insulation material, bobbins and magnetic cores. You can add to this database by creating your own magnetic parts.

21.1. DESIGN CYCLE

The design cycle consists of a series of design steps which are numbered as you progress through the design. The design steps for a DC-DC converter using the flyback topology will be presented here as an example.

The specifications are
- DC input minimum: 50 V
- DC output 12 V with <100 mV pk-pk ripple
- DC output 0.5 A with <5 mA pk-pk ripple
- switching frequency: 40 kHz
- efficiency: 75%
- maximum duty cycle: 45%

The MPE is started from the PSpice Accessories menu.

Analog Design and Simulation Using OrCAD Capture and PSpice
https://doi.org/10.1016/B978-0-08-102505-5.00021-5

21.2. EXERCISES

Exercise 1

Start the MPE from the Start menu. Start > All Programs > Cadence (OrCAD) <release number> PSpice Accessories > Magnetic Parts Editor

Step 1: Component Selection

The first design step in MPE is the selection of one of the components (topologies) to be designed (Fig. 21.1), in this case a flyback converter (discontinuous conduction mode). Select **File > New** and the Component Selection window will open as shown in Fig. 21.1.

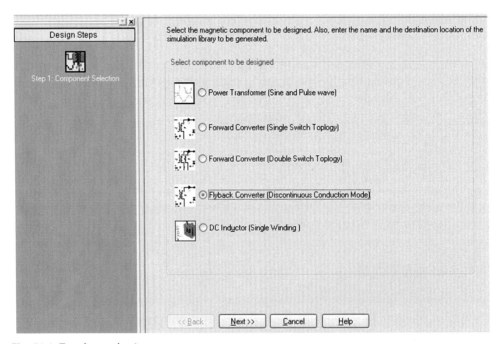

Fig. 21.1 Topology selection.

On the left-hand side of the window are the completed design steps, which allow you to keep track of which step in the design cycle you are at. You can click on any design step at any time to go back to review the parameters entered. On the right-hand side are the available design components (topologies).

Select Flyback Converter and click on Next >>.

Step 2: General Information

The second step is to enter the design specifications (Fig. 21.2) for the transformer. In this case there will be one secondary; the insulation material will be nylon with a current density of $3\,A/mm^2$. The efficiency, from the given specifications, is 75%.

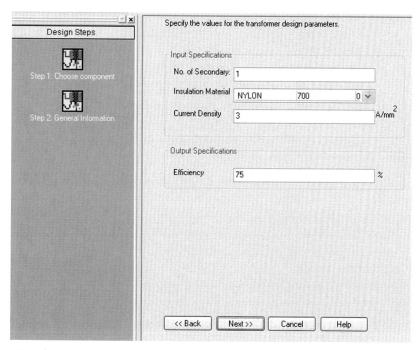

Fig. 21.2 Transformer design parameters.

The maximum number of secondary windings is nine, but for forward converters only one secondary winding is allowed. You have the option to select the transformer insulation, which by default is nylon. The other provided materials are shown in Fig. 21.3.

Insulation Material	NYLON	700	0 ⌄	
Current Density	Name	Breakdown (V/mm)	Thickness (mm)	
	MYLER	500	1,2,5,10	
	KAPTON	2000	2,4	
	NYLON	700	0.2,0.5,1	
Output Specification	TEFLON	5000	0.1,0.5,0.83	
	NONE	0	0	

Fig. 21.3 Insulation materials available in the installed database.

You can enter your own insulation materials: **Tools > Data Entry > Insulation** in the **Enter insulation material** window (Fig. 21.4).

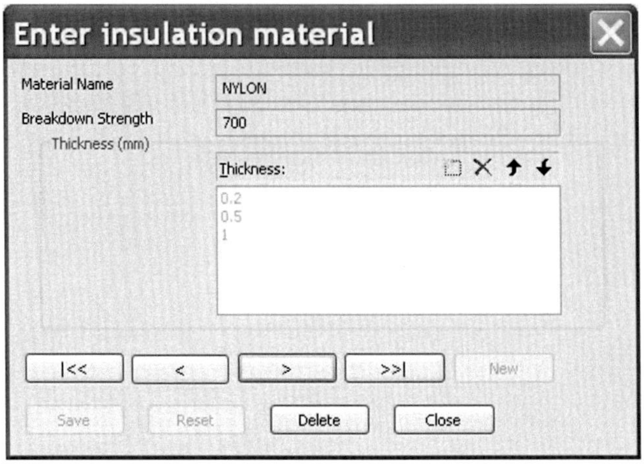

Fig. 21.4 Entering new insulation material data.

Enter the design parameters as shown in Fig. 21.2 and click on **Next >>**.

Step 3: Electrical Parameters

This is where the electrical design parameters from the specifications are entered (Fig. 21.5).

The secondary voltage and current are root mean square (RMS) values. The voltage isolation is the gap or distance between the primary and secondary windings.

Enter the electrical parameters as shown in Fig. 21.5 and click on **Next >>**.

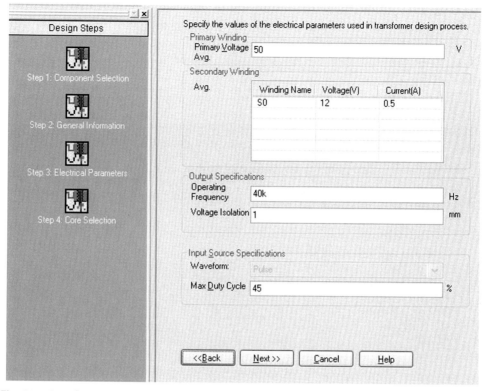

Fig. 21.5 Specifying the electrical parameters.

Step 4: Core Selection

Selection of the magnetic core for the transformer depends on the shape and material. The physical diagram shown for the core in Fig. 21.6 is updated when you select another shape such as a toroid, EE or UU.

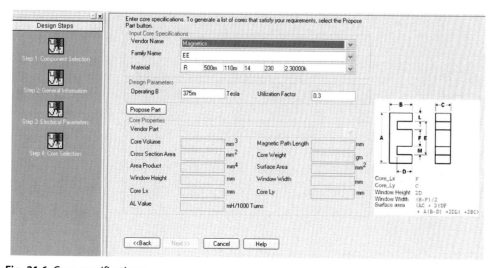

Fig. 21.6 Core specifications.

Change the Vendor name to Ferroxcube and from the **Family Name** pull-down menu, select Toroid and UU to view the respective geometrical data for the cores. Change the **Family Name** back to EE.

You can enter other manufacturer's core data by selecting **Tools > Data Entry > Core Details > Core** (Fig. 21.7).

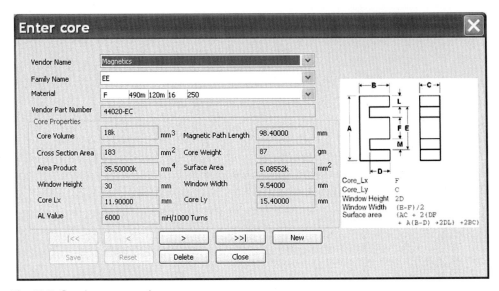

Fig. 21.7 Creating a magnetic core.

You can also enter core material data by selecting **Tools > Data Entry > Core Details > Material** (Fig. 21.8).

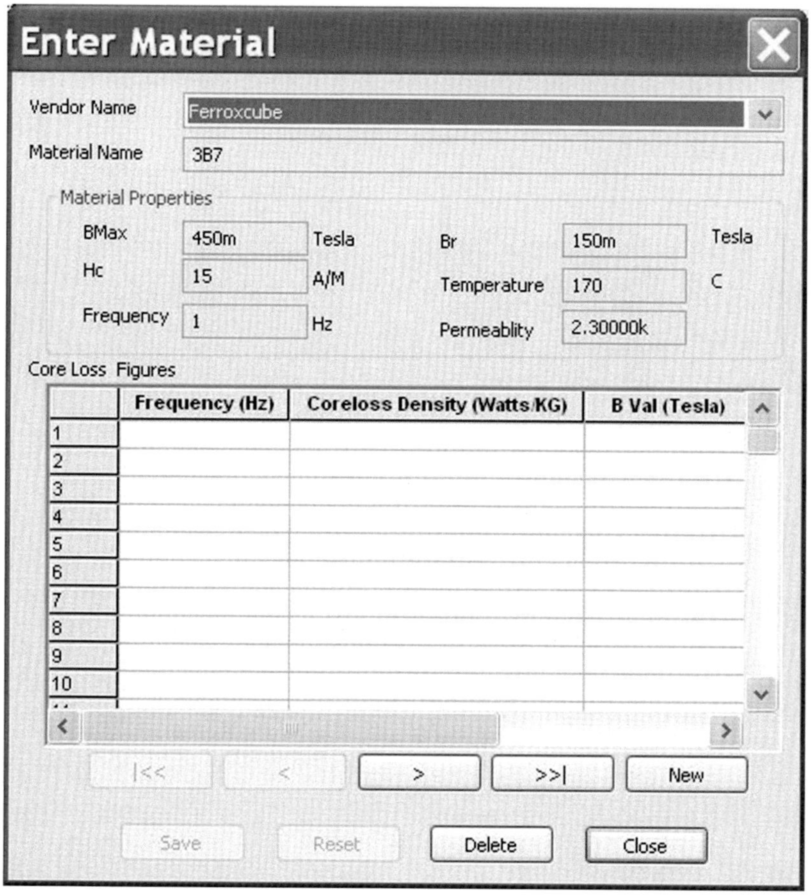

Fig. 21.8 Creating the magnetic core material.

Initially, you will specify a manufacturer's magnetic core material and then let MPE propose a magnetic core using the specified core material. The MPE will determine whether the coil windings will fit on the core.

The MPE core database includes the Ferroxcube range of magnetic cores. For this design, the Ferroxcube EE low-power (10 W) grade 3C81 material will be used as a starting point. These cores are specified for a 67 kHz switching frequency and an output voltage up to 12 V:

Set the **Vendor Name** to Ferroxcube.

Set the **Family Name** to EE.

Set the **Material** to 3C81.

Click on **Propose Part**.

The MPE will return a suitable Vendor Part and will show the respective physical dimensions for the core. The pull-down menu for **Vendor Part** contains a list of other suitable cores from the database that will also meet your specification.

In this example, the core E13_6_6 will have been selected, as shown in Fig. 21.9. Click on **Next >>**.

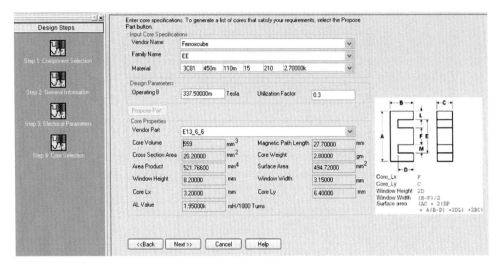

Fig. 21.9 Bobbin and wire properties.

Step 5: Bobbin-Winding Selection

This step selects the bobbin which fits on the core and on which the wire is wound. In Fig. 21.10, the Bobbin Part No. is shown as NO_NAME, which indicates that there are no bobbins in the MPE database. If no bobbin is specified, then a default bobbin wall thickness of 1 mm is used to calculate the bobbin dimensions based on the core dimensions. To create a new bobbin, select **Tools > Data Entry > Core Details > Bobbin** (Fig. 21.11).

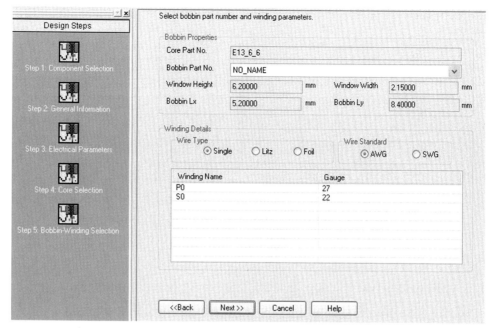

Fig. 21.10 Adding a bobbin.

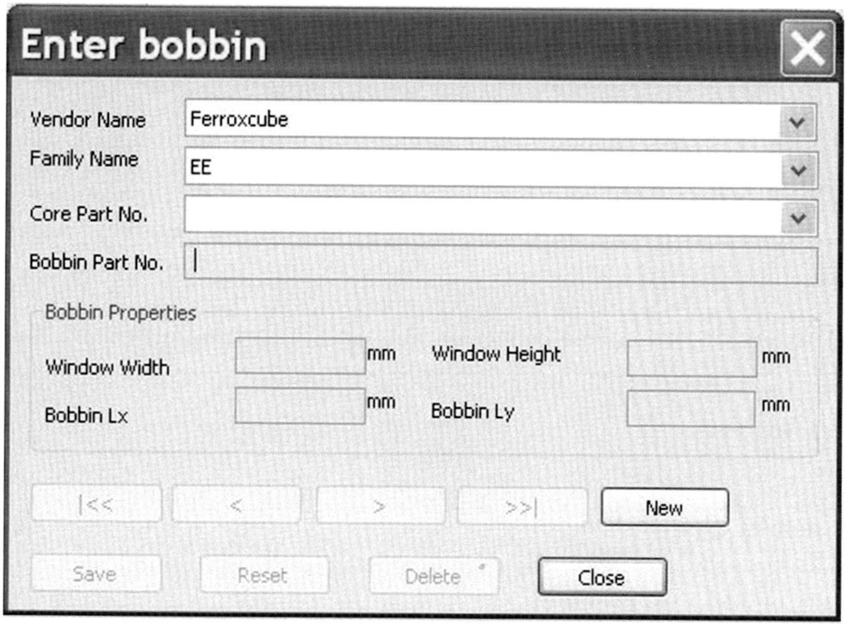

Fig. 21.11 Creating a new bobbin.

Fig. 21.12 shows the bobbin orientation and dimensions given in core vendor datasheets.

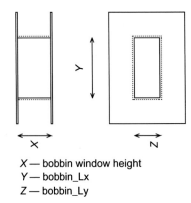

X — bobbin window height
Y — bobbin_Lx
Z — bobbin_Ly

Fig. 21.12 Bobbin Dimensions.

For the bobbin shown in Fig. 21.12, to fit into the core, the calculations are:
bobbin window height = core window height − (2 × bobbin thickness)
bobbin window width = core window width − (1 × bobbin thickness)
bobbin_Lx = core_Lx + (2 × bobbin thickness)
bobbin_Ly = core_Ly + (2 × bobbin thickness)
For this example, the default NO_NAME bobbin will be used, as the core dimensions are not yet finalized.

The Bobbin-Winding Selection window in Fig. 21.10 also shows the proposed wire types (diameters) for the primary and secondary coils. You can also select between the different AWG and SWG wire standards.

Note

You may see a warning message stating that a LITZ winding should be used instead of a single winding. In this case, select the single winding in step 5 (Fig. 21.10).

Click on **Next >>**.

Step 6: Results View

Fig. 21.13 shows the spreadsheet of results. In order to see the full spreadsheet, turn off the Steps view: **View > Steps View.**

Input Parameters		Output Parameters					
Electrical Specifications		**Winding Parameters**			**Calculated Values**		
Primary Voltage (V)	50	Winding Name	P0	S0	Core Loss (Watts)		65.13935m
Secondary Voltage 1 (V)	12	Peak Current (A)	0.7111111111111	1.818181818182	Achieved Efficiency (%)		97.28357
Power (Watts)	6	RMS Current (A)	0.2754121490636	0.7784989441615	Achieved Regulation (%)		
Frequency (Hz.)	40k	No. of Turns	81	24	Window Occupied (%)		102.5065234659
Efficiency (%)	75	Min. Inductance (H)	0.000791015625		Temperature Rise (C)		27.46545
Duty Cycle (%)	45	Wire Gauge'	27	22	Total Buildup (mm)		8.44600
Component Type	Flyback Transformer	Turns/Layer	12	7	Total Copper Loss (Watts)		102.39750m
		No of layer	7	4	Fringing Coefficient		1.22069
Design Status	Error	Inter layer Insulation (mm)'	0.2	0.2	Operating Flux Density (Tesla)'		0.3324890102445
		End Insulation (mm)'	0.2	0.2	AC Flux Density (Tesla)		
Core Details		Winding Buildup (mm)	4.042	3.404			
Vendor Name	Ferroxcube	Winding resistance (Ohm)	0.5668054726953	0.09801702558053			
Part Number	E13_6_6	Copper Loss (Watts)	0.0429932447437	0.0594042579276			
Core Type	EE	Leakage Inductance (H)	0.0002138027247611				
Core Material	3C81	Voltage Drop (V)	0.1561051133361	0.0763061509243			
Bobbin Part Number	NO_NAME						
GAP' (mm)	255.48260m						
Voltage Isolation (mm)'	1	No. Of Strands	1	1			
Maximum Flux Density (Tesla)	337.50000m	Foil Thickness (mm)					
Current Density (A/mm2)'	3	Foil Width (mm)					
Insulation Material'	NYLON						
Wire Type	AWG						

Manufacturer Report   Model View

Original

• INFO: Designing winding layout complete.
• ERROR: The P0 winding could not be fitted in the core. Try design changes to achieve success.

Fig. 21.13 Spreadsheet of results.

In the spreadsheet, the **Design Status** reports that there is an error (Fig. 21.14) and the warning message at the bottom of the results spreadsheet indicates that the P0 winding could not fit into the core.

• INFO: Designing winding layout complete.
• ERROR(ORMAGDB-1104): The P0 winding could not be fitted in the core. Try design changes to achieve success.

Fig. 21.14 Design has not been successful.

So either the wire diameter for the secondary needs to be reduced or you need to use a different core material or a larger core. You could also reduce the distance between the primary and secondary windings (Voltage isolation in Step 3).

Go back to **Step 4**: **Core selection** and select the material as 3C90.

Click on **Propose Part**.

Select the E19_8_9 core, which has a larger core size, as seen in Fig. 21.15.

Click on **Next >>** and progress to the **Results View**.

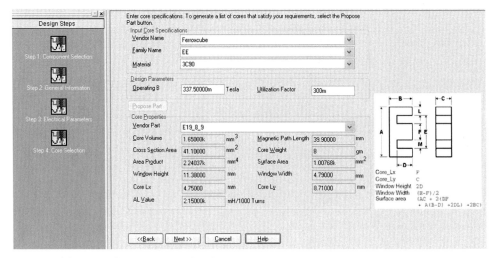

Fig. 21.15 Changing the core material and core size.

You should see the **Design Status** reporting Success (Fig. 21.16).

Now you need to create the bobbin which will fit in the core and run the calculations again.

Input Parameters					Output Parameters	
Electrical Specifications		**Winding Parameters**			**Calculated Values**	
Primary Voltage (V)	50	Winding Name	P0	S0	Core Loss (Watts)	
Secondary Voltage 1 (V)	12	Peak Current (A)	0.711111111111	1.818181818182	Achieved Efficiency (%)	
Power (Watts)	6	RMS Current (A)	0.2754121490636	0.7784989441615	Achieved Regulation (%)	
Frequency (Hz.)	50k	No. of Turns	4	2	Window Occupied (%)	
Efficiency (%)	75	Min. Inductance (H)	0.0006328125		Temperature Rise (C)	
Duty Cycle (%)	45	Wire Gauge'	27	23	Total Buildup (mm)	
Component Type	Flyback Transformer	Turns/Layer	19	6	Total Copper Loss (Watt	
		No of layer	1	1	Fringing Coefficient	
Design Status	Success	Inter layer Insulation (mm)'	0	0	Operating Flux Density (T	
		End Insulation (mm)'	0.2	0.2	AC Flux Density (Tesla)	
Core Details		Winding Buildup (mm)	0.406	0.63		
Vendor Name	Ferroxcube	Winding resistance (Ohm)	0.02388271934797	0.003140432372736		
Part Number	E19_8_9	Copper Loss (Watts)	0.001811546489802	0.001903292347112		
Core Type	EE	Leakage Inductance (H)	6.1995080260078e-008			
Core Material	3C90	Voltage Drop (V)	0.006577591061108	0.002444823286385		
Bobbin Part Number	BB01					
GAP' (mm)	1.01587m					
Voltage Isolation (mm)'	1	No. Of Strands'	1	2		
Maximum Flux Density (Tesla)	337.50000m	Foil Thickness (mm)				
Current Density (A/mm2)'	3	Foil Width (mm)				
Insulation Material'	NYLON					
Wire Type	AWG					

Fig. 21.16 Results view showing success.

Go back to the **Core selection** (**Step 4**) and select **Tools > Data Entry > Core Details > Bobbin** as shown in Fig. 21.17. Select the proposed part, which was the Ferroxcube, EE core 19_8_9. Enter a suitable Bobbin Part No., for example BB01.

The core winding area dimensions given in Fig. as 21.15 are shown again in Fig. 21.17.

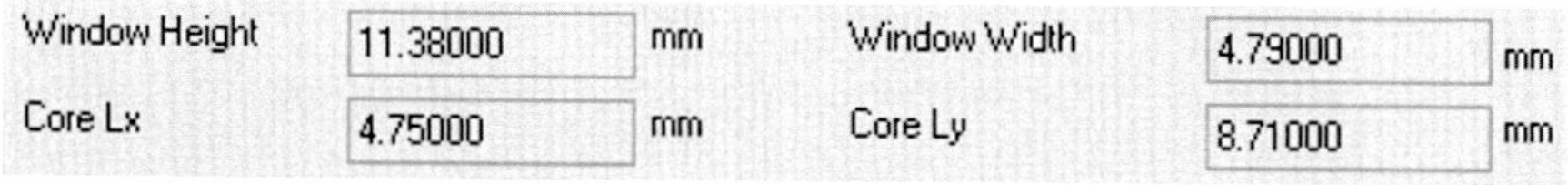

| Window Height | 11.38000 | mm | Window Width | 4.79000 | mm |
| Core Lx | 4.75000 | mm | Core Ly | 8.71000 | mm |

Fig. 21.17 Core winding area dimensions.

Referencing Fig. 21.12 and the associated calculations and using a bobbin wall thickness of 1 mm, the required bobbin dimensions are given as:

bobbin window height $= 11.38 - 2 = 9.39$ mm

bobbin window width $= 4.79 - 1 = 3.79$ mm

bobbin_Lx $= 4.75 + 2 = 6.75$ mm

bobbin_Ly $= 8.71 + 2 = 10.71$ mm

Go back to **Step 5 (Bobbin–winding selection)** and select **Tools > Data Entry > Core Details Bobbin**. Name the Bobbin BB01, enter the Bobbin data as shown in Fig. 21.18 and click on Save.

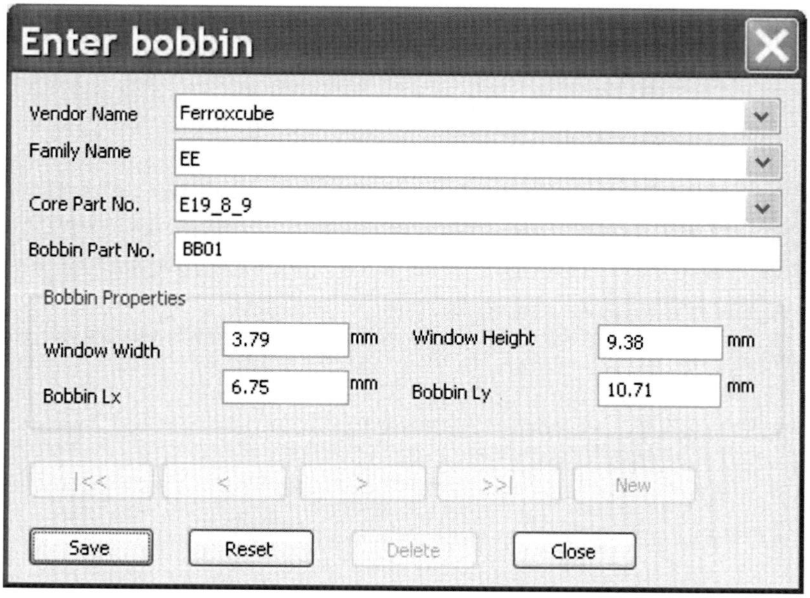

Fig. 21.18 Creating the bobbin.

When you save the new bobbin, a message will appear informing you that the record was entered successfully into the database. Click on OK.

Proceed to **Step 6** and check that the Design Status is still showing **Success**. Save the design, which will generate a flyback .mgd file.

When you save the design, a PSpice model will be created. In this example the flyback.lib file will be created. By selecting the **Model View** tab at the bottom of the **Results Spreadsheet**, you can view the PSpice transformer model as shown in Fig. 21.19.

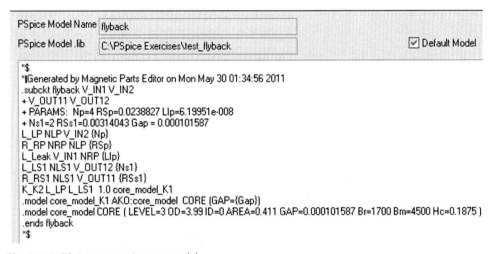

Fig. 21.19 PSpice magnetic core model.

For the flyback topology, the transformer is modeled as a subcircuit with four terminals, V_IN1, V_IN2, V_OUT11 and V_OUT12. The transformer circuit representation is shown in Fig. 21.20.

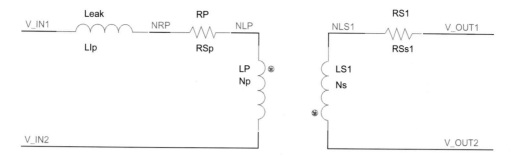

Lip   Leakage inductance referred to primary
RSp   Primary winding resistance
RSs1   Secondary winding resistance
Np   Number of turns in primary winding
Ns   Number of turns in secondary winding

Fig. 21.20 Flyback transformer model. *Llp*, leakage inductance referred to primary; *Np*, number of turns in primary winding; *Ns*, number of turns in secondary winding; *RSp*, primary winding resistance; *RSs1*, secondary winding resistance.

With the Model View open, click on Save again to save the PSpice model.

Using the Model Editor Wizard in the Model Editor, a Capture part can be generated and associated to a transformer symbol as described in Chapter 16.

Exercise 2

Creating a Transformer Model

1. Open the PSpice Model Editor from the **Start** menu.
       Start > All Programs > Cadence (or OrCAD) > PSpice > Simulation Accessories > Model Editor
2. In the Model Editor, select, **File > New**.
3. Select **File > Model Import Wizard**.
4. In the Specify Library window (Fig. 21.21), browse to the flyback.lib, accept the **Destination Symbol Library** as shown and click on **Next >**.

Fig. 21.21 Enter the flyback.lib PSpice model file.

5. In the **Associate/Replace Symbol** window (Fig. 21.22), click on **Associate Symbol**. Select the Flyback model and click on Associate Symbol.

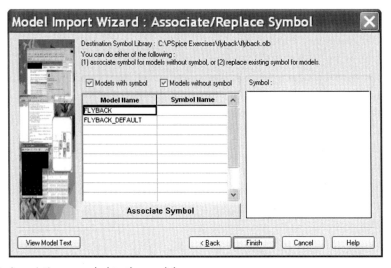

Fig. 21.22 Associating a symbol to the model.

6. In the **Select Matching** window, click on the icon [...] and select the breakout.olb library from the installed folder as **<install path> OrCAD (or Cadence) > version xx.x > Tools > Capture > library > pspice > breakout.olb.**

7. A list of matching symbols with the same number of pins as the flyback model has will be listed. Select the **XFRM_NONLINEAR** transformer as shown in Fig. 21.23 and click on **Next >**.

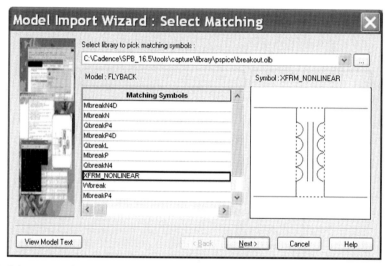

Fig. 21.23 Select a matching transformer symbol.

8. You have to associate the symbol pins with the model terminals. Select the pins as shown in Fig. 21.24 and then click on **Save Symbol**. The pins are in the same order as for the subcircuit model (Fig. 21.19).

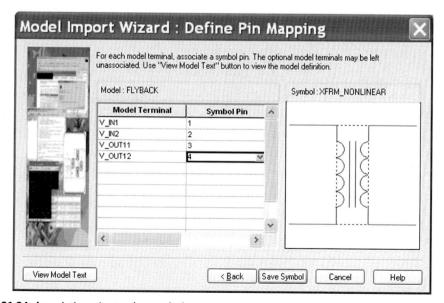

Fig. 21.24 Associating pins to the symbol.

9. You should see the model name shown with an associated symbol (Fig. 21.25).

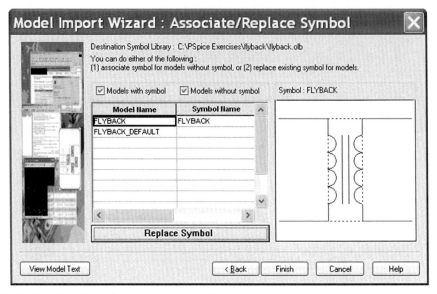

Fig. 21.25 The Flyback model now has an associated symbol.

10. Click on **Finish** and click on No in the sch2cap window. Check there are no error messages in the Summary Status window and click on OK. The Capture part and model are ready to be used.

Exercise 3

The flyback converter will be tested using the circuit in Fig. 21.26.

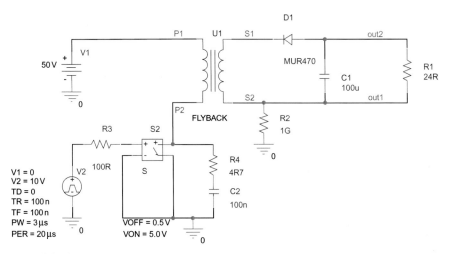

Fig. 21.26 Flyback converter.

1. Draw the circuit in Fig. 21.26. The voltage-controlled switch is from the analog library, the pulse source is from the source library and the diode is from the diode library. In the Project Manager add the newly created library for the transformer by selecting the library folder, **rmb > Add File** and browse to the **Flyback.olb** library. The flyback.olb library will appear in the Place Part menu. Place the transformer as shown in Fig. 21.26.

2. You have to make the flyback.lib PSpice library available for the design. Create a PSpice simulation profile, **PSpice > New Simulation Profile**, and set up a transient analysis for 10 ms. Select **Configuration Files > Library**, browse to the flyback.lib file and click on **Add to Design** (Fig. 21.27).

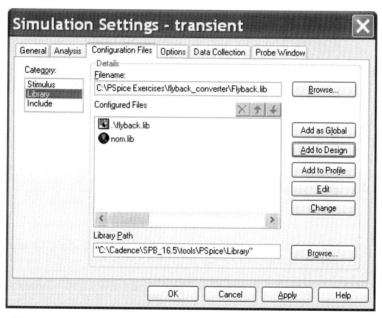

Fig. 21.27 Adding the flyback.lib model file.

3. Select **PSpice > Markers > Voltage > Differential** and place the first marker on the **out1** net. The second differential marker will appear automatically, which you need to place on the **out2** net. The output voltage of the flyback converter is negative; hence the most positive differential marker is placed on **out2**. Run the simulation.

4. You should see that the output voltage is larger than 12 V (Fig. 21.28). As the maximum duty cycle was specified as 45%, this can be decreased to reduce the output voltage. With a duty cycle of 15% (T_{on} is 3 μs) the output voltage reduces to just over 12 V with <100 mV of ripple. The current measured through R1 is just over 510 mA with <5 mA of ripple (Fig. 21.29).

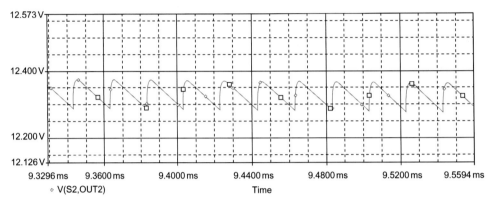

Fig. 21.28 Output voltage meets specification.

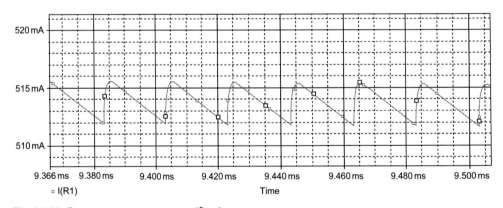

Fig. 21.29 Output current meets specification.

CHAPTER 22

Test Benches

Chapter Outline

22.1. Selection of Test Bench Parts   332
22.2. Unconnected Floating Nets   333
22.3. Comparing and Updating Differences Between the Master Design
    and Test Bench Designs   335
22.4. Exercises   336
    Exercise 1   336
    Exercise 2   343

Normally, when you run simulations on a circuit, you add, for example, voltage sources and load resistors to test the circuit. You may even remove components from a circuit in order to run simulations. However, once the simulations are complete, these add-on components will have to be removed and any deleted components restored.

Before version 16.5, you could add a PSpiceOnly property to the parts that are only used for simulation and therefore these parts would not be included, for example, in the printed circuit board (PCB) netlist. From version 16.5, you can use the Partial Design Feature, which uses test benches to allow you to define those components that are used only for simulations. You can also selectively partition designs for different simulation profiles and build up designs using circuits from other projects. Using test benches is very useful when you have a design that has been put together from a collection of circuits from other projects, as it will allow you to test the functionality of each individual circuit as you build up to the complete design.

When you create a test bench, a Test Bench folder, which contains all the design schematics, is added to the bottom of the Project Manager. All the components in all the schematics in the Test Bench folder will be grayed out. You then selectively "activate" those parts that are required for simulation and add parts such as voltage sources and load resistors. Parts can be deselected and selected either from the master design or from the created test benches.

When you create a test bench, another design folder is created in the project folder. The project folder will then contain two folders:

<project name>-PSpiceFiles
<project name>-TBFiles

The Schematic to Schematic (SVS) utility will compare test bench designs to the master design such that the master design can be updated with modified component values.

Analog Design and Simulation Using OrCAD Capture and PSpice
https://doi.org/10.1016/B978-0-08-102505-5.00022-7

22.1. SELECTION OF TEST BENCH PARTS

As mentioned previously, there will be two design folders, the master design and the test bench design. You can select those parts required for simulation from either the master design or the test bench design depending on which design you have open, but ultimately you will be simulating the test bench design. For example, Fig. 22.1 shows a hierarchical design of a digital counter which contains a clock oscillator and a modulus-3 counter.

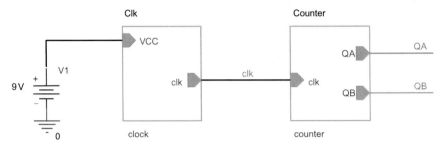

Fig. 22.1 Hierarchical design of a digital counter.

Both blocks will be tested separately using two test benches, Test_Clock and Test_Counter. Initially, all schematic parts are grayed out. From the master design you create a test bench by selecting **Tools > Test Bench > Create Test Bench** and name the test bench, which is added to the master design Project Manager. You can create multiple test benches, but only one test bench will be active, denoted by an A in front of the test bench name. You can then select parts from the master design and add to the active test bench by **rmb > Add Part(s) To Active Testbench**. In Fig. 22.2, the Test_Counter is the active test bench.

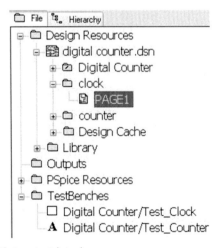

Fig. 22.2 Master design with two test benches.

From the master design you can also select and deselect parts using the hierarchy tab in the Project Manager. Fig. 22.3 shows the hierarchy tab of the master design Project Manager. In this example, only the parts in the **Clock** hierarchical block will be selected. All the other parts in the design will be grayed out.

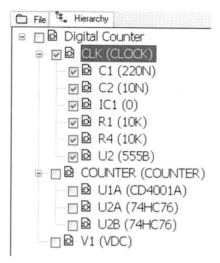

Fig. 22.3 Hierarchy selection of parts from the master design.

Alternatively, you can add and remove parts from a test bench design. In the above example, if you open the **Test_Clock/Test Bench > clock** schematic, you can select parts and then **rmb > Test Bench > Add Part(s) To Self** or **Remove Part(s) From Self** (Fig. 22.4), where **Self** refers to the active **Test_Clock/Test Bench**.

TestBench		Add Part(s) To Self
Add Part(s) To Group..	Ctrl+Shift+A	Remove Part(s) From Self

Fig. 22.4 Adding or removing test bench design parts.

22.2. UNCONNECTED FLOATING NETS

Adding and removing parts in a design can lead to unconnected wires, resulting in floating node errors. As discussed in Chapter 2, all nodes must have a DC path to ground. You can search for floating nets using the **Text to Search Box**, which presents a list of searchable objects (Fig. 22.5), one of which is floating nets.

Fig. 22.5 Searching for floating nets.

Floating nets must be resolved, otherwise the simulation will not proceed. Sometimes, all that is required is to connect a resistor between the floating net and ground to provide the DC path to ground.

In 16.6 two advanced search features have been added as shown in Fig. 22.6. **Regular Expressions** and **Property Name = Value**.

Fig. 22.6 New search features.

Property Name = Value requires the full Property name whereas for Value, the * wildcard and character? marks can be used. For example in the Digital Counter, to find all ICs, you select **Property Name = Value** as shown in Fig. 22.7.

Fig. 22.7 Select Property Name=Value.

Then enter, Part Reference=U*. Part Reference=U^x 🔍 ▾ To be more specific in searching only for digital 74 series type 76 JK flip flops ICs, regardless of the technology, i.e. LS, HC, AC etc., then you can enter; Value=74??76.

Value=74??76 🔍 ▾ The **Regular Expression** offers more flexibility in providing conditional searching for strings, i.e. you can specify a range of values or can be selective in using AND or OR (|) functions. For example, if you want to find the first resistors R1,R2 **OR** the first capacitors C1, C2 in the Digital Counter, then you could enter; Part Reference=(C|R)[1-2]. Note that both the **Regular Expression** and **Property Name = Value** are both selected. Part Reference=(C|R)[1-2] 🔍 ▾ The result of the search finds R1, C1, C2 and IC1, the initial condition part as shown in Fig. 22.8.

Reference	Value	Source Part	Source Library	Page	Page Number	Schematic
C1	220n	C	C:\CADENCE...	PAGE1	1	clock
C2	10n	C	C:\CADENCE...	PAGE1	1	clock
IC1	0	IC1	C:\CADENCE...	PAGE1	1	clock
R1	10k	R	C:\CADENCE...	PAGE1	1	clock

Fig. 22.8 Result of Regular Expression search.

22.3. COMPARING AND UPDATING DIFFERENCES BETWEEN THE MASTER DESIGN AND TEST BENCH DESIGNS

Design differences between the master design and test bench designs can be viewed in the SVS utility, which displays the differences using a color-coded system to highlight missing parts (red), unmatched parts (yellow) and matched parts (white). In Fig. 22.9, the test bench design is shown on the left-hand side and the master design on the right-hand side. The test bench clock schematic contains an extra capacitor C3, shown in red, and a different value for R4.

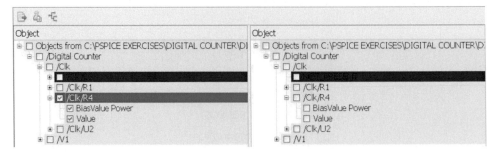

Fig. 22.9 Comparing test bench design with master design.

You can update the test bench design with modified component values to the master design by selecting the **Accept Left** icon ⤇ . However, only modified values will be updated to the master design. In the example above, if you check the Value box for R4 in the left-hand panel, the modified value for R4 will be updated from the test bench to the master design. The extra capacitor C3 will not be updated to the master design. In addition, the master design cannot be updated if there are missing parts, so removed parts in test benches will not be removed from the master design.

22.4. EXERCISES

Exercise 1

Fig. 22.10 shows the hierarchical Digital Counter design from Chapter 20, Exercise 5.

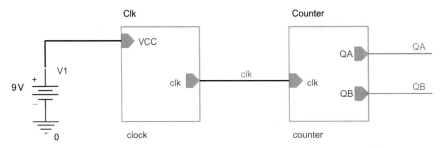

Fig. 22.10 Hierarchical digital counter design.

You will now create a Test_Clock Test Bench to simulate and verify the performance of the clock oscillator only. In the master design, you will only add clock parts to the active test bench by selecting those parts in the Hierarchy tab in the master design Project Manager.

1. Select **Digital Counter.dsn** and from the top toolbar select **Tools > Test Bench > Create Test Bench**. Name the test bench **Test_Clock**, as shown in Fig. 22.11. Click on OK and click on OK again if prompted to save the design.

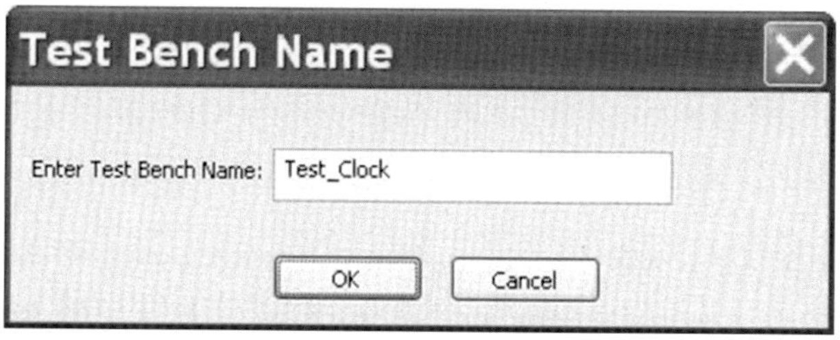

Test Bench Name   [X]

Enter Test Bench Name: Test_Clock

                        [OK]   [Cancel]

Fig. 22.11 Creating a Test_Clock test bench.

The **Test_Clock** test bench will be placed at the bottom of the Project Manager
in the **TestBenches** folder (Fig. 22.12).

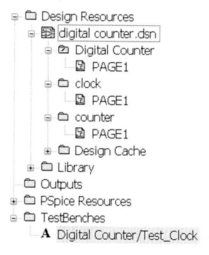

Fig. 22.12 Test_Clock test bench.

2. In the **TestBenches** folder, double click on **Digital Counter/Test_Clock**, which
   will open the **Test_Clock.dsn** design.
3. In the **Test_Clock** Project Manager, double click on **Test_Clock.dsn** and open
   the **counter** Page1 schematic. You will see that all the components are grayed out.
   Close the schematic page.
4. Still in the **Test_Clock** Project Manager, open the **clock** schematic and you will
   also see all the components grayed out.
      You will now activate the clock components in the Test_Clock test bench from
   the master design.
5. Select the master design Project Manager by either selecting the Digital Counter tab
   above the schematics (Fig. 22.13) or selecting **Window > Digital Counter**. It may
   be easier to place the master and test bench designs side by side.

Fig. 22.13 Selecting the Project Manager.

6. In the master design (Digital Counter) Project Manager, click on the **Hierarchy** tab to display the hierarchical design. Expand **Digital Counter** and **Clock** if not already expanded and check all the parts for **Clock** as shown in Fig. 22.14. Click on the File tab.

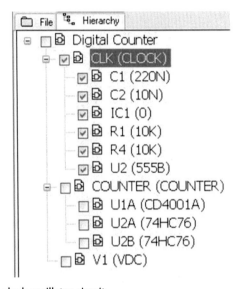

Fig. 22.14 Selecting the clock oscillator circuit.

7. In the **TestBenches** folder, double click on the **Digital Counter/Test_Clock** test bench to activate the Test_Clock test bench Project Manager and open the **clock** schematic to check that all parts are active (nongray). Close and save the schematic.

8. Still in the Test_Clock Project Manager, open the **Digital Counter** schematic page as shown in Fig. 22.15.

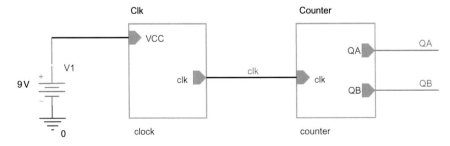

Fig. 22.15 Test_Clock > Digital Counter schematic.

The voltage source V1 is still grayed out. Draw a box around V1, the connecting wires and the 0 V symbol and **rmb > TestBench > Add Part(s) To Self** (Fig. 22.16).

TestBench		Add Part(s) To Self
Add Part(s) To Group..	Ctrl+Shift+A	Remove Part(s) From Self

Fig. 22.16 Adding the voltage source, V1, to the active test bench.

Now that only the clock oscillator circuit is active; you need to search for unconnected nets in the **Test_Clock** test bench. Make sure the Searchmenu is displayed: select **View > Toolbar > Search**.

9. Highlight the **Test_Clock.dsn** and select, from the top toolbar, the pull-down menu to the right of the Binocular ![binocular icon], which is next to the **Text to Search Box** (Fig. 22.17).

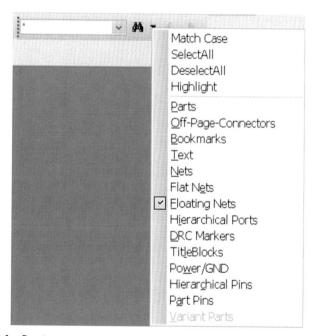

Fig. 22.17 Search for floating nets.

10. Click on **DeselectAll**. Next, select **Floating Nets** and then click on the Binocular icon ![binocular icon].

The **Find** window at the bottom of the screen, below the schematic, reports a floating net on the clk net (Fig. 22.18).

Object ID	Net Name	Page	Page Number	Schematic	Pin
clk(Wire Alias)	CLK	PAGE1	1	Digital Counter\	Counter.clk,clk.clk

Floating Nets

Fig. 22.18 Reported floating nets.

Note

If you see a 0 (Global) net reported as floating this is because there is no connecting wire between V1 and the 0 V symbol. Select the 0 V symbol and drag it down so that a length of connecting wire appears, and run the search for floating nets again so that only the clk net is floating.

---

11. Place a 1k resistor from the **clk** net to ground as shown in Fig. 22.19. This provides a DC path to ground for the clk net.

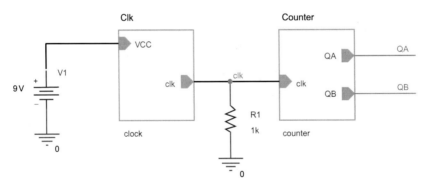

Fig. 22.19 Adding a 1k resistor from the clk net to 0 V.

12. Create a PSpice simulation profile for a transient analysis with a run to time of 20 ms. Do not exit the simulation profile.

    You will need to initialize all flip-flops to 0. Select the **Options tab > Gate-level Simulation** and initialize all flips-flops to 0.
13. Place a voltage marker on the **clk** net and run the simulation.
14. You should see the clock output waveform as in Fig. 22.20, which has a frequency of 216 Hz.

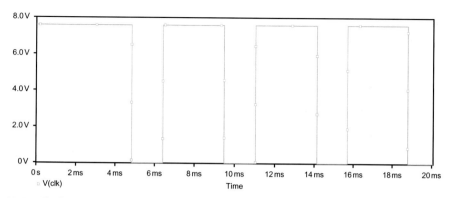

Fig. 22.20 Clock output waveform.

15. Double click on the clock hierarchical block to open the clock schematic and place another 220 nF in parallel with C1. Change the value of R4 in Fig. 20.42 to 6k8 and rerun the simulation. The clock frequency will now be 138 Hz. Close and save the clock schematic.

    Now you will verify the operation of the counter circuit, but this time you will select the active parts from an active Test Bench. The counter parts will be added to the active Test Bench and the clock parts will be removed from the active Test Bench.

16. Open the Project Manager for the Digital Counter (master design).

17. Highlight the **Digital Counter.dsn** and select **Tools > Test Bench > Create Test Bench**. Name the test bench **Test_Counter** (Fig. 22.21).

Fig. 22.21 Creating the Test_Counter test bench.

18. There will be two test benches in the master design counter, as shown in Fig. 22.22. The **A** in front of the Test Counter test bench represents the active test bench. You can make a test bench active by **rmb > Make Active**.

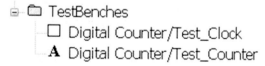

Fig. 22.22 The Test_Counter is the active test bench.

19. Double click on the **Test_Counter** test bench to open the Test_Counter design
20. Double click on **Test_Counter.dsn** and open the **Digital Counter** schematic Page1.
21. Draw a box around the **clock** hierarchical block, V1 and the 0V symbol and then **rmb > Remove Part(s) From Self** (Fig. 22.23).

TestBench		Add Part(s) To Self
Add Part(s) To Group..	Ctrl+Shift+A	Remove Part(s) From Self

Fig. 22.23 Removing the clock and associated parts from the active Test_Counter test bench.

22. Draw a box around the **counter** hierarchical block and make sure the clk, QA and QB nets are also selected; **rmb > TestBench > Add Part(s) To Self** (Fig. 22.24).

TestBench	Add Part(s) To Self
Assign Power Pins	Remove Part(s) From Self

Fig. 22.24 Adding the counter parts to the Test_Counter test bench.

23. Repeat Step 17 and check for floating nets. The QA and QB should be reported as floating. These are digital output nodes and therefore do not require a DC path to ground. Ignore the warning.
24. Add a DigClock source from the source library and set its parameters as shown in Fig. 22.25.

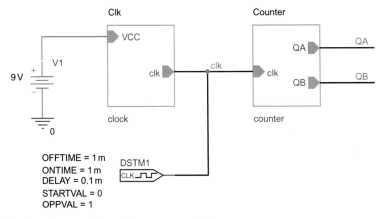

Fig. 22.25 Testing the operation of the counter circuit.

25. Create a PSpice simulation profile for a transient analysis with a run to time of 20 ms. Do not exit the simulation profile.

    Do not forget to initialize all flip-flops to 0. Select the **Options tab > Gate-levelSimulation** and initialize all flips-flops to 0.

26. Place voltage markers on QA and QB and run the simulation. You should see the circuit response as shown in Fig. 22.26.

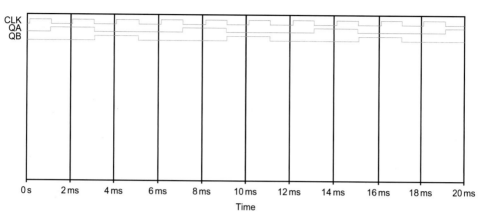

Fig. 22.26 Output of counter.

Exercise 2

After the test bench simulation testing, any differences between the master and test bench designs, component additions, removals and change in component values can be highlighted.

1. In the master design, highlight **Digital Counter.dsn** and select **Tools > TestBench > Diff and Merge** (Fig. 22.27).

| Test Bench | | Create Test Bench | Shift+B |
| Design Rules Check... | | Diff and Merge | Shift+D |

Fig. 22.27 Comparing the differences between the master and test bench designs.

2. In the SVS window, the test bench is on the left-hand side panel and the master design on the right-hand side panel. In the left panel, expand the **/Clk** as shown in Fig. 22.28.

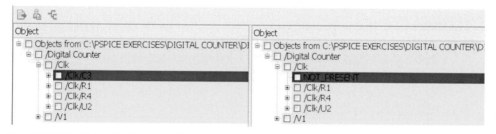

Fig. 22.28 Highlighted differences between the master design and the Test_Clock test bench.

Yellow indicates the design differences between the Test_Clock test bench and the master design, which allows you to update any modified values to the master design.

The extra capacitor in the clock circuit has been highlighted in red and shown as NOT_PRESENT in the master design.

Note

You may have to save and close the Test_Benches before any differences are detected.

3. Expand **/Clk/R4** and you will see a Value box displayed for both the Test_Clock test bench and the master design. This allows you to decide which value of R4 to accept by checking the appropriate Value box.

The modified value for R4 will be updated to the master design.

Check the /Clk/R4 box as shown in Fig. 22.29. The BiasValue Power is a measurement and is not important for this exercise.

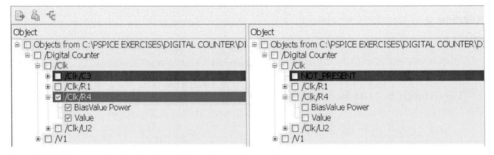

Fig. 22.29 The modified value for R4 has been detected.

4. To update the master design with the new value for R4, click on the **Accept Left** icon ⬛; as there is no longer a difference between R4 in the test bench and the master design, R4 will not be displayed in the SVS window (Fig. 22.30).

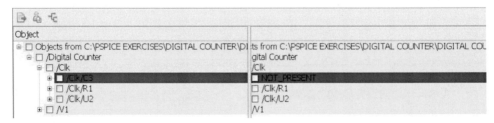

Fig. 22.30 Differences for R4 no longer detected.

If you tried to update the extra capacitor, C3, the session log will display the warning message:

```
WARNING(ORCAP-37003): Could not add object '/Clk/C3' at the target design, as
this operation is not supported
```

5. Close the SVS utility and open the Test_Clock test bench design.
6. Open the **clock** schematic.
7. Delete the capacitor C3 and modify the value of C1 to 470n.
8. Save and close the clock schematic.
9. Open the master design and highlight the Digital Counter.dsn. Check for design differences: **Tools > Test Bench > Diff and Merge**.
10. The SVS window will show only value differences between the designs highlighted in yellow. Check the **/Digital Counter** box as in Fig. 22.31 and click on the **Accept Left** icon ▶.

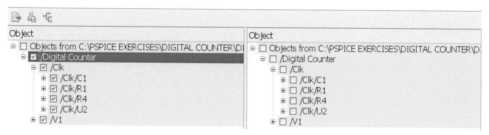

Fig. 22.31 Selecting the test bench values to be updated.

11. The message window will appear, reporting no differences (Fig. 22.32).

Information message from Difference Vie...

ⓘ Both fabrics are identical.. no difference found

OK

Fig. 22.32 No design differences.

The SVS window will appear as shown in Fig. 22.33 with no entries.

Object	Object Type/Value		Object	Object Type/Value

Fig. 22.33 Clear SVS window.

Note

You could have selected the Value box for C1 as you did for R4, but if you have a large number of values to update, it is easier to select the whole test bench design to update. When you select the whole design, warning messages will be displayed in the session log, such as:

```
WARNING(ORCAP-37003): Could not add object '/Clk/R1' at the target design,
as this operation is not supported
```

which can be ignored.

CHAPTER 23

Advanced Analysis

Chapter Outline

23.1. Introduction   347
    23.1.1 Advanced Analysis Libraries   349

23.1. INTRODUCTION

Circuits designed in PSpice may well meet design specifications, but there is always the uncertainty of circuit performance with variations in component tolerances and reliability of components operating within manufacturer's safe operating limits. Has the circuit been optimized to maximize circuit performance in terms of reliability, yield, and cost? With this in mind, Advanced Analysis was created to improve circuit performance and reliability using a suite of simulation tools that are applied to working analog PSpice circuits. Circuit performance can effectively be maximized and optimized for variations in component tolerances, temperature effects, manufacturing yields, and component stress. Fig. 23.1 shows a typical design flow using PSpice Advanced Analysis.

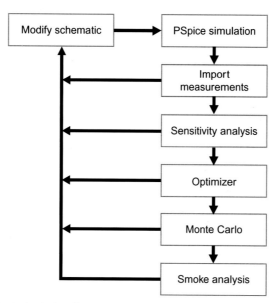

Fig. 23.1 Advanced Analysis design flow.

Analog Design and Simulation Using OrCAD Capture and PSpice
https://doi.org/10.1016/B978-0-08-102505-5.00023-9

Circuit performance can be evaluated by analyzing the circuit output waveforms in response to known stimulus inputs. Measurements applied to the output waveforms determine characteristics such as; high and low pass −3dB corner frequency, rise and fall times of a waveform, center frequency, duty cycle, slew rate, etc. A measurement definition applied to an output circuit variable is called a Measurement Expression and although PSpice provides over 50 measurement definitions, you can create your own definitions. The use of Measurement definitions and how to create custom measurement definitions was introduced in the Performance Analysis chapter (Chapter 12). A good example of using measurement expressions was also introduced in the Mixed Simulation chapter (Chapter 19, Exercise 2) where the period and duty cycle from the output of a 555 timer was evaluated using the Period and Duty Cycle measurement definitions.

Measurement expressions return single numerical values from an output waveform as a measure of circuit performance. The single value returned can be that of a single point x-y value, an expression, a multifunction expression, a mathematical expression, or a user-defined expression.

A good example of single point x-y value is to measure maxima and minima of a trace or even the output current of a voltage-controlled current source for a set input voltage. Similarly, a measurement expression can be used to characterize the output of a pulse generator in terms of its time period and a multifunctional expression can combine expressions together to measure, for example, the Ton and Toff of analog pulses. Mathematical expressions can be used to measure, gain, noise figure, etc.

Sensitivity analysis identifies components whose values are critical to the overall circuit performance and hence require tighter tolerances compared to the less critical components where the tolerances can be relaxed. Using the initial sensitivity results, a Worst Case analysis is performed by setting component tolerances to their maximum limits in order to estimate the largest possible deviations in circuit performance. Sensitivity can be calculated as relative sensitivity that represents a 1% change in component value or absolute sensitivity that represents a one unit change in component value.

Having identified the most critical components that affect circuit performance, the optimizer will adjust those component values and evaluate circuit performance against target specifications to maximize performance. The optimizer can also return nearest preferred standard component values as specified in resistor and capacitor tables such as the E48, E96, and E196 for resistors.

Using the revised component values, Monte Carlo analysis predicts circuit performance by randomly varying component values between their tolerance limits. The component values generated follow a statistical distribution and the greater the number of simulation runs, the more chance there is of all component values being used. This gives an indication of the robustness or yield of the circuit performance using component values within their tolerance limits.

The results are shown as a probability density function (PDF) histogram or as a cumulative distribution function (CDF) together with a summary of the statistical data. Unlike PSpice Monte Carlo, the Advanced Analysis Monte Carlo analysis can be run using more than one measurement expression.

Smoke analysis gives an indication as to whether components are operating within manufacturer's absolute maximum ratings. Components continually operating close to maximum ratings may stress the component resulting in early component failure. Smoke analysis allows the maximum operating conditions to be derated, to provide for a margin of safety.

The Parametric Plotter enables multiple parameters to be swept to evaluate circuit performance. These can be model or circuit design parameters. The sweep results can be displayed in the PSpice Probe waveform viewer where a trace is generated for each Parametric Plotter run or the simulation results are presented in a spreadsheet tabular format. The Parametric Plotter is not covered in this edition of the book.

23.1.1 Advanced Analysis Libraries

Although Advanced Analysis contains over 30 libraries of components that already contain MOC-defined parameters as well as standard components, you are not limited to using the parameterized components in the PSpice Advanced Analysis libraries. Smoke information is now available on almost all of the supported device types in the PSpice libraries. The passive components in the analog library are parameterized components that can be used for Smoke analysis and model parameters for supported device types can be assigned tolerances for Sensitivity and Monte Carlo analysis.

Advanced Analysis parts are found in:

C:\Cadence\SPB_17.2\tools\capture\library\pspice\advanls

The PSpice models are found in the same location as the standard PSpice models:

C:\Cadence\SPB_17.2\tools\pspice\library

CHAPTER 24

Sensitivity Analysis

Chapter Outline

Introduction   351
24.1. Absolute and Relative Analysis   352
    Absolute Sensitivity   352
    Relative Sensitivity   353
24.2. Example   353
24.3. Assigning Component and Parameter Tolerances   354
24.4. Exercises   357
    Exercise 1   357
    Exercise 2   362

INTRODUCTION

Sensitivity Analysis is used to identify components that are most sensitive to circuit performance. The sensitivity of a circuit is defined as the ratio of the change in an output measurement of a circuit, to a change in a circuit parameter value that has a defined tolerance. Typically you would analyze the gain of a circuit, the frequency response, noise figure, etc. for a change in component values up to their tolerance limits. This will enable you to tighten up on tolerances for critical components and conversely to relax tolerances for noncritical components.

Sensitivity analysis is run in conjunction with a transient, AC, or DC sweeps and uses measurement expressions on the resultant output waveforms. An initial nominal simulation run is performed using nominal component values with all tolerances set to zero. Individual component tolerances are then set in turn to 40% of their maximum positive limit while keeping all other components at their nominal values in order to determine the effect each component has on circuit performance. A Worst Case analysis is then run to determine the maximum and minimum measured values by setting component values to their respective tolerance limits. The Worst Case analysis does not take into account the interdependence of the component parameters. Worst Case analysis assumes a linear tolerance relationship between the 40% and 100% values and also for the 1% interpolated value for the relative sensitivity calculation.

Analog Design and Simulation Using OrCAD Capture and PSpice
https://doi.org/10.1016/B978-0-08-102505-5.00024-0

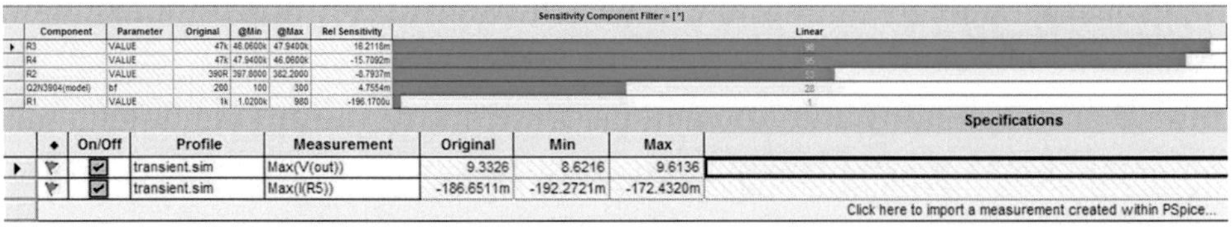

Fig. 24.1 Sensitivity analysis showing component sensitivity and Worst Case measured values.

Fig. 24.1 shows the result of the sensitivity analysis that lists the components in order that have the greatest effect or deviation from the nominal measured value. It can be seen that in the top window (Sensitivity Component Filter), R3 and R4 are the most sensitive components and R1 has the least significant effect on circuit performance. The Worst Case results shown in the Specifications window indicate the maximum and minimum measured values from the original values for V(out) and I(R5). By default, the sensitivity analysis values are calculated in the linear domain but can also be calculated in the logarithmic domain. However, the logarithmic values are not displayed in the bar graphs.

Note

Sensitivity analysis in Advanced Analysis is different to sensitivity analysis in PSpice A/D in that tolerance properties and parameters are assigned to components rather than using the simulation setting, RELTOL to set up the tolerance values.

24.1. ABSOLUTE AND RELATIVE ANALYSIS

There are two types of sensitivity analysis, absolute and relative. Absolute sensitivity is related to a unit change in a parameter value, i.e., a 1 ohm change in resistor value, whereas relative sensitivity is related to a unit percentage change in parameter value, i.e., a 1% change in resistor value. Advanced Analysis calculates both absolute and relative sensitivities but because of typically small capacitor and inductor values being used in circuits, a unit change of 1 Farad for capacitors and 1 Henry for inductors are not realistic, hence the results of relative sensitivity are typically used for capacitors and inductors. The results of the sensitivity and Worst Case analysis are written to the log file, **View > Log File > Sensitivity**.

The sensitivity calculations are based on the following equations:

Absolute Sensitivity

$$S_A = \frac{M_S - M_N}{P_N * S_V * \text{Tol}} \tag{24.1}$$

where M_S, measurement value from the sensitivity run for specified parameter; M_N, measurement value from the nominal run; P_N, nominal value of specified parameter; S_V, sensitivity variation (default 40%); and Tol, relative tolerance of the specified parameter.

Relative Sensitivity

$$S_R = \frac{M_S - M_N}{S_V {}^* \text{Tol}} \times 1\% \qquad (24.2)$$

where M_S, measurement value from the sensitivity run for specified parameter; M_N, easurement value from the nominal run; S_V, sensitivity variation (default 40%); Tol, relative tolerance of the specified parameter.

Eqs. (24.1), (24.2) are related by

$$\text{Relative sensitivity} = \text{absolute sensitivity} \times \text{Pn} \qquad (24.3)$$

24.2. EXAMPLE

For example, for the simple voltage divider resistor network in Fig. 24.2, sensitivity analysis will determine how the voltage at node **out** is affected by the resistor tolerances by using the **Max** measurement on the output voltage.

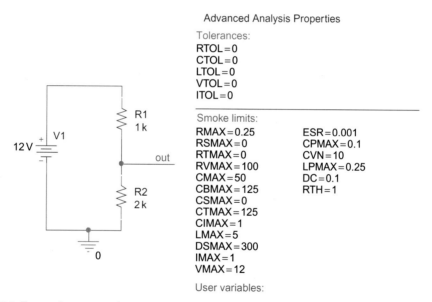

Fig. 24.2 Two resistor network.

The two resistors in Fig. 24.2 both have a tolerance of 2%. The output voltage is given by

$$V_o = V_s \frac{R2}{R1 + R2}$$

$R1 = 1000\,\Omega \pm 2\%$ or $R1 = 1000\,\Omega \pm 20\,\Omega$
$R2 = 2000\,\Omega \pm 2\%$ or $R2 = 2000\,\Omega \pm 40\,\Omega$

Using nominal values:

$$V_o = 12\frac{2000}{1000 + 2000} = 8\,V$$

Running a transient analysis and setting up a **Max** measurement (**Trace > Evaluate Measurement**), a nominal voltage of 8 V will be measured.

Advanced Analysis will initially run a nominal analysis using nominal values for the resistors and return a measured value for M_N of 8 V. The results of the simulation can be found in the log file, **View > Log File > Sensitivity**.

For absolute sensitivity analysis:

Advanced Analysis will then set the value of R1 to 40% of its maximum positive tolerance limit that equates to $0.4 \times 2 = 0.8\%$. So, R1 assumes a value of $0.8\% \times 1000 = 1008\,\Omega$ for the next simulation run. The output voltage from the voltage divider is now calculated as $M_S = 7.97872340425532$ V.

$$S = \frac{7.97872340425532 - 8}{1000 \times 0.4 \times 0.02} = -2.6596\,mV$$

The negative value for the sensitivity ties in with the fact that as R1 increases, Vout decreases from the nominal value.

Similarly for R2, the value will be set to R2 to 40% of its maximum positive tolerance limit that equates to $0.4 \times 2 = 0.8\%$. So R2 assumes a value of $0.8\% \times 2000 = 2016\,\Omega$ for the next simulation run. The output voltage from the voltage divider is now calculated as $M_S = 8.02122015915119$ V.

$$S = \frac{8.02122015915119 - 8}{2000 \times 0.4 \times 0.02} = 1.3263\,mV$$

For relative sensitivity:

The relative sensitivities for a 1% change in parameter tolerance value are calculated using Eq. (24.3):

$$\text{Relative sensitivity} = \text{absolute sensitivity} \times Pn \times 1\%$$

Therefore the relative sensitivities for the voltage divider circuit are given by

$$\text{Relative sensitivity} = -2.6595745 \times 10^{-3} \times 1000 \times 0.01 = -26.5957\,mV$$

$$\text{Relative sensitivity} = 1.3262599 \times 10^{-3} \times 2000 \times 0.01 = 26.5252\,mV$$

24.3. ASSIGNING COMPONENT AND PARAMETER TOLERANCES

The passive components in the analog library include resistors, capacitors, and inductors that have a tolerance property that allow standard symmetric tolerance values to be added to the component, i.e., a 10 k$\Omega \pm 2\%$. For asymmetric tolerances, as in the case of some electrolytic capacitors, the passive components in the **advanals > pspice_elem** library allow POSTOL and NEGTOL tolerances to be defined, i.e., $10000\mu + 20\%, -30\%$. If only POSTOL is defined, then by default, NEGTOL assumes the value of POSTOL.

The tolerance values can be set locally on individual passive components using the Property Editor or they can be set globally to apply a number of components in one go using either the **Variables** part from the **advanals > pspice_elem** library or using the **Assign Tolerance** window that is available in the later 17.2 software versions.

Fig. 24.3 shows the TOLERANCE property for a resistor set to 2% in the Property Editor. You do not have to enter the % sign in the Property editor but is useful if you want to display the component tolerance value with a % sign on the circuit diagram.

Source Part	R.Normal
TC1	0
TC2	0
TOLERANCE	2%
Value	1k
VOLTAGE	RVMAX

Fig. 24.3 Locally setting the tolerance on a passive component.

Alternatively, the component tolerance can be set globally to a number of components in one go using the **Variables** part as shown in Fig. 24.4. In the Property Editor, the value of the TOLERANCE property is set to RTOL as shown in Fig. 24.5 such that the four resistors with the RTOL value will be assigned a tolerance of 2%.

Advanced Analysis Properties

Tolerances:
RTOL = 2%
CTOL = 0
LTOL = 0
VTOL = 0
ITOL = 0

Smoke Limits:
RMAX = 0.25
RSMAX = 0
RTMAX = 0
RVMAX = 100
CMAX = 50
CBMAX = 125
CSMAX = 0
CTMAX = 125
CIMAX = 1
LMAX = 5
DSMAX = 300
IMAX = 1
VMAX = 12

ESR = 0.001
CPMAX = 0.1
CVN = 10
LPMAX = 0.25
DC = 0.1
RTH = 1

User Variables:

Fig. 24.4 Globally assigning resistor tolerances RTOL to 2% using the variables part.

TC1	0	0	0	0
TC2	0	0	0	0
TOLERANCE	RTOL	RTOL	RTOL	RTOL
Value	12k	2k	1k	1k
VOLTAGE	RVMAX	RVMAX	RVMAX	RVMAX

Fig. 24.5 Globally assigning component tolerance values using RTOL from the variables part.

Note

Global tolerance values can be overridden by setting a local tolerance value on an individual passive component.

From 17.2 onward, the **Assign Tolerance** window from **PSpice > Advanced Analysis > Assign Tolerance** can be used to assign component tolerances and distribution for Monte Carlo analysis to passive devices. Individual components can be selected and positive and negative tolerances assigned (Fig. 24.6).

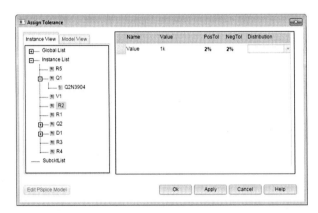

Fig. 24.6 Assign tolerance window.

The **Assign Tolerance** window also displays the PSpice model and subcircuit parameters for active components that can be assigned a tolerance and distribution for the Monte Carlo analysis. From the Assign Tolerance window, you can open the Model Editor. The Assign Tolerance window is then updated to display those model parameters as shown in Fig. 24.7.

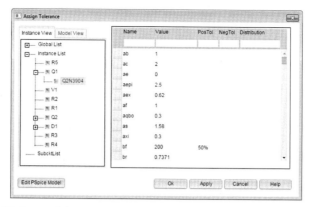

Fig. 24.7 2N3904 PSpice model parameters.

Note

For the PSpice Advanced Analysis Lite version, Sensitivity analysis can only be run on a maximum of three components with tolerances and only one measurement specification. The maximum number of runs is limited to 20.

Note

For the following exercises it is assumed that you are familiar with using the Property Editor that was introduced in Chapter 5, familiar on how to set up and run a PSpice Simulation transient analysis in Chapter 7, and are familiar with Measurement Expressions that was introduced in Chapter 12.

24.4. Exercises

Exercise 1

Assigning local Smoke parameter properties for tolerance.

1. Create a new project called for example resistor_network. After you name the project, the project folder the **Create PSpice Project** window appears. In the pull down menu, select the simple_aa.opj (Fig. 24.9) and click on **OK**.

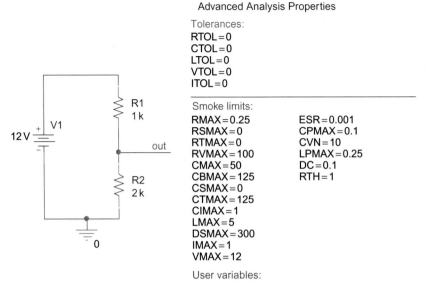

Advanced Analysis Properties

Tolerances:
RTOL=0
CTOL=0
LTOL=0
VTOL=0
ITOL=0

Smoke limits:

RMAX=0.25	ESR=0.001
RSMAX=0	CPMAX=0.1
RTMAX=0	CVN=10
RVMAX=100	LPMAX=0.25
CMAX=50	DC=0.1
CBMAX=125	RTH=1
CSMAX=0	
CTMAX=125	
CIMAX=1	
LMAX=5	
DSMAX=300	
IMAX=1	
VMAX=12	

User variables:

Fig. 24.8 Resistor network.

Note

In the latest versions of OrCAD, when you first open the schematic page, the **Variables** part will automatically appear. If you select **Empty Project**, or you have inadvertently deleted the **Variables** part, then this can be added from the **PSpice > advanals > pspice_elem** library.

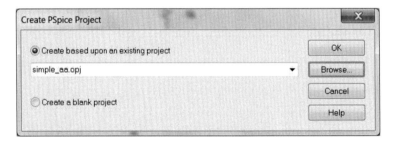

Fig. 24.9 Setting up default Advanced Analysis libraries.

2. Draw the circuit in Fig. 24.8 using resistors from the analog library.
3. You need to assign a tolerance of 2% to each of the two resistors using the Property Editor. Hold the control key down and highlight both resistors and then double click on one of the resistors or you can **rmb > Edit Properties**. The properties associated with both resistors will appear in the Property Editor (See Chapter 5).

4. For the TOLERANCE property, enter 2% for each resistor. Highlight the TOLER-
   ANCE property so that both resistor values of 2% are highlighted and click on
   the **Display** button (or rmb > Display). In the Display Properties window
   (Fig. 24.10), select **Value Only** and click on OK. Close the Property Editor. You
   should see the TOLERANCE property values of 2% displayed on the circuit diagram.

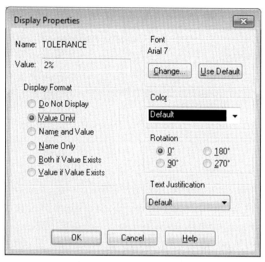

Fig. 24.10 Display properties.

5. Create a Transient analysis. PSpice > New Simulation Profile. Leave the Run to
   Time to the default value of 1000 ns. Place a voltage marker on the node "out."
6. PSpice will run and display the 8 V output in the Probe window.
7. You need to add a measurement expression (Chapter 12) to the output of the circuit.
   Select **Trace > Evaluate Measurement** that will open up the **Evaluate Measurement** window. Select **Max(1)** from the **Functions and Macros** list and then select
   V(out), see Fig. 24.11. Click on OK.

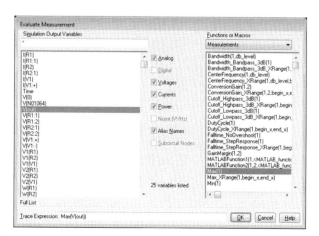

Fig. 24.11 Defining a measurement expression.

8. In Probe you should see the result of the measurement Expression as shown in Fig. 24.12.

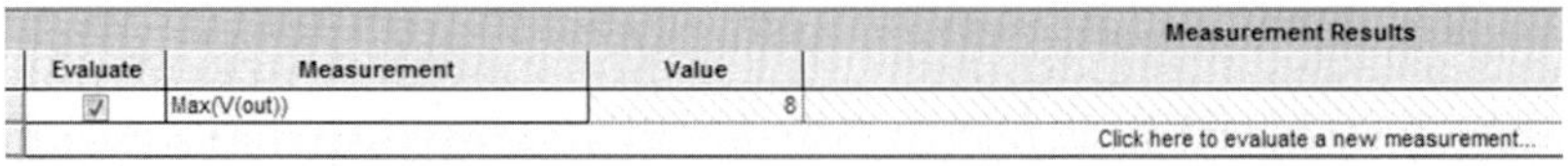

Evaluate	Measurement	Value	Measurement Results
☑	Max(V(out))	8	
			Click here to evaluate a new measurement...

Fig. 24.12 Measurement expression result.

9. Go back to Capture and run the Sensitivity analysis, **PSpice > Advanced Analysis > Sensitivity**.

Note

If a **Cadence Product Choices** menu opens, then just select PSpice Advanced Analysis.

---

10. Advanced Analysis will launch showing the Sensitivity analysis window. You need to import a Measurement Expression for the analysis. Under **Specifications** click on the text, **"Click here to import a measurement created within PSpice ..."** In the **Import Measurement(s)** window click on the **Max[V(out)]** measurement and click on OK. Run the sensitivity simulation by clicking on the green Play button.

11. Fig. 24.13 shows the results of the Sensitivity analysis showing the absolute sensitivity values for R1 and R2 and the Worst Case results, maximum and minimum for V(out).

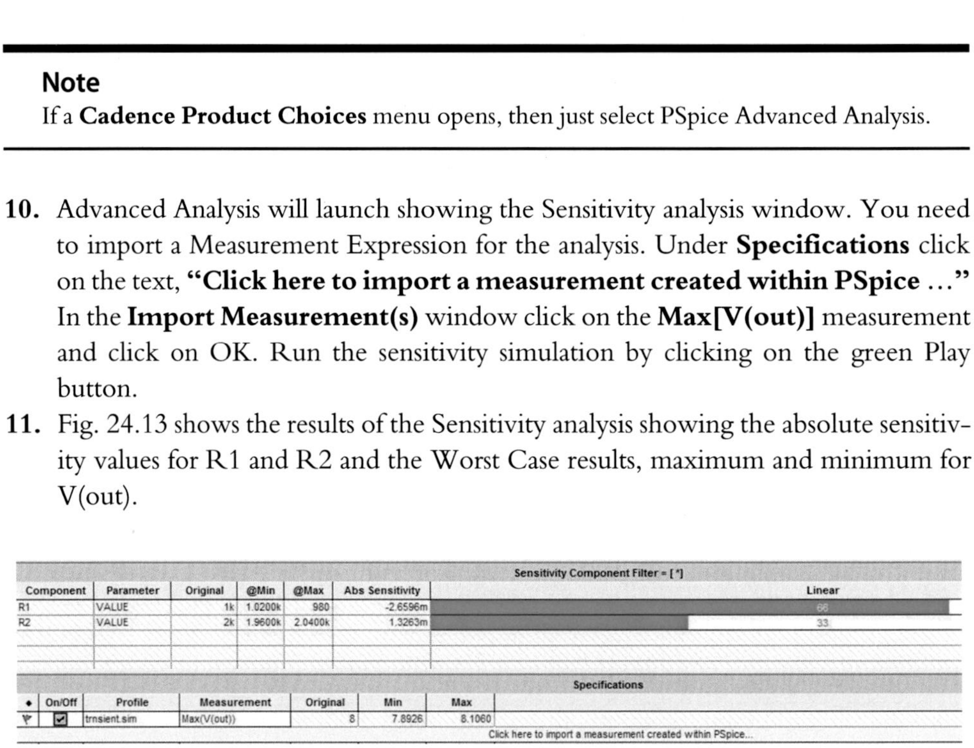

Component	Parameter	Original	@Min	@Max	Abs Sensitivity	Sensitivity Component Filter = [*] Linear
R1	VALUE	1k	1.0200k	980	-2.6596m	66
R2	VALUE	2k	1.9600k	2.0400k	1.3263m	33

•	On/Off	Profile	Measurement	Original	Min	Max	Specifications
▼	☑	trnsient.sim	Max(V(out))	8	7.8926	8.1060	
							Click here to import a measurement created within PSpice...

Fig. 24.13 Sensitivity analysis results.

12. Open the Log File, **View > Log File > Sensitivity**, to see a summary of the Sensitivity runs, Fig. 24.14.

Using Eq. (24.1) for Absolute Sensitivity for R1:

$$S = \frac{7.97872340425532 - 8}{1000 \times 0.4 \times 0.02} = -2.6596\,\text{mV}$$

```
Processing analysis specifications

Listing Profiles:
 - transient.sim

Simulation Run: 0 (Nominal Run)
Param : R1.VALUE   (R_R1.Value) = 1k
Param : R2.VALUE   (R_R2.Value) = 2k

Specs : Max(V(out)) = 8
Nominal run completed
Sensitivity runs underway.....

Simulation Run: 1
Param : R1.VALUE   (R_R1.Value) = 1.00800000000000k
Param : R2.VALUE   (R_R2.Value) = 2k

Specs : Max(V(out)) = 7.97872340425532
Sensitivity run: 1 of 2 completed

Simulation Run: 2
Param : R1.VALUE   (R_R1.Value) = 1k
Param : R2.VALUE   (R_R2.Value) = 2.01600000000000k

Specs : Max(V(out)) = 8.02122015915119
Sensitivity run: 2 of 2 completed
Sensitivity runs completed
Sensitivity bar lengths computed
Worstcase runs underway.....

Simulation Run: 3
Param : R1.VALUE   (R_R1.Value) = 1.02000000000000k
Param : R2.VALUE   (R_R2.Value) = 1.96000000000000k

Specs : Max(V(out)) = 7.89261744966443
Minimum run: 1 of 1 completed
Sensitivity minimum runs completed

Simulation Run: 4
Param : R1.VALUE   (R_R1.Value) = 980
Param : R2.VALUE   (R_R2.Value) = 2.04000000000000k

Specs : Max(V(out)) = 8.10596026490066
Maximum run: 1 of 1 completed
Sensitivity maximum runs completed
```

Fig. 24.14 Sensitivity analysis log file.

Using Eq. (24.1) for Absolute Sensitivity for R2:

$$S = \frac{8.02122015915119 - 8}{2000 \times 0.4 \times 0.02} = 1.3263 \text{ mV}$$

13. Right mouse button (rmb) anywhere in the top window pane under **Sensitivity Component Filter**, i.e., under the **Linear** field and select, **Display > Relative Sensitivity** (Fig. 24.15).

Using Eq. (24.2) for relative sensitivity for R1:

$$S = \frac{7.97872340425532 - 8}{0.4 \times 0.02} \times 0.01 = 26.5957 \text{mV}$$

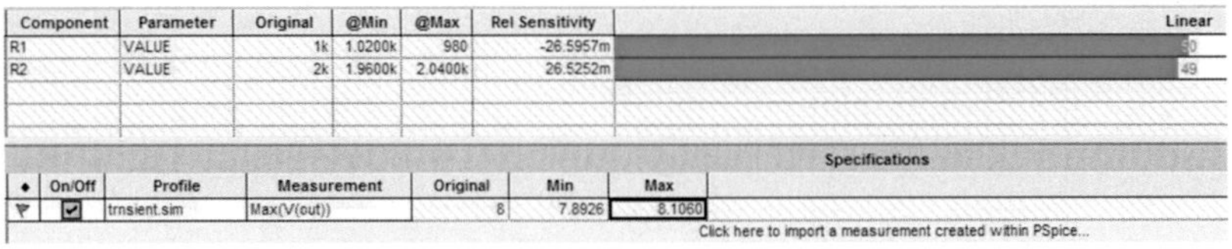

Component	Parameter	Original	@Min	@Max	Rel Sensitivity	Linear
R1	VALUE	1k	1.0200k	980	-26.5957m	0
R2	VALUE	2k	1.9600k	2.0400k	26.5252m	49

					Specifications		
●	On/Off	Profile	Measurement	Original	Min	Max	
▼	☑	trnsient.sim	Max(V(out))	8	7.8926	8.1060	

Click here to import a measurement created within PSpice...

Fig. 24.15 Sensitivity analysis results.

Using Eq. (24.2) for relative sensitivity for R1:

$$S = \frac{8.02122015915119 - 8}{0.4 \times 0.02} \times 0.01 = 26.5252 \text{mV}$$

The results of the Worst Case analysis for maximum and minimum measured values are calculated as:

sensitivity minimum run: R1 + 2%, R2 − 2% Vo = 7.89261744966443 V

sensitivity maximum run: R1 − 2%, R2 + 2% Vo = 8.10596026490066 V

The value in the green bar is calculated as a percentage of the total sensitivity.

The **Variables** symbol can be used to assign tolerances globally to a number of similar passive components such as the two resistors in Exercise 1. You need to replace the 2% tolerance with the global **RTOL** tolerance property on the **Variables** symbol and set **RTOL** to 2%. If the 2% values are displayed, double click on each 2% tolerance value and replace the text with **RTOL**. If the percentage values are not displayed, then use the Property Editor to replace the 2% value with **RTOL**.

With the later software version of 17.2, the **Assign Tolerances** window can be used to assign tolerances to individual components. **PSpice > Advance Analysis > Assign Tolerances**.

Exercise 2

Fig. 24.16 shows a series regulator circuit using voltage feedback with an opamp. The circuit is designed to have an output voltage of 9 V ± 5% with a maximum output current of 190 mA. R4 represents the load resistance that can vary by 5%.

1. Create a new PSpice project called, for example, voltage_regulator and select the project template simple_aa.opj as you did in Exercise 1.
2. Open the schematic page. In the latest versions of OrCAD, when you first open the schematic page, the **Variables** symbol will automatically appear. Delete the Variables part as you will be using the **Assign Tolerance** window to set component tolerances.
3. Draw the circuit in Fig. 24.16. Make sure you label node "out." The 2N3904 transistor can be found in the eval.olb or bipolar library, resistors are from the analog library, and the zener diode is from the phil_diode.olb or zetex.olb libraries.

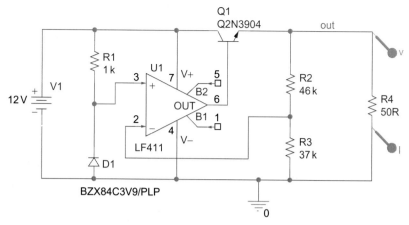

Fig. 24.16 Series regulator circuit.

Note

All the PSpice libraries are now shipped with the latest 17.2 Lite DVD. The libraries can also be downloaded for free from the OrCAD website or the local Cadence Channel Partner (CCP) website or you can copy them from someone who has the full software version installed.

---

4. Open the **Assign Tolerance** window, **PSpice > Advanced Analysis > Assign Tolerance**.
5. In the **Instance List**, select R1 and double click in the **PosTol** column for R1. Enter a value of 2%. Assign tolerances of 2% for resistors R2, R3 and 5% for R4.
6. Close the **Assign Tolerance** window. By default **NegTol** will assume the same value as **PosTol.** You will see this the next time you open the Assign Tolerances window.
7. Open the **Assign Tolerance** window.
8. Select Q1 and then Q2N3904. Click on **Edit PSpice Model** to open the Model Editor as you are going to change the transistor gain.
9. At the bottom of the Model Editor in the **Model Text** pane, change: Bf=416.4 to Bf=200 dev=50%. This adds a tolerance of 50% to the transistor current gain, see Fig. 24.17.
10. Close the Model Editor and save the changes.
11. Close the Assign Tolerances window.

Model Text

```
.model Q2N3904    NPN(Is=6.734f Xti=3 Eg=1.11 Vaf=74.03 Bf=200 dev=50% 416.4 Ne=1.259
+               Ise=6.734f Ikf=66.78m Xtb=1.5 Br=.7371 Nc=2 Isc=0 Ikr=0 Rc=1
+               Cjc=3.638p Mjc=.3085 Vjc=.75 Fc=.5 Cje=4.493p Mje=.2593 Vje=.75
+               Tr=239.5n Tf=301.2p Itf=.4 Vtf=4 Xtf=2 Rb=10)
*               National    pid=23              case=TO92
*               88-09-08 bam       creation
```

Fig. 24.17 Changing the transistor gain.

Note

When you open the Model Editor, you automatically create a copy of the transistor PSpice model file that has the same name as your project, i.e., power_supply.lib. The original transistor model is not changed.

---

12. Set up a transient analysis and use the default **Run to Time** of 1000 ns.
13. Add a PSpice voltage markers to node "out" and a current marker to the bottom pin of resistor R4. This is because current is measured into pin 1 of the resistor that is situated at the bottom end after you have rotated the resistor. If you get a negative result, it just means that the current is flowing out of pin of the resistor. Make sure the current marker is attached to the resistor pin.
14. Run the simulation.
15. As in Exercise 1, you need to add a measurement expression (Chapter 12) to the output of the circuit. Select **Trace > Evaluate Measurement** that will open up the **Evaluate Measurement** window. Select **Max(1)** from the **Functions and Macros** list and then select **V(out)** and click on OK. Repeat for load current by **Max(1)** and then **I(R4)**. You should see the same results as shown in Fig. 24.18.

	Evaluate	Measurement	Value	Measurement Results
▶	☑	Max(V(out))	8.78007	
	☑	Max(I(R4))	-175.60148m	

Fig. 24.18 Viewing the measurement results values.

Note

If you run the simulation again, then you view the previous Measurement results, **View > Measurement Result** and click in the **Evaluate** boxes to see the measurement results for the voltage and current outputs.

---

16. Go back to Capture and run the Sensitivity analysis, **PSpice > Advanced Analysis > Sensitivity**.

17. Advanced Analysis will launch showing the Sensitivity analysis window. You need to import a Measurement Expression for the analysis. Under **Specifications** click on the text, "Click here to import a measurement created within PSpice …" In the **Import Measurement(s)** window select the **Max[V(out)]** and **Max[I(R4)]** measurements and click on OK.

18. Run the sensitivity simulation by clicking on the green Play button.

19. Fig. 24.19 shows the results of the Sensitivity analysis showing the absolute sensitivity values for the circuit components and the Worst Case results, maximum and minimum for V(out) and I(R4).

Component	Parameter	Original	@Min	@Max	Abs Sensitivity	Linear
R3	VALUE	37k	37.7400k	36.2600k	-130.4708u	99
R2	VALUE	46k	45.0800k	46.9200k	105.7834u	81
R1	VALUE	1k	1.0200k	980	-66.8287u	51
R4	VALUE	50R	47.5000	52.5000	38.2815n	< MIN >
Q2N3904(model)	bf	416.4000	208.2000	624.6000	1.3592n	< MIN >

Fig. 24.19 Absolute sensitivity results for power supply circuit.

20. In the Sensitivity Component Filter window, right mouse click anywhere and select **Display > Relative Sensitivity**. Fig. 24.20 shows the results of the Sensitivity analysis showing the relative sensitivity values for the circuit components and the Worst Case results, maximum and minimum for V(out) and I(R4).

Component	Parameter	Original	@Min	@Max	Rel Sensitivity	Linear
R3	VALUE	37k	37.7400k	36.2600k	-48.2742m	99
R2	VALUE	46k	45.0800k	46.9200k	105.7834u	99
R1	VALUE	1k	1.0200k	980	-668.2870u	1
R4	VALUE	50R	47.5000	52.5000	19.1408n	< MIN >
Q2N3904(model)	bf	416.4000	208.2000	624.6000	5.6598n	< MIN >

Fig. 24.20 Relative sensitivity results for power supply circuit.

Fig. 24.21 shows the results for the Worst Case analysis for V(out) and I(R4).

	On/Off	Profile	Measurement	Original	Min	Max
	☑	transient.sim	Max(V(out))	8.7801	8.5879	8.9801
	☑	transient.sim	Max(I(R4))	-175.6015m	-189.0542m	-163.5800m

Click here to import a measurement created within PSpice…

Fig. 24.21 Relative sensitivity results for voltage regulator.

CHAPTER 25

Optimizer

Chapter Outline

Introduction   367
25.1. Optimization Engines   367
25.2. Measurement Expressions   368
25.3. Specifications   368
25.4. Exercises   369
    Exercise 1   369
    Exercise 2   376

INTRODUCTION

The Optimizer is used on existing working analog circuits to improve performance by optimizing component values or system parameters. Optimizer specifications are set using either Measurement definitions to determine circuit performance by varying circuit parameters or by using Curve-fit definitions that are applied to the profile of the output waveform. Circuit performance is evaluated using measurement expressions from transient, DC, or AC simulations, applied to the circuit output waveforms. Goal functions and constraints are then used to define circuit specifications to optimize circuit performance against measured values for a change in defined parameter values. Goal functions set "loose" target specifications that would be nice to achieve, whereas constraints set limits to which the output of the measurement expressions aim to achieve by a series of optimization steps.

Once the measurement expressions for circuit performance have been identified, you then need to define your goals or constraints that will improve circuit performance. Running a sensitivity analysis will give an indication as to which parameters have the most significant effect on circuit performance. These parameters are then imported from the results of the sensitivity analysis or can be imported directly from the circuit diagram. Component parameters are automatically set having maximum and minimum values at a factor of 10 times the nominal value. However, the values of these limits can be changed.

25.1. OPTIMIZATION ENGINES

There are three optimization engines, the Modified least square quadratic (LSQ) engine, the Random engine, and the Discrete engine. The Modified LSQ engine is a fast gradient-based engine and is used to quickly converge to an optimum solution. However, if the

Analog Design and Simulation Using OrCAD Capture and PSpice
https://doi.org/10.1016/B978-0-08-102505-5.00025-2

optimizer gets mathematically "stuck" in what is called a local minima, then the Random optimization engine is used as it can provide alternative starting points. After an optimum result has been achieved, the Discrete engine is used in conjunction with discrete tables of commercially available passive component values to provide a realistic circuit in terms of practical component values.

The efficient use of the optimization engines depends on many factors such as how the circuit behaves, the number of circuit parameters to optimize, the range of the constraints and goals, and even the number of goals and constraints. It would be sensible to minimize the number of parameter variables initially to get a feel for how the optimizer is progressing toward the optimum specifications.

25.2. MEASUREMENT EXPRESSIONS

Measurement expressions return single numerical values from an output waveform as a measure of circuit performance; for example, rise time, low pass cut off -3 dB, or a user-defined "value." The single value returned can be that of a single point trace function x-y value, a measurement expression based on a combination of functions or a user-defined optimizer expression.

25.3. SPECIFICATIONS

Circuit specifications consist of a measurement expression and a defined goal or constraint. There must be at least one goal function defined in the list of measurement expressions. For example, a circuit has been designed to deliver a rectangular current pulse of up to 1 mA with a pulse duration of 100 ms $\pm$ 5% and a pulse period 500 ms $\pm$ 5%. The Max Measurement expression can be used to measure the maximum current and since a range has not been specified, this can be set as a goal function. A Period Measurement Expression can be used for the 100 ms pulse width with constraint limits of 95–105 ms. Similarly a period measurement can be used to determine pulse period with constraint limits of 475 and 525 ms.

The progress of a measurement for a specification compared to other specifications can be emphasized in the error graph by setting up a weighting function that effectively scales up the error at each data point. For example, say you have five specifications defined. As the weight function integer is set by default to 1, each specification will contribute 20% to the total error at the start of the optimization regardless of the initial error margin from goal or constraint ranges. If you prioritize one specification, specification A, with a weight of 6 leaving all other specification weights set at 1, then the total contribution to the specification A is $(6 + 1 + 1 + 1 + 1) = 10$. Specification A will then contribute, $6/10 \times 100 = 60\%$ to total error at the start of optimization with all other specifications contributing 10%. In the error graph, you can see the progress of the optimization measurements and any effects a weighted specification has on measurements converging to a final solution.

The idea of the optimization process is to reduce the difference or error between a calculated measurement and defined goals or constraints by varying component parameters. The error graph gives an indication of the progress of the optimizer to converge toward a solution that meets the specified goals and constraints and also provides a history of the simulation data. The error is determined using the process of least mean squares and is displayed as normalized % error against the number of runs. The error calculation is based on

$$\%\text{Error}_{\text{RMS}} = 100 \times \frac{\text{ME}_c}{\text{ME}_o} \times \frac{W}{\Sigma W}$$

where

ME_c—difference between current measurement value and nearest edge of desired range (i.e., constraint).

ME_o—difference between original measurement value and nearest edge of desired range.

W—weighting integer.

Σ—sum of weighted integers.

Note

It is recommended to define nonlinear parameters as constraints rather than goals and there has to be at least one goal function set and the number of goals should not exceed the number of constraints.

Note

For the PSpice Advanced Analysis Lite version, only two component values and one measurement function can be optimized using the Modified LSQ and Random engines.

25.4. EXERCISES

Exercise 1

Following on from the results of the sensitivity analysis in Chapter 24, the power supply circuit performance will be optimized for R1, R2, and R3. The circuit diagram is shown in Fig. 25.1.

1. Open the voltage regulator circuit from the Sensitivity analysis in Chapter 24.

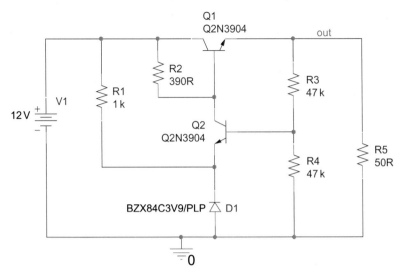

Fig. 25.1 Power supply circuit.

2. Highlight Q1 and **rmb > Edit PSpice Model** and delete the dev=50% and leave Bf=200.
3. Run a transient analysis using the default **Run To Time** of 1000 ns.
4. Measurement Expressions were set up in Sensitivity analysis and can be seen in PSpice, **View > Measurement Results**. If the expressions were not set up, then select **Trace > Evaluate Measurement** that will open up the **Evaluate Measurement** window. Select **Max(1)** from the **Functions and Macros** list and then select **V(out)**, see Fig. 25.2. Click on OK.

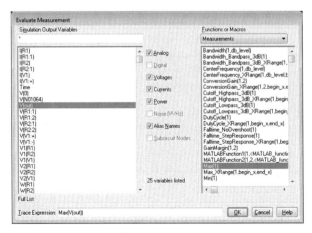

Fig. 25.2 Defining a measurement expression.

5. Create another Measurement Expression using the **Max** function for the current through R4, **Max(I(R4))**.

6. In PSpice select **View > Measurement Results** and click on the Evaluate to see the measurement results (Fig. 25.3).

	Evaluate	Measurement	Value	Measurement Results
▶	✓	Max(V(out))	8.78007	
	✓	Max(I(R4))	-175.60148m	

Fig. 25.3 Measurement expression result.

7. In Capture select **PSpice > Advanced Analysis > Optimizer**.

8. In the **Parameters [Next Run]** window, click on, **Click here to import a parameter from the design property map....**

9. In the **Parameters Selection Component Filter [*]** window, hold down the **Control Key** and select **R2 Value, R3 Value, and R4 Value**. The parameters will appear as shown in Fig. 25.4.

Parameters [Next Run]							
♦	On/Off	Component	Parameter	Original	Min	Max	Current
♈	☑ 🔥	R3	Value	37k	3.7000k	370k	
♈	☑ 🔥	R2	Value	46k	4.6000k	460k	
♈	☑ 🔥	R1	Value	1k	100	10k	
			Click here to import a parameter from the design property map...				

Fig. 25.4 Imported component parameters.

10. Component parameters are automatically adjusted to a decade either side of nominal value. Change the parameter ranges to that shown in Fig. 25.5.

Parameters [Next Run]							
♦	On/Off	Component	Parameter	Original	Min	Max	Current
♈	☑ 🔥	R3	Value	37k	27k	47k	
♈	☑ 🔥	R2	Value	46k	27k	56k	
▶	♈ ☑ 🔥	R1	Value	1k	910	2k7	
			Click here to import a parameter from the design property map...				

Fig. 25.5 Revised component parameter values.

11. You need to set up the Measurement Specifications. Click on, **Click here to import a measurement created within PSpice....** You should see the two previous Measurement Expressions as shown in Fig. 25.6. If not, open PSpice and select **View > Measurement Results** and make sure the **Evaluate** button is checked for both measurements.

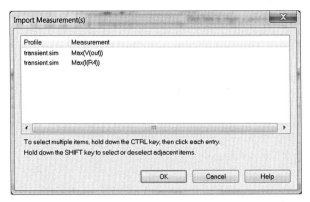

Fig. 25.6 Importing measurement expressions.

12. The circuit specifications are Vout = 9 V ± 5% and I(R4) = 190 mA. So Vout will be constrained between 8.55 and 9.45 V and I(R4) will be set as a goal. In the **Specifications [Next Run]**, for **Max(V(out))**, enter a Min value of 8.55 and a Max value of 9.45. From the **Type** pull down menu, change **Goal** to **Constraint**. Leave the Weight as 1.

13. For **Max(I(R4))**, enter a Max value of 190 m and leave the **Type** as **Goal**. Change the Weight factor to 3 (see Fig. 25.7).

Specifications [Next Run]							
Measurement	Min	Max	Type	Weight	Original	Current	Error
Max(V(out))	8.5500	9.4500	Constraint	3			
Max(I(R4))		190m	Goal	1			
Click here to import a measurement created within PSpice							

Fig. 25.7 Setting up goals and constraints.

14. From the top tool bar, the Modified LSQ engine will be shown by default. Click on the run button.

15. You should see the optimized results shown in Fig. 25.8.

Specifications [Next Run]							
Measurement	Min	Max	Type	Weight	Original	Current	Error
Max(V(out))	8.5500	9.4500	Constraint	1	8.7801	9	0%
Max(I(R4))		190m	Goal	1	-175.6015m	-180m	0%
Click here to import a measurement created within PSpice							

Fig. 25.8 Modified LSQ optimization results.

The minus sign in front of the original −175.6015 mA refers to the direction of current flow associated with pin 1 of resistor R4. By convention, current flow into pin 1 is positive. When you place R4 in the circuit and rotate once, pin 1 is at the bottom of the resistor so current flows out of pin1 and is recorded as negative.

Note

If an optimized value is not being achieved, check how close the parameter values are to their limits. If they are too close, expand the limits. Also run the optimizer again.

16. The results in the Parameters [Next Run] (Fig. 25.9) show that the current component values that are not commercially available.

♦	On/Off		Component	Parameter	Original	Min	Max	Current
▸	☑	🖊	R3	Value	37k	27k	47k	27.3067k
	☑	🖊	R2	Value	46k	27k	56k	35.8008k
▸	☑	🖊	R1	Value	1k	910	2k7	1.9822k
					Click here to import a parameter from the design property map...			

Parameters [Next Run]

Fig. 25.9 Parameter values for Modified LSQ results.

17. From the top tool bar, select **Edit > Profile Settings** to open the **Profile Settings** window and select **Discrete** from the **Engine** pull down menu. Click on the square-dotted ion next to the red cross, **New (Insert)** as shown in Fig. 25.10. If the **discretetables\resistance** folder is not shown, then click on the three ellipses (…) and navigate to the **discretetables\resistance** folder that can be found in, <install path>\tools\pspice\library\discretetables\resistance.

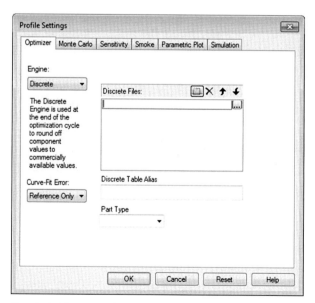

Fig. 25.10 Adding discrete tables.

18. Select the **res1%.table** and in **Part Type**, select **Resistance** from the pull down menu, see Fig. 25.11. Click on OK to close the **Profile Settings** window.

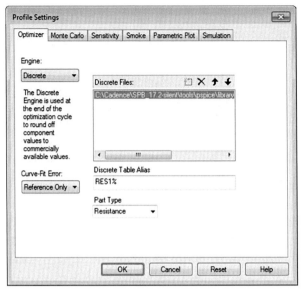

Fig. 25.11 Resistance 1% table values selected.

19. Change the Modified LSQ engine to the **Discrete** engine from the pull down menu.
20. In the **Parameters [Next Run]** window, click in the **Discrete Table** box for R4 and from the pull down menu, select **Resistors—1%**. Repeat for R3 and R2, see Fig. 25.12.

♦	On/Off	Component	Parameter	Discrete Table	Original	Min	Max	Current
▼ ☑ ♦		R3	Value	Resistor - 1%	37k	27k	47k	27.3067k
▼ ☑ ♦		R2	Value	Resistor - 1%	46k	27k	56k	35.8008k
▼ ☑ ♦		R1	Value	Resistor - 1%	1k	910	2k7	1.9822k

Click here to import a parameter from the design property map...

Fig. 25.12 Adding 1% resistor tables.

21. Rerun the Discrete simulation.
22. Fig. 25.13 shows the specification results.

Measurement	Min	Max	Original	Current	Error
Max(V(out))	8.5500	9.4500	8.7801	8.9677	0%
Max(I(R4))		190m	-175.6015m	-179.3543m	0%

Click here to import a measurement created within PSpice...

Fig. 25.13 Specification results using 1% resistors.

23. Fig. 25.14 shows the discrete 1% resistor parameter values.

	On/Off	Component	Parameter	Discrete Table	Original	Min	Max	Current
☑	☑	R3	Value	Resistor - 1%	37k	27k	47k	27.4000k
	☑	R2	Value	Resistor - 1%	46k	27k	56k	35.7000k
	☑	R1	Value	Resistor - 1%	1k	910	2k7	2k
				Click here to import a parameter from the design property map...				

Fig. 25.14 Commercially available 1% resistor values.

24. Right mouse click in the Error Graph and the cursor will appear. Move the cursor to previous Run Numbers and you will see the Current results and Error change based on previous Measurement data.

25. Right mouse click in the Error Graph and select Clear History.

26. Change the simulation engine to Random and run the simulation. You should see similar results but bear in mind you still have to run the Discrete engine.

27. Replace the resistor values for R1, R2, and R3 with the 1% resistor values and re-run the PSpice transient simulation to confirm the optimized circuit outputs. For circuits with a large component count, a useful feature is to rmb on a resistor and select **Find in Design**.

The revised resistor values will be used for the Monte Carlo analysis and Smoke analysis.

Note

When starting a new optimization simulation from new, it is best to place the cursor in the Error Graph and **rmb > Clear History** and also to delete the Parameters [Next Run] **Current** values.

The optimized power supply circuit is shown in Fig. 25.15.

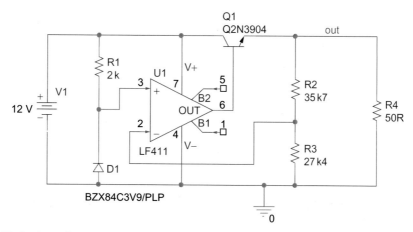

Fig. 25.15 Optimized power supply circuit.

Exercise 2

Curve fitting can be used to optimize model parameters to match characteristic graphs from datasheets or from empirical measured data. Curve-fit definitions can also be applied to the profile of a circuit's output waveform that is measured against a reference waveform where single data points can be measured using the trace expression, YatX. A curve fit specification consists of a reference file of measured data points and a corresponding measurement. Optimizer specifications are set using measurement expressions to determine circuit performance by varying circuit parameters. Circuit performance is evaluated using measurement expressions from transient, DC, or AC simulations, applied to the circuit output waveforms.

1. **File > Open > Demo Designs** and select the **Optimization using Curve-Fitting** design.
2. The circuit is that of a 4–pole low pass filter.
3. An AC logarithmic sweep from 100 Hz to 1 kHz has been set up.
4. Run the simulation and you will see the magnitude and phase response waveforms for the circuit.
5. From the top tool bar, select **PSpice > Advanced Analysis > Optimizer**. The Optimizer should be set up as shown in Fig. 25.16. Select the Curve Fit tab if not already selected.

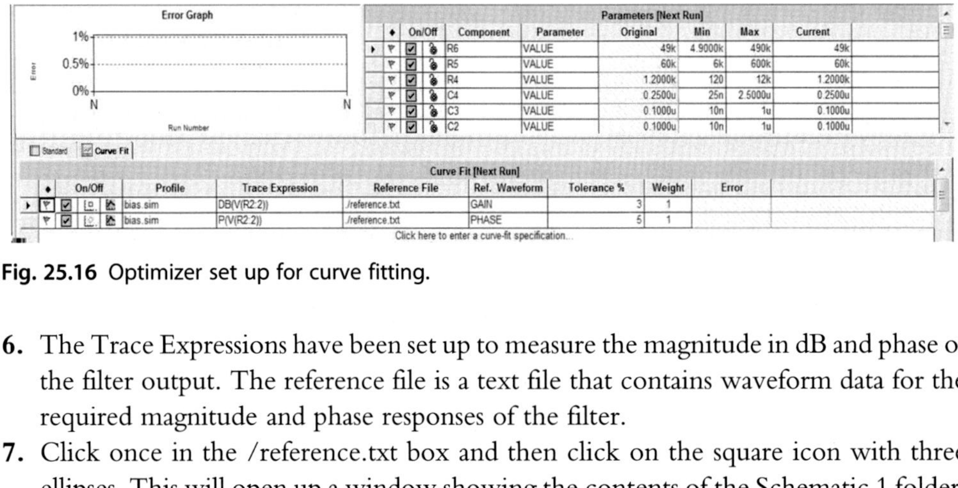

Fig. 25.16 Optimizer set up for curve fitting.

6. The Trace Expressions have been set up to measure the magnitude in dB and phase of the filter output. The reference file is a text file that contains waveform data for the required magnitude and phase responses of the filter.
7. Click once in the /reference.txt box and then click on the square icon with three ellipses. This will open up a window showing the contents of the Schematic 1 folder; ac and bias folders and the reference file.
8. Select the reference file and rmb > Open with > Wordpad. The reference file contains data columns for Frequency, Phase, and Gain. Close Wordpad and Cancel the Open window.
9. The **Ref. Waveform** fields incorporate a pull down menu that lists the GAIN and PHASE names from the reference file. The tolerance value is used in conjunction

with the method for error calculation using optimizer parameters known as Gears that is set in Profile Settings under **Curve-Fit Error**. From the top tool bar, select **Edit > Profile Settings > Optimizer**.

10. Close **Profile Settings**.

11. More information regarding Gears can be found in the PSpice Advanced Analysis User Guide.

12. Make sure the Modified LSQ engine is selected and run the Optimizer simulation. You should see two curves in the Error Graph, one for gain and one for phase. The results of the simulation show that optimization has been achieved. The optimized component values are shown in the Parameters [Next Run] window pane (see Fig. 25.17).

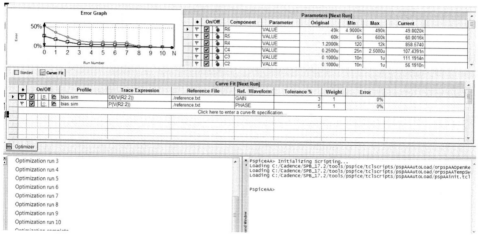

Fig. 25.17 Optimizer curve fitting results.

13. Go back to PSpice and you will see the optimized curves displayed for gain and phase.

CHAPTER 26

Monte Carlo

Chapter Outline

26.1. Introduction   379
26.2. Exercise   382
    Exercise 1   382

26.1. INTRODUCTION

The Monte Carlo analysis was introduced in Chapter 10. In summary, the Monte Carlo analysis is used to estimate the statistical performance of a circuit by randomly varying component tolerances and model parameter tolerances between their specified tolerance limits. The generated component values are based upon statistical distributions. The circuit analysis (DC, AC, or transient) is repeated a number of specified times using newly generated component and model tolerance values. Increasing the number of simulation runs will increase the spread of component tolerance values used for each simulation. The statistical results will give an indication on the robustness or yield of circuit performance to a range of different component values within their tolerance limits. However, using regular PSpice A/D, you can only run one Monte Carlo analysis for each defined measurement whereas Advanced Analysis lets you define multiple measurements for a single Monte Carlo analysis. You can also add tolerances to any SPICE model and subcircuit parameter. This will enable you to run Monte Carlo analysis on third party models or models downloaded from manufacturer's website.

The Monte Carlo results can be shown as either a probability density function (PDF) or as a cumulative distribution function (CDF), together with a summary of the statistical data. Monte Carlo yield results can also be updated for alternate specifications by adjusting and defining the appropriate limits using maximum and minimum cursors on the PDF display. Fig. 26.1 shows the results of a Monte Carlo simulation being displayed in PDF format. The number of graphical bins can be increased for finer statistical resolution and different data can be generated by changing the default random seed value. This is useful if you want to compare different data sets using the same number of Monte Carlo runs. The raw data from the simulations can also be accessed from the Raw Measurements tab seen in Fig. 26.1.

Analog Design and Simulation Using OrCAD Capture and PSpice
https://doi.org/10.1016/B978-0-08-102505-5.00026-4

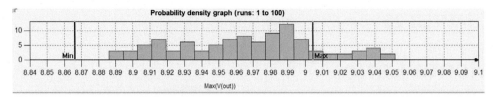

Fig. 26.1 Probability distribution function.

Only components or model parameters that have a tolerance value defined will be included in a Monte Carlo simulation. Monte Carlo starts with a nominal run (number one) by setting component tolerances to zero. Successive runs are numbered such that you can selectively rerun Monte Carlo for component values of interest without having to rerun the simulation again. Monte Carlo uses a default random seed number of 1 when running a simulation. If you change the seed number, then this will enable you to collect a different set of data with the same number of runs previously specified. The maximum number of runs depends on the how much system memory you have available.

The component and model parameter deviations from the nominal values up to the tolerance limits are determined by a probability distribution curve. By default, the distribution curve is flat or uniform, that is, each value has an equal chance of being used. The other option is the Gaussian distribution which is the more familiar bell-shaped curve commonly used in manufacturing. Component values are more likely to take on values found near the center of the Gaussian distribution compared to the outer edges of the tolerance limits. The distribution is characterized by the mean value (μ) and the standard deviation that is called sigma (σ). The standard deviation is a measure of the spread in measured values about the mean value. For a spread of one sigma (1σ) deviation, there is a probability that 68.26% of all the measured values are within that range. Advanced Analysis Monte Carlo calculates 3σ (99.73% probability) and 6σ (99.999998%) as well as the Mean and Median. The mean is the sample mean that is calculated from all the measured values divided by the total number of runs, whereas the Median is the measured value that represents the middle value of all the measured values.

The PDF gives an indication as to the number of measured values that fall into small specific ranges displayed as histograms or bins. The outline of the histograms then gives an indication to the distribution of measured values. The CDF is obtained from the integration of the PDF. The CDF in effect accumulates the probabilities from the PDF so that the probability of a measured value being less than or equal to a specified range can be determined by the cumulative number of runs on the y-axis of the CDF (Fig. 26.2).

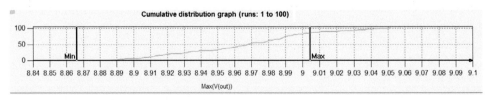

Fig. 26.2 Cumulative distribution function.

Monte Carlo uses five standard tolerance value distributions; Flat, Gauss, Gauss0.4, BIMD4.2, and Skew4.8. These distributions are defined in text files and can be found in:

   <install path>\tools\pspice\library\distribution

You can create your own distributions in the form of a text file that you place in the distribution folder. These distributions are readily available with the Advanced Analysis parameterized components. Fig. 26.3 shows the PSpice model for the 2N3904 from the advanls\bjn library. Once the tolerances have been set, the distribution is selected from the pull down menu.

Simulation Parameters

Property Name	Description	Value	Default	Unit	Distribution	Postol	Negtol	Editable	
IS	Saturation current	1 728E-16	0 1f	A				✓	
BF	Maximum forward beta	254 395	100		FLAT ▼	40		✓	
NF	Ifwd emission coef	0 85	1		FLAT			✓	
VAF	Fwd early voltage	10	100MEG	V	bimd 4 2			✓	
IKF	Hi cur beta rolloff	0 0163741	10	A	gauss			✓	≡
ISE	B-E leakage cur	9 97446E-15	1E-13	A	gauss0 4			✓	
NE	B-E leak emis coef	1 20863	1 5		skew 4 8			✓	
BR	Max reverse beta	0 1	1					✓	
NR	Irev emission coef	0 891964	1					✓	
VAR	Rev early voltage	7 74046	100MEG	V				✓	
IKR	Hi Irev beta rolloff	0 163741	100MEG	A				✓	
ISC	B-C leakage cur	9 97446E-15	1E-15	A				✓	
NC	B-C leak emis coef	2 84343	2					✓	
RB	Zero bias Rbase	28 3394	0	Ohm				✓	
IRB	Rbase cutoff current	0 01	100MEG	A				✓	
RBM	Min base resistance	0 01	0	Ohm				✓	
RE	Emitter resistance	0 00133911	0	Ohm				✓	
RC	Collector resistance	2 08353	0	Ohm				✓	
XTB	Beta temp exponent	1 20888	0					✓	

Fig. 26.3 2N3904 parameterized PSpice model.

If you are not using parameterized models, you can enter the distribution and tolerance to the model text parameter in the PSpice model editor text. For example,

   BF = 200 dev/gauss0.4 =40%

   BF = 200 dev/BIMD4.2=20%

   BF = 200 Skew4.8Q= 20%

Note

You can use the TOL_ON_OFF property to exclude a component from a Monte Carlo simulation even if a tolerance value has been defined for that component. Use the Property Editor to add the TOL_ON_OFF property and assign a value of either OFF or ON accordingly.

26.2. EXERCISE

Exercise 1

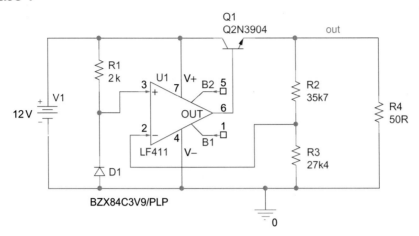

Fig. 26.4 Optimized voltage regulator circuit.

Fig. 26.4 shows the optimized power supply circuit from Chapter 25. You will need to have set up a transient analysis and created measurement expressions as in steps 1 to 6 of Exercise 1 of Chapter 25.

1. From the Optimizer results, resistor R1, R2, and R3 values were optimized with 1% resistor values. Assign a tolerance of 1% to resistors R1, R2, and R3 if not already assigned. Assign a 5% tolerance to the load resistor R4.
2. From Capture, select **PSpice > Advanced Analysis > Monte Carlo**.
3. From the top tool bar, select **Edit > Profile Settings**.... Set the **Number of Runs** to 100, the Number of Bins to 20 (Fig. 26.5), and click on OK.

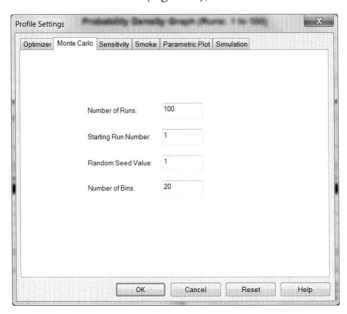

Fig. 26.5 Setting number of runs to 100 and Number of Bins to 20.

4. In the **Statistical Information** window, click on, **Click here to import a measurement created within PSpice ....** Select V(out) and I(R4) and click on OK.
5. Run the simulation by clicking on the play button.
6. You should see similar results to that in Fig. 26.6. The distribution for V(out) shows that there are possible output voltages down toward 8.870 V and the median is 8.9678 V. Ideally the median should have a value of 9 V with an equal distribution either side. The distribution is not symmetrical about the mean but the values are within the specified 9 V$\pm$5%.

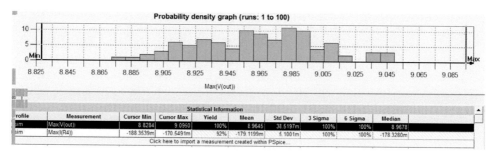

Fig. 26.6 Probability distribution function.

7. To take a measurement over a range of values, click once on the Min or Max cursor that will change from black to orange and then move and click to new position. If you mess up, **rmb > Zoom Fit** to rescale the PDF display.
8. Right mouse click in the PDF display and select MC Graph (PDF/CDF) to display the CDF distribution as shown in Fig. 26.7.

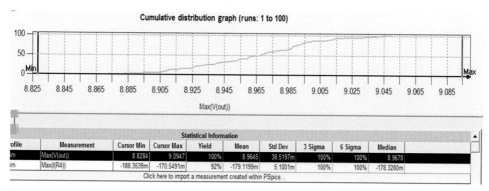

Fig. 26.7 Cumulative distribution function.

CHAPTER 27

Smoke Analysis

Chapter Outline

27.1. Passive Smoke Parameters   387
    27.1.1 Resistor Smoke Parameters   387
    27.1.2 Inductor Smoke Parameters   389
    27.1.3 Capacitor Smoke Parameters   391
27.2. Active Smoke Parameters   392
    27.2.1 Bipolar Transistors   393
27.3. Derating Files   395
27.4. Example 1   398
27.5. Exercise   401
    Exercise 1   401
27.6. Example 2   404
    Exercise 2   406
27.7. Example 3   414
    Exercise 3   415
27.8. Example 4   419

Smoke analysis gives an indication as to whether components are operating within manufacturer's absolute maximum ratings. Components that are continually operating close to maximum ratings can result in electrical and thermal stress leading to premature component failure. To provide for a margin of safety, manufacturer's maximum operating conditions (MOCs) can be derated to provide safe operating limits for components.

Smoke Analysis reports on stress-related parameters such as breakdown voltages, operating currents, device temperatures, junction temperatures, and power dissipations.

Fig. 27.1 shows the results of a Smoke analysis where the red bars indicate that the component Smoke limits are exceeding the MOCs. Orange bars indicate that the limits are >90% of the MOCs and the green bars indicate that parameters are operating within 90% of the MOC.

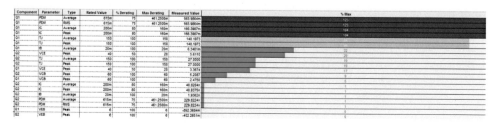

Component	Parameter	Type	Rated Value	% Derating	Max Derating	Measured Value	% Max
Q1	PDM	Average	615m	75	461.2500m	565.9864m	123
Q1	PDM	RMS	615m	75	461.2500m	565.9864m	123
Q1	IC	Average	200m	80	160m	166.3967m	104
Q1	IC	Peak	200m	80	160m	166.3967m	104
Q1	TJ	Average	150	100	150	140.1973	
Q1	TJ	Peak	150	100	150	140.1973	
Q1	IB	Average	20m	100	20m	6.3461m	
Q2	VCE	Peak	40	50	20	5.6110	32
Q2	TJ	Average	150	100	150	27.0000	28
Q2	TJ	Peak	150	100	150	27.0000	18
Q1	VCE	Peak	40	50	20	3.3674	16
Q2	VCB	Peak	60	100	60	5.2087	17
Q1	VCB	Peak	60	100	60	2.4750	9
Q2	IC	Average	200m	80	160m	46.8204m	
Q2	IC	Peak	200m	80	160m	40.8375n	
Q2	IB	Average	20m	100	20m	1.9362n	
Q2	PDM	Average	615m	75	461.2500m	229.6224m	
Q2	PDM	RMS	615m	75	461.2500m	229.6224m	
Q1	VEB	Peak	6	100	6	-892.3684m	
Q2	VEB	Peak	6	100	6	-402.2851m	

Fig. 27.1 Smoke analysis simulation results.

Analog Design and Simulation Using OrCAD Capture and PSpice
https://doi.org/10.1016/B978-0-08-102505-5.00027-6

Smoke analysis parameters allow MOCs to be defined for a component, the parameter values being those given in manufacturer data sheets. Advanced analysis contains libraries of components that already have Smoke parameters defined. However, in the latest Cadence 17.2 software release, Smoke parameters can be added to almost every device. For active components, Smoke parameter values can be entered via the PSpice Model Editor whereas for passive components, Smoke parameter values can be added to individual components or they can be added globally using the **Design Variables** part (Fig. 27.2).

Advanced Analysis Properties

Tolerances:
RTOL = 0
CTOL = 0
LTOL = 0
VTOL = 0
ITOL = 0

Smoke Limits:

RMAX = 0.25	ESR = 0.001
RSMAX = 0.0125	CPMAX = 0.1
RTMAX = 200	CVN = 10
RVMAX = 100	LPMAX = 0.25
CMAX = 50	DC = 0.1
CBMAX = 125	RTH = 1
CSMAX = 0.005	
CTMAX = 125	
CIMAX = 1	
LMAX = 5	
DSMAX = 300	
IMAX = 1	
VMAX = 12	

User Variables:

Fig. 27.2 Design Variables part table showing Smoke Analysis parameters and their default global limits for passive components.

The Assign Tolerance window, that displays all circuit components with tolerances and components with associated models, can also be used to open the PSpice Model Editor (see Fig. 27.3).

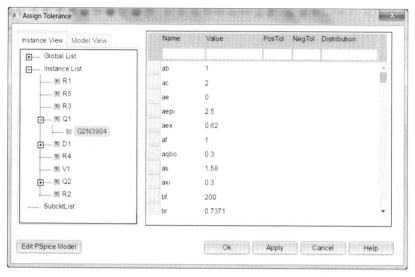

Fig. 27.3 Assign Tolerance window displays components with associated PSpice models as well as components with defined tolerances.

27.1. PASSIVE SMOKE PARAMETERS

The passive components consist of resistors, capacitors, inductors, and power supplies placed from either the analog or pspice_elem libraries. All of these components are specified to operate over a given rated temperature range beyond which the output parameters have to be derated in order for the components to operate reliably with an increase in ambient temperature.

27.1.1 Resistor Smoke Parameters

Smoke Analysis calculates and displays the maximum power dissipation, maximum body temperature, and maximum voltage drop across a resistor.

A current flowing through a resistor will generate heat resulting in an increase in temperature that will affect the normal operation of the resistor. The maximum specified power rating for a resistor relates to the heat that can be dissipated up to a specified ambient temperature limit above which the operation of the resistor will degrade or fail. Beyond the temperature limit, the power that can be dissipated as heat decreases as a percentage of the maximum power rating. Fig. 27.4 shows the Derating Power curve for a resistor used in Smoke analysis.

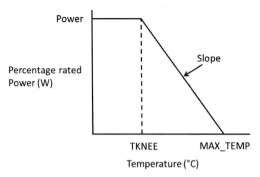

Fig. 27.4 Resistor power derating curve.

The maximum power rating (POWER) for a resistor is specified over its normal operating ambient temperature range up to its rated temperature, TKNEE. Above this temperature, the power rating decreases at a rate given by the SLOPE, eventually falling to zero at the maximum temperature rating for a resistor, MAX_TEMP.

For example, a 0.25 W resistor is capable of dissipating a maximum power of 0.25 W up to a typical ambient temperature of 70°C. Exceeding the 0.25 W power rating of the resistor will result in excess heat that cannot be efficiently dissipated resulting in an increase in resistor temperature. Beyond 70°C, the effective power dissipation will decrease up to the maximum temperature typically 150–200°C at which point the resistor will fail. Premature resistor failure can be avoided by increasing the power rating of the resistor, increasing airflow across the resistor, using an equivalent parallel combination of resistors, using ventilated cases or to avoid placing resistors close to heat sources such as power transistors. Table 27.1 summarizes the resistor Smoke Analysis parameters.

Table 27.1 Resistor parameters used in Smoke Analysis

Maximum operating condition	Smoke parameter property	Smoke parameter value	Default value	Parameter name displayed in Smoke window
Maximum power dissipation	POWER	RMAX	0.25 W	PDM
Slope of power dissipation versus temperature	SLOPE	RSMAX	0.0125 W/°C	Not displayed
Maximum temperature	MAX_TEMP	RTMAX	200°C	TB
Voltage rating	VOLTAGE	RVMAX	100V	RV

In Smoke analysis, the resistor derating curve is defined by the KNEE temperature and the MAX temperature both of which are related by the SLOPE of the power derating curve. You can enter the SLOPE parameter value using the Variables symbol from which the KNEE value will be calculated. Alternatively, you can add the TKNEE property to the resistor using the Property Editor. See Example 1 at the end of the chapter.

27.1.2 Inductor Smoke Parameters

Smoke Analysis displays the maximum DC current rating, maximum dielectric strength, maximum power loss due to equivalent series resistance (ESR) and inductor temperature rise.

Inductors are made from insulated coils of wire wound in air or wound on a ferrite magnetic core. When current flows through the wire windings, average real power is dissipated in the form of heat associated with the inherent DC resistance of the wire windings that make up the coils. Exceeding the inductor maximum DC current ratings can lead to saturation effects. The self-heating losses due to winding resistance will also cause a temperature rise in the inductor resulting in the wire insulation material to fail. Similarly, if the inductor maximum voltage ratings are exceeded, the insulation can also break down resulting in windings to short circuit and overheat. For core wound inductors and transformers, the current flowing through the inductor windings will generate heat resulting in an increase in temperature that will affect the magnetic characteristics of ferrite core materials. Core saturation levels decrease with an increase in temperature resulting in inductance values to decrease.

Subsequently, manufacturers specify the maximum inductor current rating (DC) and maximum voltage rating over a maximum ambient temperature range for reliable operation. Smoke analysis does not calculate the derating of inductor current with temperature.

The power dissipated by the inherent DC resistance of the inductor is attributed to the average current flowing through the inductor resulting in a temperature rise above ambient. Manufacturers specify the average current as an Irms current that causes a maximum allowable rise in temperature by a specified value above the ambient temperature. Hence, the maximum ambient temperature and maximum allowable temperature rise define the inductor maximum temperature MAX_TEMP.

Inductors also specify a rated saturation current (Isat) that is associated with the saturation of the inductor core resulting in the inductance value to decrease. Smoke analysis supports four parameters associated with inductors: DC current for saturation, temperature rise, breakdown voltage, and MAX current. See Example 2 at the end of the chapter. Table 27.2 lists the inductor Smoke parameters.

Table 27.2 Inductor Smoke Analysis parameters

Maximum operating condition	Smoke parameter property	Smoke parameter value	Default value	Parameter name displayed in Smoke window
Current rating	CURRENT	LMAX	5 A	LI
DC current value	DC_CURRENT	DC	0.1 A	LIDC
Dielectric strength	DIELECTRIC	DSMAX	300	LV
Maximum power loss due to series resistance	POWER	LPMAX	0.25 W	PDML
DC_RESISTANCE	DC_RESISTANCE	ESR	0.001 Ω	Not displayed
Rise in temperature	RTH	THERMR	1°C	TJL
Maximum temperature	MAX_TEMP	LTMAX	125°C	Not displayed

Note

The PDML parameter is the power related to the ESR property. The default value shown is from the Variables part table.

The ESR property is not available on the standard inductor from the analog.olb library. You have to use the L_t from the analog.olb.

Tip

When you run Smoke analysis, the Parameter names are shown in short form. To see the complete description, **rmb > Parameter Descriptions** as shown in Fig. 27.5.

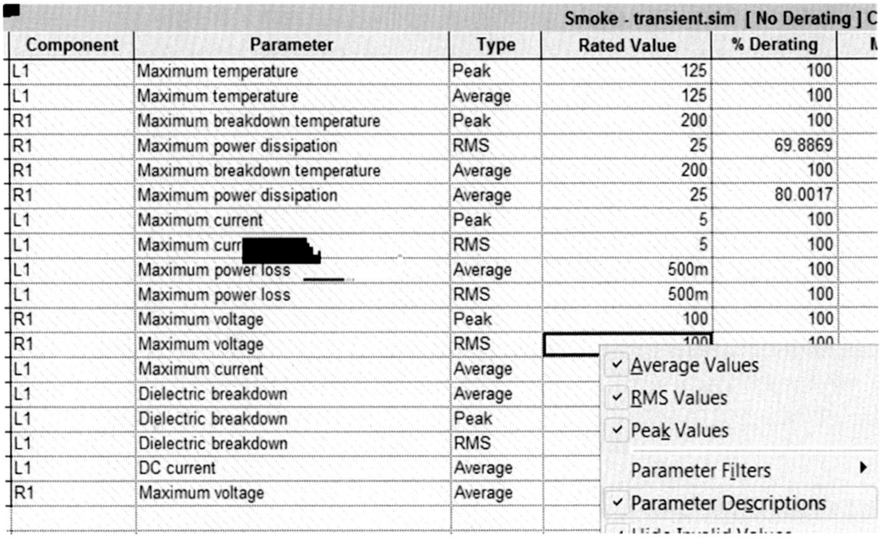

Fig. 27.5 Displaying parameter descriptions.

27.1.3 Capacitor Smoke Parameters

Smoke Analysis calculates and displays the maximum rated voltage, maximum ripple current, maximum reverse voltage, maximum ESR power dissipation and capacitor temperature rise.

Applied voltage as well as temperature affects the reliability of a capacitor and its initial power on performance. The range of ambient temperature over which a capacitor can continuously operate is known as the category temperature range and is defined by upper and lower temperature limits. The rated temperature (upper category limit) is defined as the maximum ambient temperature that a capacitor can operate continuously without exceeding the rated voltage. Exceeding the capacitors maximum voltage rating will cause the dielectric to break down. Subsequently, manufacturers provide voltage derating guidelines to improve the long-term reliability of their capacitors. Fig. 27.6 shows the voltage-temperature derating curve for a capacitor used in Smoke analysis.

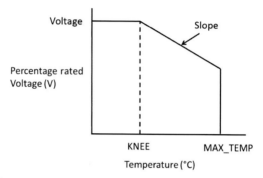

Fig. 27.6 Capacitor voltage derating curve.

The rated VOLTAGE is the maximum peak DC operating voltage that can be applied to a capacitor for continuous operation, also known as the working voltage that appears printed on the capacitor. The rated temperature, KNEE is the maximum ambient temperature over which the capacitor can continuously operate at the rated VOLTAGE. The maximum ripple current is the AC rms current that flows in and out of a capacitor and it is this current that produces heat and hence power dissipation (PDML) in the ESR. Smoke analysis supports electrolytic and nonelectrolytic capacitors. The CVN parameter is used to report on short reverse transient voltages. The maximum ambient temperature at which the capacitor can operate is given by MAX_TEMP. Table 27.3 summarizes the capacitor parameters used in Smoke analysis.

Table 27.3 Capacitor parameters used in Smoke analysis

Maximum operating condition	Smoke parameter property	Smoke parameter value	Default value	Parameter name displayed in Smoke window
Maximum ripple	CURRENT	CIMAX	1 A	CI
Voltage Rating	VOLTAGE	CMAX	50 V	CV
Maximum reverse voltage	NEGATIVE_VOLTAGE	CVN	10 V	CVN
Maximum power loss due to series resistance	POWER	CPMAX	0.1 W	PDML
Temperature derating slope	SLOPE	CSMAX	0.005 V/°C	Not displayed
Breakpoint temperature	KNEE	CBMAX	125°C	Not displayed
Rise in temperature	RTH	THERMR	1°C	TJL
Maximum temperature	MAX_TEMP	CTMAX	125°C	Not displayed
Equivalent series resistance	ESR	ESR	0.001 Ω	Not displayed

The rated temperature is different for capacitors with different dielectrics. Electrolytic aluminum and tantalum capacitors have a working voltage dependence on temperature compared to nonelectrolytics such as ceramic, polyester, and other plastic film capacitors. Capacitance and leakage current are also dependent on temperature. Smoke analysis supports electrolytic and nonelectrolytic capacitors and will test for reverse transient voltages. The CVN parameter defines the negative voltage ratings.

Note

The ESR property is not available on the standard capacitor from the analog.olb library. You have to use the C_t or C_elect capacitors from the analog.olb. See Example 3 at the end of the chapter.

27.2. ACTIVE SMOKE PARAMETERS

Smoke analysis incorporates parameters for semiconductor devices that include diodes, diode bridge rectifiers, zener diodes, bipolar transistors, JFET, MESFET, MOSFET, power MOSFET, dual MOSFET, IGBT, LED, optocouplers, varistors, and thyristors.

27.2.1 Bipolar Transistors

Bipolar transistors used in power amplifiers or power switching circuits should be operated within their safe operating area in order to avoid excessive power dissipation that can lead to premature failure. Manufacturers provide derating curves that plot maximum continuous power dissipation against collector junction temperature also known as case temperature. For silicon transistors, the range is normally specified from room temperature at 25°C to 150°C.

The power derating curve for a transistor is shown in Fig. 27.7. The transistor case temperature is related to the collector-base junction temperature (TJ). The transistor junction thermal (ΘJA) resistance is composed of the junction-to-case thermal resistance (RJC) that is a fixed value and the case-to-ambient thermal resistance (RCA). The case-to-ambient thermal resistance can be decreased by using a heatsink that effectively increases the surface area of the transistor case. In the case of power transistors, the collector-base junctions have large areas in order to dissipate heat quickly away from the junction. Table 27.4 lists the Smoke analysis parameters used for bipolar (BJT) transistors.

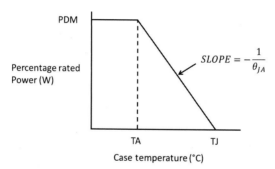

Fig. 27.7 Transistor power derating curve.

Table 27.4 Bipolar parameters used in Smoke analysis

Semiconductor component	Maximum operating condition	Smoke parameter name property and symbol property name	Parameter name displayed in Smoke window
BJT	Maximum base current (A)	IB	IB
BJT	Maximum collector current (A)	IC	IC
BJT	Maximum power dissipation (W)	PDM	PDM
BJT	Thermal resistance, Case-to-Ambient (°C/W)	RCA	Used for calculating junction temperature

Continued

Table 27.4 Bipolar parameters used in Smoke analysis—cont'd

Semiconductor component	Maximum operating condition	Smoke parameter name property and symbol property name	Parameter name displayed in Smoke window
BJT	Thermal resistance, Junction-to-Case (°C/W)	RJC	Not displayed
BJT	Secondary breakdown intercept (A)	SBINT	Not displayed
BJT	Derated percent at TJ (secondary breakdown)	SBMIN	Not displayed
BJT	Secondary breakdown slope	SBSLP	Not displayed
BJT	Temperature derating slope (secondary breakdown)	SBTSLP	Not displayed
BJT	Maximum junction temperature (°C)	TJ	TJ
BJT	Maximum collector-base voltage (V)	VCB	VCB
BJT	Maximum collector-emitter voltage (V)	VCE	VCE
BJT	Maximum emitter-base voltage (V)	VEB	VEB

Semiconductor smoke analysis parameters are defined in the semiconductor PSpice models. The smoke parameters can be viewed using the PSpice Model Editor that can be accessed either from the circuit diagram or from the **Start Menu.** In the circuit diagram you highlight the transistor part and **rmb > PSpice Model Editor**. See Example 4 at the end of the chapter. Fig. 27.8 shows the Smoke parameters and their values for a Q2N3904 transistor.

Smoke Parameters

These are Device Maximum Operating condition parameters
required for Smoke Analysis

Device Max Ops	Description	Value	Unit
IB	Max base current		A
IC	Max collector current		A
VCB	Max C-B voltage	60	V
VCE	Max C-E voltage	40	V
VEB	Max E-B voltage	6	V
PDM	Max pwr dissipation		W
TJ	Max junction temp	150	C
RJC	J-C thermal resist	83.3	C/W
RCA	C-A thermal resist	116.7	C/W
SBSLP	Second brkdown slope		
SBINT	Sec brkdwn intercept		A
SBTSLP	SB temp derate slope		%/C
SBMIN	SB temp derate at TJ		%

Fig. 27.8 Bipolar transistor Smoke Analysis parameters.

The other semiconductor types and their associated Smoke parameters can be found in the online PSpice Advanced Analysis User Guide.

Note

Smoke analysis will only be performed using parameters that have defined parameter values. The Log File (View > Log File > Smoke) will report which Smoke tests have not been performed. For example in Fig. 27.7, no Smoke tests will be done using IB, IC, and SB.

Note

You can exclude components or hierarchical blocks from Smoke analysis simulations. In the **Property Editor**, in the **Filter by**: box, select **Capture PSpiceAA** and set the value of SMOKE_ON_OFF to OFF.

27.3. DERATING FILES

As well as providing MOCs, many manufacturers provide derating guidelines for their components so they can operate within safe operating limits (SOL) that is given by

$$SOL = MOC \times \text{derating factor} \qquad (27.1)$$

For example, using a derating of 75% on the Vmax of a 200 V capacitor, Smoke Analysis will indicate any excursions >150 V in the form of a red bar in a bar graph display.

You can also define different deration specifications for each of the supported device types. This allows you to define a unique deration margin for each specification and associate this to an instance in the circuit diagram. For example, you can have two different derating definitions for MOSFET drain currents. A derating margin of 50%, say for use in high reliability products otherwise a derating margin of 70% for all other designs. These derating specifications can be stored in global files for reuse in other circuit designs.

Smoke Analysis comes with standard passive and active derating factors that can be found in the standard.drt file in the **tools > PSpice > Library** folder:

C:\Cadence\SPB_17.2\tools\pspice\library

The standard.drt file can be opened with a text editor. However, it is easier to use Advanced Analysis using the **Profile Settings** window, **Edit > Profile Settings.** You can then browse to the default standard.drt derating file by clicking on **New (Insert)** as shown in Fig. 27.9. Cadences also supply a custom_derating_template file that can be modified to create new derating files.

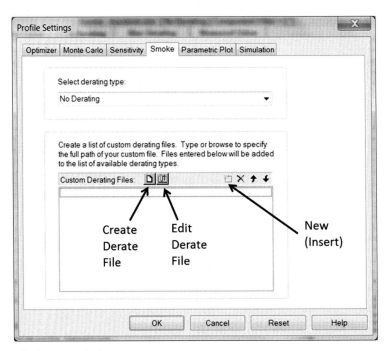

Fig. 27.9 Profile Settings window.

Note
Profile Settings can also be opened by clicking anywhere in Smoke Analysis worksheet and **rmb > Derating > Custom Derating Files.**

Selecting either **Create Derate File** or **Edit Derate File** opens up the **Edit Derate File** window (Fig. 27.10) where you can enter the smoke parameters. You can build up a collection of derating files that will be displayed in the Profile Settings windows. The files will then appear in a pull down box under **Select derating type.**

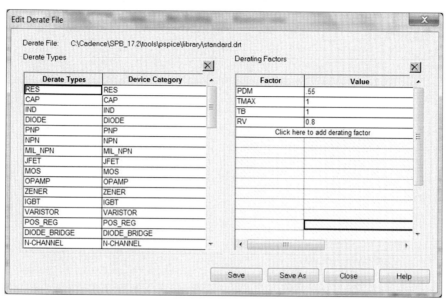

Fig. 27.10 Edit Derate File window where you can enter smoke parameters.

Fig. 27.10 shows the default derating factors using the Smoke Analysis "standard derating" file. When you select "Standard Derating" in Smoke Analysis, all the derating factors will be applied to the components. For example, a derating factor of 0.55 will be applied to the maximum power (PDM) rating of the resistors. PDM is the same as RMAX. If the calculated power exceeds the derated value then a red bar indicates that the power has exceeded the derated value and by how much (see Fig. 27.11).

♦	Component	Parameter	Type	Rated Value	% Derating	Max Derating	Measured Value	% Max
▼	R1	PDM	Average	250m	52.8661	132.1652m	136.1702m	104
▼	R1	PDM	RMS	250m	52.8661	132.1652m	136.1702m	104
▼	R1	TB	Average	155	100	155	73.2979	48
▼	R1	TB	Peak	155	100	155	73.2979	48
▼	R1	RV	Average	400	80	320	8	3
▼	R1	RV	Peak	400	80	320	8	3
▼	R1	RV	RMS	400	80	320	8	3

Smoke - transient.sim [Standard Derating] Component Filter = [*]

Fig. 27.11 Resistor Smoke Analysis showing derated power.

Note

If after a SMOKE analysis, the PDM shows zero, this is because the maximum operating temperature has been exceeded therefore the PDM is automatically derated to zero.

27.4. EXAMPLE 1

Fig. 27.12 shows the power-temperature derating curve for a 470 kΩ metal film resistor. The corresponding specifications from the manufacturer's data sheet are given as:

Rated Power @70°C (W) 0.25 W

Max overload voltage 400 V

Operating temperature range −55°C to +155°C

The resistor body temperature, TB is given by

$$TB = Tambient + \Delta T \qquad (27.2)$$

where ΔT is the temperature rise due to the self-heating effect of the ESR and subsequent power dissipation.

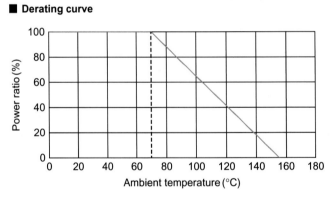

Fig. 27.12 Resistor power-temperature derating curve.

$$TB = Tambient + \frac{Pdis}{POWER} Rth \qquad (27.3)$$

where

Tambient—ambient temperature that has a default value of 27°C set by TNOM

Rth—resistor to ambient thermal resistance

Pdis—calculated power dissipation

POWER—maximum rated power dissipation

Rth = 1/SLOPE and Pdis is calculated from the result of running a transient analysis.

$$TB = Tambient + \frac{Pdis}{POWER}\frac{1}{SLOPE} \tag{27.4}$$

As manufacturer's derating curves show unit power derating, the slope RSMAX is given by

$$SLOPE = \frac{1}{MAX\_TEMP - TKNEE} \tag{27.5}$$

where MAX_TEMP is the maximum operating temperature.

From Eq. (27.5), TKNEE can be calculated as:

$$TKNEE = MAX\_TEMP - \frac{1}{SLOPE} \tag{27.6}$$

Combining Eqs. (27.4), (27.5):

$$TB = Tambient + \frac{Pdis}{POWER}(MAX\_TEMP - TKNEE) \tag{27.7}$$

If TKNEE is not known, the derating slope is used to calculate TB using Eq. (27.3) else if TKNEE is known then Eq. (27.7) can be used to calculate TB. The TKNEE property can be added to individual resistors.

Therefore,

$$SLOPE = \frac{1}{155 - 70}$$

$$SLOPE = \frac{1}{85} = 0.01176\,W/^{\circ}C$$

Therefore the Smoke parameters defined in Table 23.1 are
POWER, RMAX = 0.25
MAX_TEMP, RTMAX = 155
SLOPE, RSMAX = 0.01176
VOLTAGE, RVMAX = 400

Fig. 27.13 shows the resistor circuit with Smoke parameters added. From the results of a transient analysis, power dissipation Pdis was calculated as 136.17 mW.

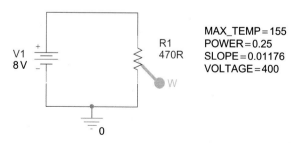

Fig. 27.13 Resistor Smoke Analysis.

From Eq. (27.3), the resistor body temperature is given by

$$TB = 27 + \frac{0.13617}{0.25} \frac{1}{0.01176}$$

$$TB = 73.3163\,^{\circ}C$$

Note

Using 1/SLOPE = 85 results in a more accurate calculation to give TB = 73.2978°C.

Derating Factor

From the derating curve in Fig. 27.12, if the resistor body temperature exceeds TKNEE, the power dissipation will be dynamically derated by a Power Derating Factor, PDF given by

$$PDF = SLOPE\,(MAX\_TEMP - TB) \qquad (27.8)$$

From which the derated power is given by

$$Pdis = PDF \times POWER \qquad (27.9)$$

The standard derating file for a resistor in Smoke Analysis is 0.55. Therefore the derated power is given by

$$Pdis = 0.55 \times 0.13617 = 0.07489 = 74.89\,mW$$

Note

The derating curve is dynamically derated for resistors and capacitors and not for inductors.

Note

With the Cadence demo Lite version, Smoke Analysis is limited to resistors, capacitors, diodes, and transistors.

27.5. EXERCISES

Exercise 1

1. Create a new PSpice project called for example Smoke_Resistor and in the **Create PSpice Project** window, select **simple_aa.opj** from the pull-down menu (Fig. 27.14).

Fig. 27.14 Advanced Analysis libraries.

2. Delete the AA Variables part. Draw the resistor circuit in Fig. 27.15.

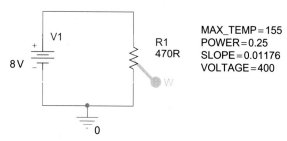

Fig. 27.15 Resistor Smoke Analysis circuit.

3. Double click on the resistor to open the **Property Editor** and add the following Smoke Limit parameters values:
   MAX_TEMP = 155
   SLOPE = 0.01176
   POWER=0.25
   VOLTAGE = 400
   No need to enter the SI units.
   To display the Smoke limits, highlight each Smoke parameters by holding down the control key and clicking on each property. Select **Display** (or rmb > Display) and in the **Display Properties** window, select **Name and Value** as shown in Fig. 27.16.

Fig. 27.16 Displaying inductor Smoke parameter properties on the schematic.

4. Set up a transient analysis (PSpice > New Simulation Profile) and leave the default **Run To Time** as 1000 ns.
5. Place a Power marker on the body (middle) of the resistor. **PSpice > Markers > Power Dissipation** or click on the icon 🖋 .
6. Run the simulation.
7. The Probe window in PSpice window will appear displaying a power dissipation trace of 136.170 mW.
8. Go back to Capture and select, **PSpice > Advanced Analysis > Smoke**.
9. PSpice Advanced Analysis will open up showing the results of the Smoke Analysis. Invalid values will be displayed as gray bar lines. For example, only average and peak values are calculated for TB, the RMS value is invalid. Right mouse click anywhere and select **Hide Invalid Values**. Fig. 27.17 shows the results of the Smoke Analysis with a power dissipation calculated as 136.1702 mV.
   % Max is given by: $100 \times 136.1702\,m/240.3004\,m = 57\%$.

Smoke - transient.sim [No Derating] Component Filter = [*]								
◆	Component	Parameter	Type	Rated Value	% Derating	Max Derating	Measured Value	% Max
▼	R1	PDM	Average	250m	96.1202	240.3004m	136.1702m	57
▼	R1	PDM	RMS	250m	96.1202	240.3004m	136.1702m	57
▼	R1	TB	Average	155	100	155	73.2979	48
▼	R1	TB	Peak	155	100	155	73.2979	48
▼	R1	RV	Average	400	100	400	8	3
▼	R1	RV	Peak	400	100	400	8	3
▼	R1	RV	RMS	400	100	400	8	3

Fig. 27.17 Smoke Analysis results for a single resistor.

Note

Invalid values indicate the type of measurement that does not apply to that specific Smoke test. For example, the breakdown voltage of a diode is calculated as the more meaningful peak value rather than the average value.

10. Using Eq. (27.3) the resistor body temperature is calculated as $TB = 73.3163°C$. Obviously Smoke Analysis uses a greater number of significant digits in the calculations. A more accurate "hand" calculation would be to use $1/SLOPE = 85$ rather than use $SLOPE = 0.01176$. This gives the same result as Smoke Analysis as $73.2978°C$.

11. The resistor body temperature of $73.2978°C$ is greater than the TKNEE temperature of $70°C$ and so the power dissipation is dynamically derated. The value of the derating factor (% Derating) is given by Eq. (27.8) as:

$$PDF = SLOPE \left(MAX\_TEMP - TB \right)$$

For a more accurate calculation:

$$PDF = \frac{(155 - 73.2978)}{85} = 0.9612 = 96.12\%$$

Therefore using Eq. (23.9), the power dissipation (Max Derating) is given by

$$Pdis = PDF \times POWER$$
$$Pdis = 0.9612 \times 0.25 = 0.2403\,W$$
$$Pdis = 240.3\,mW$$

This is the value of PDM shown in Fig. 27.17.

12. Right mouse click anywhere in Smoke Analysis and select **Derating > Standard Derating** and run the simulation by clicking on the play button.

13. You should see two red bars that indicate that the derated values for PDM (RMAX) have been exceeded by 104% Fig. 27.18. The resistor is not operating within the defined safe operating limits.

	Component	Parameter	Type	Rated Value	% Derating	Max Derating	Measured Value	% Max
▼	R1	PDM	Average	250m	52.8661	132.1652m	136.1702m	104
▼	R1	PDM	RMS	250m	81.6521	104.1302m	136.1702m	104
▼	R1	TB	Average	155	100	155	73.2979	48
▼	R1	TB	Peak	155	100	155	73.2979	48
▼	R1	RV	Average	400	80	320	8	3
▼	R1	RV	Peak	400	80	320	8	3
▼	R1	RV	RMS	400	80	320	8	3

*Smoke - transient.sim [Standard Derating] Component Filter = [*]*

Fig. 27.18 Power dissipation safe operating limits have been exceeded.

14. In Capture, change the value of POWER to 0.5.

15. Re-run the transient simulation and then run Smoke Analysis. You will see that using a 0.5 W resistor, the power dissipation of the resistor (Fig. 27.19) meets the defined safe operating limits of operation. In fact you could even use a 0.4 W resistor.

	Component	Parameter	Type	Rated Value	% Derating	Max Derating	Measured Value	% Max
▼	R1	PDM	Average	500m	55	275m	136.1702m	50
▼	R1	PDM	RMS	500m	55	275m	136.1702m	50
▼	R1	TB	Average	155	100	155	50.1582	33
▼	R1	TB	Peak	155	100	155	50.1582	33
▼	R1	RV	Average	100	80	80	8	10
▼	R1	RV	Peak	100	80	80	8	10
▼	R1	RV	RMS	100	80	80	8	10

Smoke - transient.sim [Standard Derating] Component Filter = [*]

Fig. 27.19 Resistor operating within defined safe operating limits.

27.6. EXAMPLE 2

Fig. 27.20 shows the temperature rise characteristics curve for the Murata 1.5 μH Inductor LQM2HPN1r5MG0, respectively.

Specification for a Murata 1.5 μH Inductor are given as:

Temperature range: $-55°C$ to $+125°C$

Rated temperature $T_R = 85°C$

1.5 A @ambient temperature 85°C

1.1 A @ambient temperature 125°C

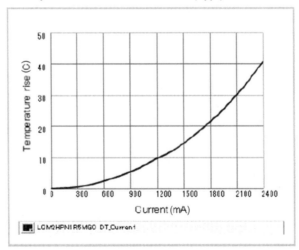

Fig. 27.20 Inductor temperature rise characteristics.

The inductor temperature, TJL is given by the sum of the ambient temperature and inductor temperature rise ΔT:

$$TJL = Tambient + \Delta T \qquad (27.10)$$

For example, an inductor specifies the maximum current rating up to a typical ambient temperature of 85°C and a maximum allowable temperature rise of 40°C. The maximum operating temperature is therefore 125°C.

The temperature rise ΔT associated with the power dissipation of the inductor DC resistance is given by

$$\Delta T = Pdis \times RTH \qquad (27.11)$$

RTH is the inductor thermal resistance and Pdis is related to the average real power dissipated by the inductor resistance given by

$$Pdis = (Irms)^2 ESR \qquad (27.12)$$

Therefore (23.10) can be written as:

$$TJL = Tambient + (Pdis \times RTH) \qquad (27.13)$$

$$TJL = Tambient + \left[(Irms)^2 ESR \times RTH\right] \qquad (27.14)$$

Inductor thermal resistances (RTH) are not normally specified by manufacturers as the value of RTH is dependent on how the inductors are mounted on circuit boards. However, some manufacturers specify different current ratings for surface mount inductors using different PCB mountings and land pattern areas. Thermal resistances are not usually specified for frame core inductors and transformers as the thermal resistance paths are not clearly defined. However, the thermal resistance can be approximated using the manufacturer's specified Irms current.

From Eqs. (23.11), (23.12), the inductor thermal resistance can be calculated as:

$$RTH = \frac{\Delta T}{(Irms)^2 ESR} \qquad (27.15)$$

where ΔT is a specified temperature rise at a given Irms current. From the graph of Temperature Rise versus Current (Fig. 27.19), Irms=2.4 A @40°C. The inductor thermal resistance is therefore calculated as:

$$RTH = \frac{\Delta T}{(Irms)^2 ESR} \qquad (27.16)$$

$$RTH = \frac{40}{(2.4)^2 0.088} = 78.91°C/W$$

Fig. 27.21 shows the inductor circuit with added Smoke parameters. The result of a transient analysis calculates the Irms as 1.0606 A. From Eq. (27.12):

$$Pdis = (1.0606)^2 0.088 = 98.9888\,mW$$

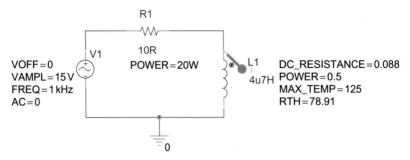

Fig. 27.21 Smoke Analysis inductor circuit.

Using Eq. (27.14), the inductor temperature is

$$TJL = 27 + \left[(1.606)^2 0.088 \times 78.91\right] = 34.8812^\circ C$$

Exercise 2

1. Create a new PSpice project called Smoke_Inductor and in the **Create PSpice Project** window, select **simple_aa.opj** from the pull-down menu (Fig. 27.22).

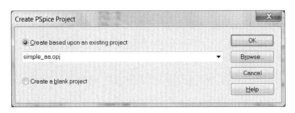

Fig. 27.22 Advanced Analysis libraries.

2. Draw the inductor circuit in Fig. 27.23. Use an L_t inductor from the analog library as this has the ESR property attached. Place a VSIN source from the source library.

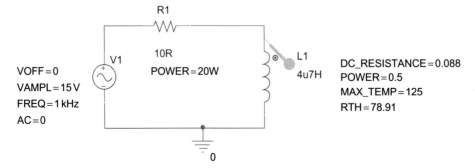

Fig. 27.23 Inductor.

3. Double click on the resistor and in the Property Editor for POWER, set RMAX to 20 W. To display the power value 20 W, click on **Display > Name and Value**. Close the Property Editor.

4. Double click on the inductor to open the Property Editor and add the following Smoke Limit parameters values:
DC_RESISTANCE = 0.088
MAX_TEMP = 125
RTH = 78.91
POWER = 0.5
Highlight each Smoke parameters by holding down the control key and clicking on each property. Select Display and in the Display Properties Window select Name and Value (see Fig. 27.24).

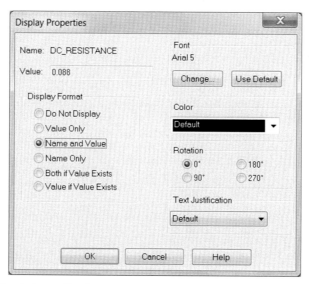

Fig. 27.24 Displaying inductor Smoke parameter properties.

5. Set up a transient analysis with a **Run To Time** of 4 ms.

6. Place a Current marker on pin1 of the inductor that is marked with the dot symbol. Run the simulation.

7. The Probe window in PSpice window will appear displaying the inductor current with a max value (LI) of 1.4942 A.

8. Go back to Capture and select, **PSpice > Advanced Analysis > Smoke**.

9. PSpice Advanced Analysis will open up showing the results of the Smoke Analysis for the resistor R1 and inductor L1. Invalid values will be displayed as gray bars. For example, only average and peak values are calculated for TJL. Right mouse click any-where and select **Hide Invalid Values** (Fig. 27.25).

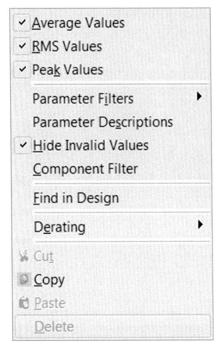

Fig. 27.25 Hide Invalid Values.

You should see the results in Fig. 27.26.

Component	Parameter	Type	Rated Value	% Derating	Max Derating	Measured Value	% Max
R1	PDM	RMS	20	17.6132	3.5226	13.7774	392
R1	PDM	Average	20	30.2796	6.0559	11.2441	186
R1	TB	Peak	200	100	200	250.2553	126
R1	TB	Average	200	100	200	139.4407	70
L1	LI	Peak	5	100	5	1.4942	30
L1	TJL	Peak	125	100	125	36.4033	30
L1	TJL	Average	125	100	125	34.8080	28
L1	LI	RMS	5	100	5	1.0604	22
L1	PDML	Average	500m	100	500m	98.9479m	20
L1	PDML	RMS	500m	100	500m	98.9479m	20
L1	LI	Average	5	100	5	21.6296u	1
L1	LV	Average	300	100	300	6.6546u	1
L1	LV	Peak	300	100	300	44.5608m	1
L1	LV	RMS	300	100	300	31.5756m	1
L1	LIDC	Average	1	100	1	21.6296u	1

Fig. 27.26 Murata inductor Smoke Analysis results.

10. It is always a good idea to have a look at the Smoke Analysis log file. **View > Log File > Smoke** as shown in Fig. 27.27.

```
Smoke Analysis Run : Mon Jul 10 17:47:06 2017


   Reference Designator = R1
   Info:
     Smoke test RV will not be done.
     The Maximum Operating Value is not defined.

   Reference Designator = L1
   Info:
     Smoke test LIDC will not be done.
     The Maximum Operating Value is not defined.

   Reference Designator = R1
   Warning: Deration
   INFO(ORPSPAA-7028): TBreak (Tknee)is less than Simulation Temprature.
   Check the slope[RSMAX] or maximum temperature[RTMAX].

   Reference Designator = R1
   Info: Deration
     Tbrk Calculated:0.000000

   Reference Designator = R1
   Warning: Deration
   INFO(ORPSPAA-7028): TBreak (Tknee)is less than Simulation Temprature.
   Check the slope[RSMAX] or maximum temperature[RTMAX].

   Reference Designator = R1
   Info: Deration
     Tbrk Calculated:0.000000

   Reference Designator = R1
   Warning: Deration
   INFO(ORPSPAA-7028): TBreak (Tknee)is less than Simulation Temprature.
   Check the slope[RSMAX] or maximum temperature[RTMAX].

   Reference Designator = R1
   Info: Deration
     Tbrk Calculated:0.000000


   ***** Analysis Summary *****

   Reference Designator = R1
     The following parameter(s) had undefined
     Maximum Operating Value(s)
            RV
   Reference Designator = L1
     The following parameter(s) had undefined
     Maximum Operating Value(s)
            LIDC
```

Fig. 27.27 Smoke Analysis log file.

11. The log file in Fig. 27.27 reports undefined CV parameter for R1 and an undefined parameter LIDC for L1. A warning message reports on TKNEE being less than the simulation temperature and to check the values for the SLOPE and MAX_TEMP. These values were not added on the resistor and so default values were used.

12. In Capture double click on the resistor to open the Property Editor. Set the limit of MAX_TEMP (RTMAX) to 155 and the VOLTAGE limit (RVMAX) to 200. Do not close the Property Editor.

    Rather than calculate the SLOPE, you can set the TKNEE limit. Select **New Property** and add TKNEE as shown in Fig. 27.28.

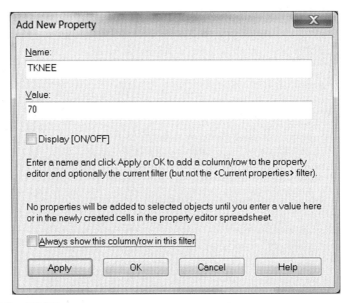

Fig. 27.28 Adding TKNEE property.

13. Re-run the simulation in Capture.

14. Re-run Advanced Analysis and look at the Smoke Analysis log file (Fig. 27.29).

```
Smoke Analysis Run : Mon Jul 10 18:40:25 2017


Reference Designator = L1
Info:
        Smoke test LIDC will not be done.
        The Maximum Operating Value is not defined.


***** Analysis Summary *****
                      659x526
Reference Designator = R1
        The following parameter(s) exceeded the
        Maximum Operating Value(s) specified
              PDM ( PEAK )


Reference Designator = L1
        The following parameter(s) had undefined
        Maximum Operating Value(s)
              LIDC
```

Fig. 27.29 Smoke Analysis log file.

15. Select the inductor and set the DC_Current limit (DC) to 1 A. As shown in Table 27.2, Smoke Analysis displays the DC_Current as LIDC.
16. Re-run the PSpice simulation in Capture.
17. Re-run Advanced Analysis.
18. The results for the resistor and inductor are displayed. Right mouse click anywhere and select Component Filter. The default wildcard setting (*) displays all components. As shown in Fig. 27.30, enter L1 and click on OK.

Component Filter	X
Find what:	L1
OK	Cancel

Fig. 27.30 Component Filter.

19. Fig. 27.31 shows the results of the Smoke Analysis for the inductor only. The current LIrms is 1.0604 A

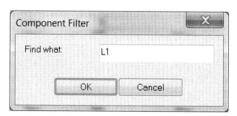

Component	Parameter	Type	Rated Value	% Derating	Max Derating	Measured Value	% Max
L1	LI	Peak	5	100	5	1.4942	30
L1	TJL	Peak	125	100	125	36.4033	30
L1	TJL	Average	125	100	125	34.6000	28
L1	LI	RMS	5	100	5	1.0604	22
L1	PDML	Average	500m	100	500m	98.9479m	20
L1	PDML	RMS	500m	100	500m	98.9479m	20
L1	LI	Average	5	100	5	21.6296u	1
L1	LV	Average	300	100	300	6.6546u	1
L1	LV	Peak	300	100	300	44.5608m	1
L1	LV	RMS	300	100	300	31.5756m	1
L1	LIDC	Average	1	100	1	21.6296u	1

Fig. 27.31 Smoke Analysis showing only results of Murata inductor.

Therefore from Eq. 27.12 the approximate value for Pdis is

$$Pdis = (1.0606)^2 0.088 = 0.09895 = 98.95\,mW$$

The average inductor temperature is given by

$$TJL = Tambient + \left[(LIrms)^2 \times ESR \times RTH\right]$$

$$TJL = 27 + \left[(1.060\mu)^2 \times 0.088 \times 78.91\right] = 34.8083\,^\circ C$$

However, the values are shown with no derating.

20. In Smoke Analysis, select **Edit > Profile Settings** and in the **Profile Settings** window, click on the **New (Insert)** icon as shown in Fig. 27.32. Click on three ellipses (…) and browse to the default standard.drt derating file in the; <install path> \tools\pspice\library. For example:

C:\Cadence\SPB_17.2\tools\pspice\llbrary

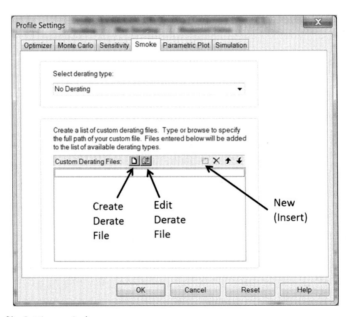

Fig. 27.32 Profile Settings window.

21. Click on the **Edit Derate File** to display the Derate Types and Derating Factors (Fig. 27.33). Maximum power dissipation (PDM) is derated to 0.55 and voltage rating (RV) is 0.8.

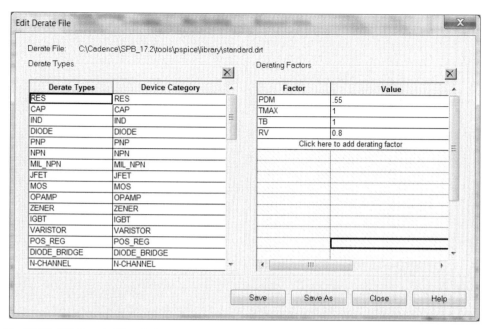

Fig. 27.33 Edit Derate File window.

22. Select the inductor (IND) and you will see that LI, LV, and LIDC all have a derating factor of 0.9.

23. Close both the **Edit Derate File** and **Profile Settings** windows.

24. Right mouse click anywhere in Smoke Analysis and select **Component Filter** and enter * to select all components.

25. Again, right mouse click anywhere in Smoke Analysis but this time select **Derating > Standard Rating**. Run Smoke analysis again. You will see two red lines indicating that the derated resistor power has been exceeded.

26. Change the resistor power in Capture to 30 W and re-run the PSpice simulation and then the Smoke Analysis. The final Smoke Analysis is shown in Fig. 27.34.

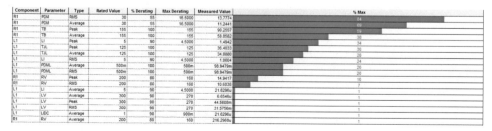

Fig. 27.34 Final Smoke Analysis results.

27.7. EXAMPLE 3

The datasheet for a 1 μF plastic film, polyethylene terephthalate (PET) capacitor gives

Rated temperature T_R—85°C

Derating factor—1.25% per °C

Maximum temperature—125°C

Dissipation factor of 0.004 @ 1 kHz

Rated voltage—50 VDC

The ESR is not given in the datasheets, so it can be calculated from

$$\text{ESR} = \text{Xc} \times \tan\delta = \frac{\tan\delta}{2\pi f C} \tag{27.17}$$

where

$\tan\delta$—dissipation factor, DF

f—test frequency, typically 120 Hz

C—capacitance

Therefore the equivalent series resistance is

$$\text{ESR} = \frac{0.004}{2\pi \times 10^3 \times 10^{-6}} = 0.159\,\Omega$$

The capacitor temperature, TJL is given by

$$\text{TJL} = \text{Tambient} + \Delta T \tag{27.18}$$

where ΔT is the temperature rise due to the self-heating effect of the ESR and subsequent power dissipation and is given by

$$\Delta T = \text{Pdis} \times \text{Rth} \tag{27.19}$$

Rth is the capacitor thermal resistance and Pdis is related to the average real power dissipated by the capacitor's ESR. Power dissipation is attributed to the AC Irms ripple current that flows in and out of a capacitor producing internal heating effects. Ignoring the effects of the leakage current, the power dissipation in the ESR is given by

$$\text{Pdis} = \text{I}^2\text{rms} \times \text{ESR} \tag{27.20}$$

Therefore the rise of capacitor temperature is given by

$$\text{TJL} = \text{Tambient} + \left[\left(\text{I}^2\text{rms} \times \text{ESR} \right) \times \text{Rth} \right] \tag{27.21}$$

The thermal resistance of the capacitor, Rth, is given by the reciprocal of the SLOPE from the voltage derating curve and I^2rms is calculated from the result of running a transient analysis.

$$\text{TJL} = \text{Tambient} + \frac{\left(\text{I}^2\text{rms} \times \text{ESR} \right)}{\text{SLOPE}} \tag{27.22}$$

If the capacitor thermal resistance is not specified in manufacturer's datasheets, the SLOPE can be calculated from

$$SLOPE = \frac{1}{CTMAX - TKNEE} \qquad (27.23)$$

Therefore Eq. (27.22) can be written as:

$$TJL = Tambient + \frac{(I^2 rms \times ESR)}{CTMAX - TKNEE} \qquad (27.24)$$

Where CTMAX is the maximum operating temperature and TKNEE is the rated temperature that a capacitor can operate continuously without exceeding the rated voltage, also known as the upper category temperature.

For this example both the slope and TKNEE are given. Only the TKNEE value will be used.

TKNEE, CBMAX = 85
MAX_TEMP, CTMAX = 125
SLOPE, CSMAX = 0.0125
VOLTAGE, CMAX = 50
ESR = 0.159

Exercise 3

The voltage tripler circuit in Fig. 27.35 is used to demonstrate how Smoke analysis can be used to improve circuit performance by detecting components that do not meet the derated MOCs.

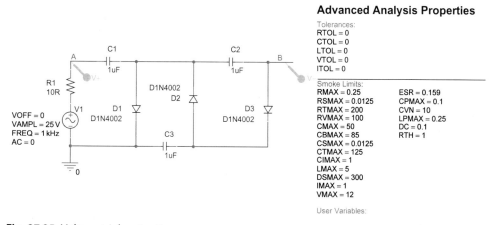

Fig. 27.35 Voltage tripler circuit.

1. Create a new project called, for example, voltage_tripler. After you name the project, the **Create PSpice Project** window appears. In the pull-down menu select the simple_aa.opj and click on **OK** (Fig. 27.36).

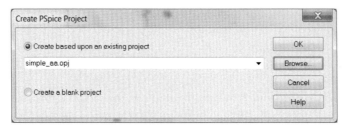

Fig. 27.36 Setting up default Advanced Analysis libraries.

Note

In the latest versions of OrCAD When you first open the schematic page, the **Variables** part will automatically appear. If you select **Empty Project**, or you have inadvertently deleted the **Variables** part, then this can be added from the **PSpice > advanals > pspice_elem** library.

---

2. Draw the voltage_tripler circuit in Fig. 27.35 Use a C_t capacitor from the analog library as this has the ESR property attached.
3. Add SMOKE parameter values to the Variables part as shown in Fig. 27.35.
4. Set up a transient analysis with a **Run To Time** of 50 ms.
5. Place a differential voltage marker on node A and B, **PSpice > Markers > Voltage Differential**. When you place the first V + marker, the second V − marker appears automatically. Alternatively, click on the icon   .
6. Run the simulation.
7. The Probe window in PSpice window will appear displaying the differential voltage between nodes A and B.
8. In PSpice, select **View > Measurement Results** and in the **Measurement Results** window, check on the box for **Max(V(A)-V(B))** that will display a value of 73.25649 V, see Fig. 27.37.

	Evaluate	Measurement	Value	Measurement Results
▶	☑	Max(V(A)-V(B))	73.25649	
				Click here to evaluate a new measurement...

Fig. 27.37 Differential voltage measurement result.

9. Go back to Capture and select, **PSpice > Advanced Analysis > Smoke**.

10. PSpice Advanced Analysis will open up showing the results of the Smoke Analysis for the voltage_tripler circuit. Invalid values will be displayed as gray bar lines. For example, only average and peak values are calculated for TJL. Right mouse click anywhere and select **Hide Invalid Values** Fig. 27.38.

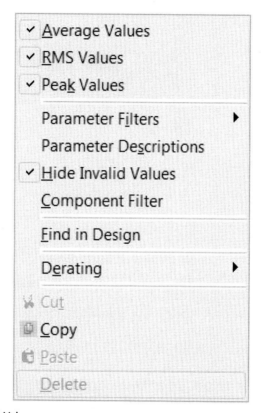

Fig. 27.38 Hide Invalid Values.

11. Right mouse click anywhere again and select **Component Filter** and enter **C*** as shown in Fig. 27.39.

Fig. 27.39 Selecting results for capacitors only.

12. Fig. 27.40 shows the results of the Smoke Analysis for the voltage_tripler circuit. The capacitor ripple current is given by PDML (rms) for capacitor C1 and is shown as 43.5728 μA. Therefore from Eq. (27.20) the approximate value for Pdis is

$$\text{Pdis} = \left(43.5728 \times 10^{-6} \times 0.159\right)^2 = 0.302\text{nW}$$

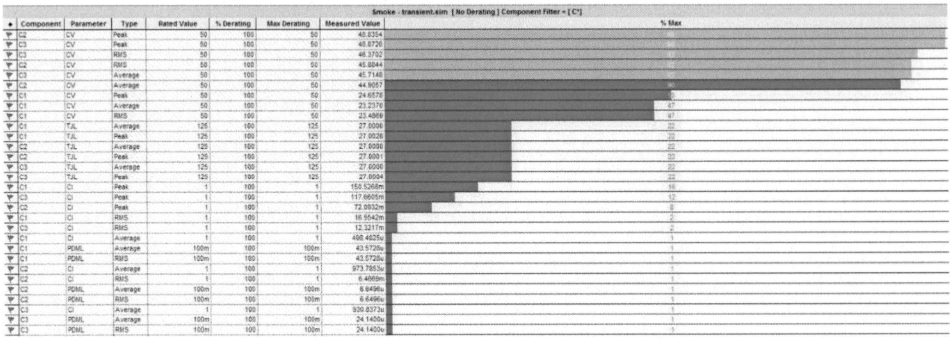

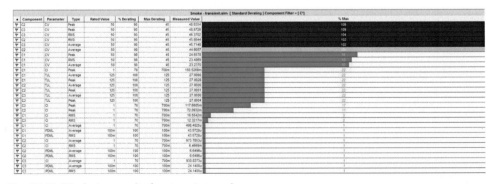

Fig. 27.40 Smoke Analysis showing only results of the voltage tripler circuit.

This small power dissipation is attributed to the small ESR value and is not going to change the capacitor temperature significantly. What is of more interest is the voltage rating for the capacitors. The yellow bar lines indicate that the voltage ratings are within 90% of their maximum ratings.

13. Right mouse click anywhere and select **Derating > Standard Derating** and run the Smoke analysis simulation.

14. You should see the results as shown in Fig. 27.41 where the red bars indicate that the derated MOCs have been exceeded.

Fig. 27.41 Displaying results for capacitors only.

15. Capacitors with a higher working voltage of 63 V can be used instead. Globally change the value of CMAX from 50 to 63 V and run the PSpice simulation and then the Smoke analysis simulation. You should see that the derated MOCs for the capacitors have not been exceeded.

27.8. EXAMPLE 4

Transistor datasheets normally give two values for device power dissipation, one for ambient temperature and one for case temperature. Two values are also given for thermal resistances (Rth): junction to ambient and junction to case. If you use a heatsink, then the thermal resistance from junction to ambient is given by the sum of the transistor junction to case and heatsink thermal resistances. With no heatsink, the thermal resistance is given by the value of the transistor junction to ambient (see Fig. 27.42).

MAXIMUM RATINGS

Rating	Symbol	Value	Unit
Collector – Emitter Voltage	V_{CEO}	40	Vdc
Collector – Base Voltage	V_{CBO}	60	Vdc
Emitter – Base Voltage	V_{EBO}	6.0	Vdc
Collector Current – Continuous	I_C	200	mAdc
Total Device Dissipation @ T_A = 25°C Derate above 25°C	P_D	625 5.0	mW mW/°C
Total Device Dissipation @ T_C = 25°C Derate above 25°C	P_D	1.5 12	W mW/°C
Operating and Storage Junction Temperature Range	T_J, T_{stg}	−55 to +150	°C

THERMAL CHARACTERISTICS (Note 1)

Characteristic	Symbol	Max	Unit
Thermal Resistance, Junction–to–Ambient	$R_{\theta JA}$	200	°C/W
Thermal Resistance, Junction–to–Case	$R_{\theta JC}$	83.3	°C/W

Fig. 27.42 2N3904 datasheet.

With no heatsink, transistor thermal resistance is given by Rja.
Transistor temperature is given by

$$TJ = Tambient + \Delta T$$

where ΔT is given by

$$\Delta T = Pdis \times RTH$$

Hence

$$TJ = Tambient + (Pdis \times RTH)$$

where Pdis is the calculated power dissipation from the result of a transient analysis. If there is no heatsink, the thermal resistance RTH is given by RJC + RCA.

where

RJC—junction-to-case thermal resistance

RCA—case-to-ambient thermal resistance

Therefore,

$$TJ = Tambient + [(Pdis \times (RJC + RCA)]$$

1. Use the voltage regulator circuit that was used in Sensitivity analysis. Otherwise draw the circuit in Fig. 27.43.

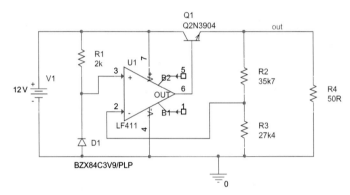

Fig. 27.43 9V Series voltage regulator.

2. You need to edit the Smoke parameters for Q1 using the PSpice Model Editor. In the later 17.2 versions, you can open and edit the Smoke parameters via the Assign Tolerance window. Either way, you will be using the PSpice Model Editor to assign Smoke parameters. Either:

   (a) Highlight **Q1 > 2N3904** and **rmb > Edit PSpice Model**.
   OR
   (b) Select **PSpice > Advanced Analysis > Assign Tolerance**, select **Q1 > 2N3904** then click on Edit PSpice Model.

3. Enter the Smoke Parameter values as shown in Fig. 27.44. If the Smoke Parameters are not shown in the Model Editor, then select **View > Model**.

Smoke Parameters

These are Device Maximum Operating condition parameters
required for Smoke Analysis

Device Max Ops	Description	Value	Unit
IB	Max base current	0.02	A
IC	Max collector current	0.2	A
VCB	Max C-B voltage	60	V
VCE	Max C-E voltage	40	V
VEB	Max E-B voltage	6	V
PDM	Max pwr dissipation	0.625	W
TJ	Max junction temp	150	C
RJC	J-C thermal resist	83.3	C/W
RCA	C-A thermal resist	116.7	C/W
SBSLP	Second brkdown slope		
SBINT	Sec brkdwn intercept		A
SBTSLP	SB temp derate slope		%/C
SBMIN	SB temp derate at TJ		%

Fig. 27.44 Smoke parameter values for 2N3904.

Note

If you are using the circuit from Sensitivity Analysis, then you will already have changed
the Bf of the transistor to 200. If not, then in the PSpice Model Editor change the Bf from
Bf = 416.4 to 200.

Save the file and close the PSpice Model Editor.
4. Place a voltage marker on node "out" and a current marker on the bottom pin of load
resistor R4.
5. Run a transient analysis using the default **Run To Time** of 1000 ns.
6. In PSpice set up two measurement expressions for V(out) and I(R4). **Trace >
Evaluate Measurement**.
Select **Max(1)** from the Functions or Macros and then select V(out).
Select **Max(1)** from the Functions or Macros and then select I(R4).
See Fig. 27.45.

	Evaluate	Measurement	Value	Measurement Results
	☑	Max(V(out))	8.96771	
▶	☑	Max(I(R4))	-179.35427m	

Fig. 27.45 The Measurement Results.

7. Return to Capture and select **PSpice > Advanced Analysis > Smoke**. RMB anywhere in the window and select **> Hide Invalid Values**. You should see the results in Fig. 27.46.

◆	Component	Parameter	Type	Rated Value	% Derating	Max Derating	Measured Value	% Max
▼	R4	PDM	Average	250m	0	0	1.6084	< MAX >
▼	R4	PDM	RMS	250m	0	0	1.6084	< MAX >
▼	R4	TB	Average	200	100	200	1.3137k	657
▼	R4	TB	Peak	200	100	200	1.3137k	657
▼	Q1	TJ	Average	150	100	150	134.1865	90
▼	Q1	TJ	Peak	150	100	150	134.1865	90
▼	Q1	IC	Average	200m	100	200m	175.6343m	88
▼	Q1	IC	Peak	200m	100	200m	175.6343m	88
▼	Q1	PDM	Average	615m	100	615m	535.9327m	86
▼	Q1	PDM	RMS	615m	100	615m	535.9327m	86
▼	U1	VMMIN	Peak	-3	100	-3	-3.8941	71
▼	U1	VPMIN	Peak	200	100	-3	-3.8941	71
▼	U1	VCCMAX	Peak	18	100	18	12	67
▼	U1	VSMAX	Peak	36	100	36	12	34
▼	R1	TB	Average	200	100	200	53.2825	27
▼	R1	TB	Peak	200	100	200	53.2825	27
▼	Q1	IB	Average	20m	100	20m	3.8621m	20
▼	D1	TJ	Average	150	100	150	27	19
▼	D1	TJ	Peak	150	100	150	27	19
▼	R1	PDM	Average	250m	73.3588	183.3969m	32.8531m	18
▼	R1	PDM	RMS	250m	73.3588	183.3969m	32.8531m	18
▼	U1	IOUT	Peak	25m	100	25m	3.8621m	16
▼	R3	TB	Average	200	100	200	27.4427	14
▼	R3	TB	Peak	200	100	200	27.4427	14
▼	R2	TB	Average	200	100	200	27.5768	14
▼	R2	TB	Peak	200	100	200	27.5768	14
▼	Q1	VCE	Peak	40	100	40	3.0323	8
▼	D1	IRMX	Peak	61.0500m	100	61.0500m	4.0530m	7

Smoke - dc_sweep.sim [No Derating] Component Filter = [*]

Fig. 27.46 Smoke analysis with no derating applied.

The results of the Smoke analysis give an average power dissipation for Q1 of 535.9327mW. There is no heatsink so using:

$$TJ = Tambient + [(Pdis \times (RJC + RCA)]$$
$$TJ = 27 + [(0.5359327 \times (83.3 + 116.7)] = 134.1865°C$$

In Smoke analysis, the Rated Value of PDM is shown as 615 mW although the Smoke PDM parameter for the 2N3904 transistor was set at 625 mW in the Model Editor. Using the PDM value from the 2N3904 datasheet, the transistor junction thermal resistance is given by

$$RJA = \frac{TJ - Ta}{PDM}$$
$$RJA = \frac{150 - 25}{0.625} = 200°C/W$$

where
  TJ—maximum operating junction temperature
  Ta—ambient temperature
  RJA—junction thermal resistance (RJC + RCA)

However, the nominal simulation temperature TNOM that represents Ta is set at 27°C. Therefore using $RJA = 200°C/W$, the PDM displayed in Smoke analysis is given by

$$PDM = \frac{150 - 27}{200} = 615\,mV$$

Smoke analysis will use the smallest PDM value.

8. From Fig. 27.46, the transistor temperature is within 90% of its maximum value.
9. Click anywhere in the window and **rmb > Derating > Standard Derating**. Run the Smoke simulation again. You should see the results in Fig. 27.47.

•	Component	Parameter	Type	Rated Value	% Derating	Max Derating	Measured Value	% Max
▼	R4	PDM	Average	250m	0	0	1.6064	< MAX >
▼	R4	PDM	RMS	250m	0	0	1.6064	< MAX >
▼	R4	TB	Average	200	100	200	1.3137k	657
▼	R4	TB	Peak	200	100	200	1.3137k	657
▼	Q1	PDM	Average	615m	75	461.2500m	535.9327m	117
▼	Q1	PDM	RMS	615m	75	461.2500m	535.9327m	117
▼	Q1	IC	Average	200m	80	160m	175.6343m	110
▼	Q1	IC	Peak	200m	80	160m	175.6343m	110
▼	Q1	TJ	Average	-150	100	150	134.1865	90
▼	Q1	TJ	Peak	615.0	100	150	134.1865	90
▼	U1	VMMIN	Peak	-3	100	-3	-3.8941	71
▼	U1	VPMIN	Peak	-3	100	-3	-3.8941	71
▼	U1	VCCMAX	Peak	18	100	18	12	67
▼	U1	VSMAX	Peak	36	100	36	12	34
▼	R1	PDM	Average	250m	40.3473	100.8683m	32.8531m	33
▼	R1	PDM	RMS	250m	40.3473	100.8683m	32.8531m	33
▼	R1	TB	Average	200	100	200	53.2825	27
▼	R1	TB	Peak	200	100	200	53.2825	27
▼	Q1	IB	Average	20m	100	20m	3.8621m	20
▼	D1	TJ	Average	150	100	150	27	19
▼	D1	TJ	Peak	150	100	150	27	19
▼	Q1	VCE	Peak	40	50	20	3.0323	16
▼	U1	IOUT	Peak	25m	100	25m	3.8621m	16
▼	R3	TB	Average	200	100	200	27.4427	14
▼	R3	TB	Peak	200	100	200	27.4427	14
▼	R2	TB	Average	200	100	200	27.5768	14
▼	R2	TB	Peak	200	100	200	27.5768	14
▼	D1	PDM	Average	250m	75	187.5000m	15.7825m	9

Smoke - dc_sweep.sim [Standard Derating] Component Filter = [*]

Fig. 27.47 Smoke analysis with standard rating applied.

You need to use a transistor with a higher power dissipation specification, for example, a ZTX450 from the Zetex library. The standard derating is 0.8 for collector current and 0.75 for power dissipation. Alternatively, you can add a heatsink to the transistor. R4 is the load resistance and is not part of the circuit. However, you can assign a value of 5 W to the resistor POWER property via the Property Editor just to dismiss the red bar.

10. Replace the Q2N3904 with a ZTX450 from the Zetex library.
11. Run the transient analysis.
12. Run the Smoke analysis.
13. In Smoke, **rmb > Component Filter > Q1**. You should see that with standard derating, the ZTX450 is operating within safe limits, see Fig. 27.48.

•	Component	Parameter	Type	Rated Value	% Derating	Max Derating	Measured Value	% Max
▼	Q1	PDM	Average	1	75	750m	541.2641m	73
▼	Q1	PDM	RMS	1	75	750m	541.2641m	73
▼	Q1	IC	Average	1	80	800m	178.1531m	23
▼	Q1	IC	Peak	1	80	800m	178.1531m	23
▼	Q1	TJ	Average	200	100	200	27	14
▼	Q1	TJ	Peak	200	100	200	27	14
▼	Q1	VCE	Peak	45	50	22.5000	3.0323	14
▼	Q1	VCB	Peak	60	100	60	2.2483	4
▼	Q1	VEB	Peak	5	100	5	-783.9703m	0

Smoke - ransient.sim [Standard Derating] Component Filter = [Q1]

Fig. 27.48 Smoke analysis results for Q1 with standard rating applied.

APPENDIX

PSpice Measurement Definitions

Bandwidth	Bandwidth of a waveform (you choose dB level)
Bandwidth_Bandpass_3dB	Bandwidth (3 dB level) of a waveform
Bandwidth_Bandpass_3dB_XRange	Bandwidth (3 dB level) of a waveform over a specified X-range
CenterFrequency	Center frequency (dB level) of a waveform
CenterFrequency_XRange	Center frequency (dB level) of a waveform over a specified X-range
ConversionGain	Ratio of the maximum value of the first waveform to the maximum value of the second waveform
ConversionGain_XRange	Ratio of the maximum value of the first waveform to the maximum value of the second waveform over a specified X-range
Cutoff_Highpass_3dB	High-pass bandwidth (for the given dB level)
Cutoff_Highpass_3dB_XRange	High-pass bandwidth (for the given dB level)
Cutoff_Lowpass_3dB	Low-pass bandwidth (for the given dB level)
Cutoff_Lowpass_3dB_XRange	Low-pass bandwidth (for the given dB level) over a specified range
DutyCycle	Duty cycle of the first pulse/period
DutyCycle_XRange	Duty cycle of the first pulse/period over a range
Falltime_NoOvershoot	Fall time with no overshoot
Falltime_StepResponse	Fall time of a negative-going step response curve
Falltime_StepResponse_XRange	Fall time of a negative-going step response curve over a specified range
GainMargin	Gain (dB level) at the first 180-degree out-of-phase mark
Max	Maximum value of the waveform
Max_XRange	Maximum value of the waveform within the specified range of X
Min	Minimum value of the waveform
Min_XRange	Minimum value of the waveform within the specified range of X
NthPeak	Value of a waveform at its nth peak
Overshoot	Overshoot of a step response curve
Overshoot_XRange	Overshoot of a step response curve over a specified range
Peak	Value of a waveform at its nth peak
Period	Period of a time domain signal
Period_XRange	Period of a time domain signal over a specified range
PhaseMargin	Phase margin
PowerDissipation_mW	Total power dissipation in milliwatts during the final period of time (can be used to calculate total power dissipation, if the first waveform is the integral of V(load)

Pulsewidth	Width of the first pulse
Pulsewidth_XRange	Width of the first pulse at a specified range
Q_Bandpass	Calculates Q (center frequency/bandwidth) of a bandpass response at the specified dB point
Q_Bandpass_XRange	Calculates Q (center frequency/bandwidth) of a bandpass response at the specified dB point and the specified range
Risetime_NoOvershoot	Rise time of a step response curve with no overshoot
Risetime_Step	Response rise time of a step response curve
Risetime_StepResponse_XRange	Rise time of a step response curve at a specified range
SettlingTime	Time from <begin_x> to the time it takes a step response to settle within a specified band
SettlingTime_XRange	Time from <begin_x> to the time it takes a step response to settle within a specified band and within a specified range
SlewRate_Fall	Slew rate of a negative-going step response curve
SlewRate_Fall_XRange	Slew rate of a negative-going step response curve over an X-range
SlewRate_Rise	Slew rate of a positive-going step response curve
SlewRate_Rise_XRange	Slew rate of a positive-going step response curve over an X-range
Swing_XRange	Difference between the maximum and minimum values of the waveform within the specified range
XatNthY	Value of X corresponding to the nth occurrence of the given Y-value, for the specified waveform
XatNthY_NegativeSlope	Value of X corresponding to the nth negative slope crossing of the given Y-value, for the specified waveform
XatNthY_PercentYRange	Value of X corresponding to the nth occurrence of the waveform crossing the given percentage of its full Y-axis range; specifically, nth occurrence of Y=Ymin+ (Ymax-Ymin)*Y_pct/100
XatNthY_Positive	Slope value of X corresponding to the nth positive slope crossing of the given Y-value, for the specified waveform
YatFirstX	Value of the waveform at the beginning of the X-value range
YatLastX	Value of the waveform at the end of the X-value range
YatX	Value of the waveform at the given X-value
YatX_PercentXRange	Value of the waveform at the given percentage of the X-axis range
ZeroCross	X-value where the Y-value first crosses zero
ZeroCross_XRange	X-value where the Y-value first crosses zero at the specified range

INDEX

Note: Page numbers followed by *f* indicate figures, and *t* indicate tables.

0–9

0V node symbol, 9, 11–12, 29, 30*f*, 342, 342*f*
$D_HI digital symbol, 9, 12
$D_LO digital symbol, 9, 12
* (comment lines), 126, 219–220, 228–229, 235

A

A Kind Of (AKO) models, definition, 220–221
Absolute sensitivity, 352, 354
Accept Left, 336, 344–345, 345*f*
AC Markers, 72
AC sources, 69–70, 69*f*, 73
AC Sweep, 70–78, 170, 170*f*
active smoke parameters, 392–395
Add File, 32, 329
Add Library, 32
Add New Row, 255, 255*f*
Add New Row, 80–81, 82*f*
Add Part(s) To Active Testbench, 332, 332*f*
Add Part(s) To Self, 333, 333*f*, 342, 342*f*
Add to Design, 328–330
Add to Design, 111
Add to Profile, 107
Add Trace, 75–76, 75*f*, 264
Advanced Analysis, 347–349, 360, 365
   circuit performance, 347–348
   design flow, 347, 347*f*
   libraries, 349
   measurement definitions, 348
   Monte Carlo, 381
   Monte Carlo analysis, 348–349
   Parametric Plotter, 349
   reliability, 348
Advanced Analysis libraries, 406, 406*f*
air core transformers, 141–149. *See also* transformers
All markers on open schematic, 160
analog behavioral models (ABMs), 133, 185–191
analog library, 11, 16
Analog Operators and Functions, 75*f*, 76
analog power supplies, templates, 19
analog to digital (AtoD), 260
Analysis type, 36, 48, 52, 54, 96, 97*f*, 123, 123*f*, 136–137

AND gates, 259. *See also* NAND gates
area scaling factor, 202
Ascend Hierarchy, 294
Assign Tolerance window, 386, 387*f*, 420
Assign Tolerance window, 355–357, 356*f*, 362–363
Associate/Replace Symbol, 228, 326–328, 326*f*, 328*f*
AutoConverge, 135, 135*f*
Auto Wire, 30, 30*f*, 261
Auto Wire Two Points, 30
Available Sections, 257
AWG wire standard, 320
Axis Settings, 66–67
Axis Settings, 114*f*
Axis Variable, 216, 216*f*

B

bandpass filters, 152, 152*f*, 156–161, 157*f*, 160*f*
Bandwidth. *See also* Measurement Definitions Performance Analysis, 161
Base parasitic resistance thermal noise, 202
Base shot and flicker noise currents, 202
batteries, 11
Bias Point, 136–137
bias point analysis, 38–45, 53, 117–118, 132
   displays, 43–44
   Load Bias Point, 45
   Netlist Generation, 38–42
   Save Bias Point, 44–45
   significant digits displayed, 43
   suppression option, 35–36, 41
Bias Point Node Voltages, 172–173
bias simulation profile, 108*f*, 136–137
binary bus signal displays, 264
binocular button, 339, 339–340*f*
bipolar transistors, 393–395
Bobbin, 318–320, 318*f*, 322–323, 323*f*
bobbins, 311–330
bobbin-winding selection MPE step, 318–320, 323
Bode and Nydquist plots, 72
Boltzmann's constant, 197

bottom-up designs, 283–310. *See also* Hierarchical
    Symbols
breakout library, 16–18
Browse, 6
Browse File, 32–33, 32*f*
Bus Entry, 261–273
buses. *See also* digital simulations
    custom-made buses, 271*f*
    signal connections, 261–262
    signal displays, 263–264, 268–273

C

C (capacitor) prefix, PSpice implementation
    definitions, 219, 219*t*
Cadence, 7
capacitors, 391–392
CAPSYM library, 9, 11, 47
CenterFrequency, 160–164, 163*f*, 177.
    See also Measurement Definitions
Change Project Type, 26, 26*f*
characteristic impedance, transmission lines,
    243–258
Chebyshev filter characteristics, ABMs, 185–186
checkpoints. *See also* Check Points
    transient analysis, 118–124, 126–129
circuit errors, 131–139. *See also* errors
Cleanup Cache, 304
Clock Oscillator, 279–281, 332–346
Clock Oscillator, 306*f*, 307
coaxial cable models, 244
coils of wire, 313–330. *See also* inductors
Collating Functions
    Monte Carlo simulations, 154–155, 155*f*
    worst-case analysis, 165, 168–169, 171, 171*f*
Collector parasitic resistance thermal noise, 202
Collector shot noise current, 202
Color Settings, 57, 59
column format **Property Editor**, 80, 81*f*
comment lines (*), 126, 219–220, 228–229, 235
comparators, ABMs, 186, 187*f*, 275–281
Component Filter, 411, 411*f*, 413, 417
component-selection MPE step, 312–313
component tolerances, worst-case analysis,
    165–175
conditional statements, ABMs, 186, 187*f*
conductor resistance, 243–258
Configuration Files, 126, 233, 329
Connect to Bus, 261–262

constraints, 367
control system parts ABM devices, 185–191
convergence problems, 117, 119. *See also* Simulation
    Profile Settings
    AutoConverge, 135, 135*f*
Core, 220*t*
Core Details, 316, 316–317*f*, 318, 319*f*, 322–323,
    323*f*
coupling devices, 141–149. *See also* transformers
Create a blank project, 6
Create based upon an existing project, 6
Create Derate File, 397
Create parts for Library, 239
Create PSpice Project, 6, 6*f*, 357
Create PSpice Project, 401, 406, 416
Create Test Bench, 332, 336
creating a PSpice project, 217–242
creating PSpice models, 217–242, 311–330
creating test benches, 336
cumulative distribution function (CDF), 349,
    379–380, 380*f*, 383, 383*f*
curly braces, 79, 219–220, 254. *See also* global
    parameters
current markers, 54–67
current sources, 11
cursors, 11, 29, 80, 86, 86–87*f*, 104
    icons, 86, 87*f*
    pen cursors, 114
Curve-fit definitions, 367, 376–377
Curve-Fit Error, 376
custom-made buses, 271*f*
Cutoff_Lowpass_3dB, 183
CVN parameter, 391–392

D

D (diode) prefix, PSpice implementation
    definitions, 219, 219*t*
data converters, 15
DATACONV library, 277, 277*f*
DB(), 76
dB level down for measurement, 162–164, 162*f*
dB Magnitude of Voltage, 98
DC analysis, 54, 54–59, 59–62, 63–67, 85, 79,
    141–142, 333 *See also* parametric…
    DC Sweep, 54–67, 232
    Markers, 54–67, 86
    missing DC path to ground errors, 131, 141–142,
        333–334, 340, 342

DC bias point analysis, 38–53
    Load Bias Point, 45
    Netlist Generation, 38–42
    Save Bias Point, 44–45
DC Sweep, 54–67, 232, 232*f*
DC Sweep, 83–89, 85*f*, 117–118
Decade, 90
delay circuit diagram, ABMs, 193–195,
    193–194*f*
delay lines, 259–273
Delete All Traces, 76
Demo Designs, 20–21
derating curve
    capacitor, 391, 391*f*
    resistor, 387, 388*f*, 398, 398*f*
    transistor, 393, 393*f*
derating factor, 400
Descend Hierarchy, 284, 294
DeselectAll, 339, 340*f*
Design Cache, 304, 305*f*
design cycle, MPE, 311–330
design opening, 24
Design Resources, 26
Design Status, 321–322, 321–322*f*, 324
Destination Part Library, 235
Destination Part Library, 296
Destination Symbol Library, 325, 326*f*
dev, 168
Devices and Printers, 21–22
dielectric capacitance, 243–258. *See also* RLCG
dielectric conductance, 243–258. *See also* RLCG
Diff and Merge, 343–346
Differential, 328–330
differential ABMs, 185–191
digital circuits, 259–273, 275–281
Digital Counter, 332–346
Digital Counter, 306–307, 306–307*f*, 309
digital devices, 259–273
digital power supplies, templates, 19
digital pull-up resistors, 260–273, 275–281
digital simulations, 30. *See also* buses; gates
    displaying digital signals, 263–264
    profile, 262–263, 276–281
    timing hazards, 262
    timing violations, 259, 262
    warning messages, 262
digital sources, **Stimulus Editor**, 100–115
digital symbols, 8–9
digital to analog (DtoA), 259

digital to analog converters (DACs), 276–281
dig_misc library, 260
DigSTIM, 103, 103*f*
DINPUT, 220*t*
discontinuous conduction mode, flyback
    converters, 311–330
Discrete engine, 374
Discrete simulation, 374
dispersion losses, transmission lines, 243–258
Display, 80–81, 83*f*
Display Properties, 158, 158*f*, 247, 248*f*, 255–256,
    255*f*, 359, 359*f*
Display Properties, 303, 304*f*, 401–402
displays, 72, 74–75, 145
    bias point analysis, 43–44
    bus signals, 263–264, 268–273
    Property Editor, 81, 83*f*
distributed models, transmission lines, 246–258
dot convention, 142
DOUTPUT, 220*t*
downloading models from vendors, 228–229
dual opamps, 95–96, 95*f*

E

E device, 185–186, 186*f*. *See also* voltage-controlled
    voltage sources (VCVSs)
E13_6_6 core, 318, 318*f*
E19_8_9 core, 321, 322*f*
Edit Derate File, 397, 397*f*, 412–413
Edit Part, 238–239
Edit Properties, 170
Edit Properties, 358
Edit Property Values, 157, 158*f*, 170
Edit PSpice Model, 142–143, 210
Edit PSpice Model, 363, 370
EditStimulus Editor, 113
EFreq ABM, 192, 192*f*
Emitter parasitic resistance thermal noise, 202
Empty Project, 416
encryption, PSpice models, 229–230
Encrypt Library, 229–230
End Mode, 29
Enter bobbin, 318–320
Enter Insulation Material, 313, 314*f*
equivalent inductor circuits, 165, 166*f*
equivalent input noise
    concepts, 197–200
    definition, 200–202
Error Graph, 375, 377

errors, 41, 259, 262, 333, 342. *See also* mistakes; warning messages
  common error messages, 131–139
  convergence problems, 119, 131–139
  floating node errors, 62*f*, 333, 342
  less than two connections at node errors, 132
  missing DC path to ground errors, 131, 141–142, 333–334, 340, 342
  Output File, 41
  timing hazards, 262
  timing violations, 259, 262
  voltage source or inductor loop errors, 132, 138, 138*f*
ESR, 390–392, 414
eval library, 17
Evaluate Measurement, 182–183, 279–281
Evaluate Measurement, 359, 364, 370
EXP Attributes, 106
Export, 106*f*
Exporting Capture designs, 21–22, 23*f*
Export to Capture Part Library, 224, 238
Extract Model, 142–143
Extra library work, 32–33

F
F devices, 185
FALL_EDGE, 169
Family Name, 316–318, 318*f*
Ferroxcube, 316–318, 317–318*f*
filter circuits, 151
F5 key, refreshed displays, 44
flat designs. *See also* hierarchical
  definition, 283, 283*f*
flicker noise, 198 *See also* noise; semiconductors
flip-flops, 259–273, 276–281, 335. *See also* registers
  initialization, 262
  mistakes, 262
Floating Nets, 333–335
floating node errors, 62*f*, 333, 342
Flyback Converter, 312–330
flyback converters
  MPE, 311–330
  testing, 328–330
footprint, 81, 83*t*
forward converters, 20, 20*f*
frequency-related signal losses, 243–258
frequency responses
  ABMs, 188, 189–190*f*, 190
  AC analysis, 70–78, 96, 97*f*

frequency tables, ABMs, 192, 192*f*
Functions, 212
Functions and Macros list, 359, 364, 370

G
G device, 185–191. *See also* voltage-controlled voltage sources (VCVSs)
gain and phase responses, 70
Gaussian distribution, 380
Gaussian normal distributions, 153–154, 156
Generate Part, 229, 235
Generate Part, 288, 289*f*, 296
global parameters, 79–98, 85*f*. *See also* parametric analysis
  curly braces, 79, 219–220, 254
  definition, 79
  mathematical expressions, 79
GMIN, 118
goal functions, 367
ground
  missing DC path to ground errors, 131, 141–142, 333–334, 340, 342
  symbols, 61–62
gyrators, 165, 166*f*

H
H devices, 185
henry units, 141–142
Heterogeneous parts per package, 95–96, 96*f*
hexadecimal bus signal displays, 264
Hide Invalid Values, 402, 407, 408*f*, 417, 422
Hierarchical Blocks, 21, 287–288. *See also* top-down designs
hierarchical designs
  passing parameters, 289
  power supplies, 291
  Project Manager, 283–284, 284*f*, 292–294, 296–297, 300–301, 300*f*, 304, 307, 333, 333*f*, 336–343, 338–339*f*
  schematic folders, 283–284, 283*f*, 288, 300*f*, 307
  test benches, 332–346
  testing oscillators, 300–302
hierarchical netlists, 290
Hierarchical Pin, 284, 288
Hierarchical Symbols, 287–288. *See also* bottom-up designs
  definition, 288
  saving to libraries, 287
Hierarchy, 333, 333*f*, 336–343, 338*f*

Hierarchy, 284–285, 294, 296, 300–302
Histogram Divisions, 164
histograms, 152, 161, 163, 163*f*
HI symbols, 260
Homogeneous parts per package, 95–96, 96*f*
hysteresis curves, 142–143, 143*f*

I

I (current source) prefix, PSpice implementation
 definitions, 219, 219*t*
I/V Source, 199
IBIS Translator, 231
ideal transmission lines, 243–244.
 See also transmission lines
Implementation, 235
Implementation name, 235
Implementation name, 287, 293
Implementation type, 287, 288*f*, 293
Import Measurement(s), 360, 365
inductance. *See also* RLCG
 transmission lines, 243–258
inductors, 2, 18, 18*f*, 132, 219, 219–220*t*, 311–330,
 389–390
 concepts, 132
 definition, 132, 141
 MPE, 311–330
 parameters, 18
 single switch forward converter topology, 20*f*,
  311–330
 temperature analysis, 209–210
 voltage source or inductor loop errors,
  132, 138
inductor thermal resistance, 405
initial conditions (ICs), 117, 125, 181
Initialize all flip-flops, 264–272, 277–279, 278*f*
input noise. *See also* noise
 concept, 197–200
install path, 7–8, 32–33, 235, 327
Instance List, 363
Insulation, 313–314
insulation materials, transformers, 313–314
interface nodes, 276–281. *See also* mixed simulations
intervals, noise analysis, 200
inverters, 259–273
IO_LEVEL, 260
IPWL, 120, 121*f*, 122. *See also* PWL (piecewise
 linear), Stimulus Editor
IRF.lib, 226

ISWITCH, 220*t*
ITL, 118

J

JFET, 220*t*
Johnson noise, 197

K

K_Linear coupling devices, 141–149.
 See also transformers

L

L (inductor) prefix, PSpice implementation
 definitions, 219, 219*t*
Laplace ABMs, 189
LEN property, 246–247
less than two connections at node errors, 132
library, 11, 16, 32–33, 157, 231, 325–330
Library Encryption, 229–230
Linear sweep, 70–72
Linear sweep, 60, 65
linear transformers, 141–149. *See also* transformers
LITZ winding, 320
Load Bias Point, 45
load resistors, 83–89
load terminations, transmission lines, 247–258
Logarithmic sweep, 70–78
Log File, 360, 361*f*
look-up tables, ABMs, 185–191
lossy transmission lines, 243–247.
 See also transmission lines
LO symbols, 260–261
lot, 168
lumped line segment models, 243–244, 245*f*

M

M (MOSFET) prefix, PSpice implementation
 definitions, 219, 219*t*
Macros, 212
magnetic, 142
magnetic core models, 311–330.
 See also transformers
Magnetic Parts Database, 311–330
Magnetic Parts Editor (MPE), 311–330
 bobbin-winding selection MPE step, 318–320,
  323
 component-selection step, 312–313
 core-selection step, 315–318, 321–322

Magnetic Parts Editor (MPE) *(Continued)*
   creating a transformer model, 325–328
   design cycle, 311–330
   electrical-parameters step, 314
   errors, 321, 321*f*
   results-view step, 320–325
   testing flyback converters, 328–330
Major Grid, 67
Major Spacing, 67
Make Active, 341
Mark Data Points, 123–124
Markers, 54–67, 86, 328–330, 343
master designs, test benches, 331–346
mathematical expressions
   ABMs, 185–191
   global parameters, 79
mathematical functions, 186, 187*t*
MAX, 154–155, 169
Max function, 371
maximum operating conditions (MOCs),
   385–386
Maximum step size, 124
Max measurement, 353–354
MC Load/Save, Monte Carlo simulations, 154
mean value (μ), 380
Measurement Definitions
   list, 177, 178*f*, 178*t*
   Performance Analysis, 177–184
Measurement Expression, 161, 348
Measurement expression, 359–360, 359–360*f*
Measurement Results, 280, 280*f*
Measurement Results, 370–371, 416, 421, 421*f*
Measurements, 177–184, 280, 280*f*
MHz/mHz confusions, mistakes, 71
microphones, 83–89
Min, 155
Minimum Severity level, 273
Minor Grid, 67
Mirror, 96
Mirror Vertically, 96
missing DC path to ground errors, 131, 141–142,
   333–334, 340, 342
missing PSpice Template, 139, 140*f*
mixed-mode power supplies, templates, 19
mixed simulations, 275–281
MLSQ engine, 377
Model Editor, 217–218, 223–231, 363
Model Import Wizard, 224–228, 238–239,
   325–328

Model parameter values, 172–173
Models List, 234, 238
Model Text pane, 363
Model View, 324
modified component values, test benches, 331, 336
modified LSQ engine, 372
modulus 3 synchronous counters, 264–267,
   306–310, 307*f*, 332–346
Monte Carlo, 379–383
   analysis, 348–349, 375, 379
   component and model tolerance values, 379
   distribution curve, 380
   tolerance value distributions, 381
Monte Carlo simulations, 168, 379–380
   adding tolerance values, 168
   Collating Functions, 154–155, 155*f*
   filter circuits, 151
   Number of runs, 153
   Performance Analysis, 152
   Simulation Settings, 152–155, 159*f*, 160
Monte Carlo/Worst-Case, 158, 159*f*, 171
More Settings
   Monte Carlo simulations, 154–155
   worst-case analysis, 171
Murata inductor, 408–411, 408*f*, 411*f*
myIRF540, 226
myTransistors.lib, 228, 235, 237–238, 238*f*

N
Name and Value, 81, 83*t*
Name of trace to search, 161
NAND gates, 259–273. *See also* AND gates
NegTol, 363
nested sweeps, 53*f*, 63, 65, 66–67*f*
Net Alias, 31, 60, 60*f*, 265
NE555 timer, 279–281
netlists, 218, 229, 235, 290, 331
Netlist/source file, 229, 235
Netlist/source file, 288, 296
New Column, 84, 84*f*
New Model, 223–224, 224*f*, 240–242
New Page, 292, 300
New Project, 2–3, 3*f*, 27
New Property, 410–411
New Row, 80–81, 81–82*f*, 255, 255*f*
New Schematic, 292, 300
New Simulation Profile, 36, 47–48, 50, 54, 60, 65,
   328–330
New Stimulus, 104

New Stimulus, 104, 104*f*
Newton-Raphson iteration method, 131
NFIB, 198, 198–199*t*
NJF, 220*t*
NOBIAS, 41, 49–50
No Connect, 260
nodes, 54–67, 333, 342
    floating node errors, 62*f*, 333, 342
    interface nodes, 276–281
    less than two connections at node errors, 132
    Markers, 54–67
    0V node symbol, 9, 11–12, 29, 30*f*
noise current spectral density, 200
noise power spectral density, 201
noise voltage spectral density, 201
non-linear transformers, 141–149.
        See also transformers
normalized line length, transmission lines,
        243–258
notch filters, 70–78
NRB, 198, 198–199*t*
NRC, 198, 198–199*t*
NRE, 198, 198–199*t*
NSIB, 198, 198–199*t*
NSIC, 198, 198–199*t*
NTOT, 198, 198–199*t*
Number of Runs, 382, 382*f*
Number of runs, Monte Carlo simulations, 151–153

O

olb files, 32–33, 224, 327, 329
open circuit transmission lines
    RL replaced with an open circuit, 252–253
    SWR, 258
Open Demo Designs, 20
operational amplifiers (opamps), 15, 92–94, 221,
        224–226
Optimizer, 367–377
    circuit performance, 367
    circuit specifications, 368–377
    curve fitting, 376–377
    goal function, 367
    measurement expressions, 368
    optimization engines, 367–368
Options, 5, 31, 57, 59, 65, 158, 164.
        See also ABSTOL; CHGTOL; GMIN; ITL.;
        RELTOL; TNOM; VNTOL
OrCAD, 2, 7, 21, 24

OrCAD Capture Marketplace, 15
osc125Hz, 290, 294*f*, 296–298, 297–302*f*,
        300–306, 305*f*
oscillators, 117
Output File, 41
    errors, 41
    noise analysis, 207, 207*f*
Output File Options, 42*f*, 49*f*
output noise, 198–199. *See also* noise
Outputs folder, 297
Output variable
    Monte Carlo simulations, 153
    worst-case analysis, 168, 171
Output Window, 132, 132*f*

P

Packaging, 95–96
PARAM, 83–98, 255
Parameter Descriptions, 390, 390*f*
Parameter name, 97, 97*f*
PARAMETERS, 70–71
Parameters Selection Component Filter, 371
parametric analysis, 83–98
Parametric Plotter, 349
Parametric Sweep, 97, 97*f*
2N3904 parameterized PSpice model, 381, 381*f*
Part, 93, 93*f*
Part Editor, 298–300, 304
Partial Design, 332–346. *See also* test benches
Partial Encryption, 230–231
parts, 7, 10–12, 16, 31, 31*f*, 79–98, 304
Part Search, 93, 93*f*
Parts per Package, 95–96
passing hierarchical parameters, 289, 290*f*
passive inductor model, 18*f*
passive smoke parameters, 387–392
    capacitor, 391–392
    inductor, 389–390
    resistor, 387–389
Password protection, 25
Path and filename, 293, 293*f*, 309
PCB footprints, 96–98
PC Board Wizard, 3
PCB projects, 3
PDML parameter, 390
Peak Detector, 125, 127
peak detector circuits, transient analysis, 125–127,
        125*f*, 128*f*

pen cursors, 114
Performance Analysis
  Bandwidth, 161
  Monte Carlo simulations, 152
Period and Duty Cycle, 348
phase responses, AC analysis, 70–78
Pin Numbers Visible, 299
Pin Properties, 299, 299*f*
pins, 11, 44, 54–67, 219, 219*t*, 221–223, 226, 228, 228*f*, 230–231
PJF, 220*t*
Place, 60, 62, 212
Place Net, 269
Place Part, 8, 10–12, 10–11*f*, 29, 32–33, 34*f*, 93, 93*f*
Place Part menu, 33, 33–34*f*
play button, 37
Plot, 273
polyethylene terephthalate (PET) capacitor, 414
PORTRIGHT-R, 286, 286*f*, 292, 294, 308
PosTol, 363
Postscript (PS), 21–22
Postscript to PDF converter (PDD), 21–22
pot, 96
potential dividers, 59, 61, 292, 292*f*
potentiometers, 17*t*, 92–98. *See also* notch filters; resistors
Power, 76, 80, 96
Power Derating Factor (PDF), 400, 403
Power Dissipation, 83–88, 84*f*
  curves, 86
power supplies, hierarchical designs, 291, 294
power symbols, 80, 83–89
power transformers, MPE, 311–330
predefined transformers, 144. *See also* transformers
Pressure sensors, 15
primary coils, 141–149, 314, 320–321
Primary Sweep, 65, 97, 97*f*
printed circuit boards (PCBs), 3–4, 11–12, 26, 96–98, 96*f*
probabilities, 151–164
probability density function (PDF), 349, 379–380, 380*f*, 383*f*
Probe, 106, 111, 153–154, 161, 164, 275
Probe, 54–57, 59, 59*f*, 61
Probe Cursor, 75, 75*f*
Probe Window, 160
Profile Settings, 373–374, 377, 396–397, 412–413
programmable logic arrays, 259
Project Manager, 6–7, 7*f*, 328–330

project name, 4, 7
propagating velocities, transmission lines, 243–258
Properties, 57
Property Editor, 141–149, 152, 155–157, 163, 359. *See also* parametric analysis
Property Editor, 83–98, 198–199, 401. *See also* parametric analysis
Property Name=Value, 334–335, 334–335*f*
Propose Part, 317, 321
PSpice Advanced Analysis, 402, 417
PSpice digital parts, 17*t*
PSpice discrete parts, 16, 17*t*
PSpice Environment Window, 39–40
PSpiceFiles, 331
PSpice Model, 142–143, 217–242
PSpice Model Editor, 394, 420
PSpice Modeling Application, 18–19
PSpice Model Library, 229, 235–238
PSpice models, 142–143, 210, 217–242, 311–330
  adding, 217–242
  copying existing models, 239
  creating, 217–242, 311–330
  directories, 234–237
  downloading models from vendors, 228–229
  edits, 217–218, 223–242
  encryption, 229–230
  exercises, 231–242
  implementation, 311–330
  Model Editor, 217–218, 223–242
  Model Import Wizard, 224–228, 325–328
  MPE, 311–330
  subcircuits, 218, 219*t*, 221–222, 324, 325*f*, 327
  types, 311–330
2N3904 PSpice model parameters, 356–357, 357*f*
PSpiceOnly, 331
PSpice Probe waveform, 349
PSpice Runtime Settings, 133, 133*f*
PSpice source parts, 17*t*
PSpice Template, 218, 221
PULSE
  Attributes, 105, 106*f*, 113
  Stimulus Editor, 104
PWL (piecewise linear), Stimulus Editor, 104

Q

Q (transistor) prefix, PSpice implementation definitions, 238
quadratic temperature coefficients, 209, 214, 214*f*
Quick Place of PSpice Components, 16–17, 17*t*

R

R, 29, 155

R (resistor) prefix, PSpice implementation definitions, 219, 219*t*

Random number seed, Monte Carlo simulations, 154

Rbreak, 155–156, 210

RC, 244

rectifier, ABMs, 193, 193*f*

REFDES, 218–219, 221

Reference, 287, 293, 293*f*

reflected signals, 243–258

registers, 259–273. *See also* flip-flops

Regular Expression, 334–335, 334–335*f*

Relative sensitivity, 353–354, 361, 362*f*, 365

Relaxed limit, 135

RELTOL, 118, 134, 166–167, 172–173, 172–173*f*

Remove Library, 33

Remove Part(s) From Self, 333, 333*f*, 342, 342*f*

repeats forever. *See also* IPWL; VPWL transient analysis, 129

Replace Cache, 304

resistor body temperature, 398–400, 403

resistors, 7, 27, 27*f*, 387–389

Restart At, checkpoints, 119

Restart Simulation, 119

Restore, 214

Results Spreadsheet, 320–321, 321*f*, 324, 324*f*

results-view MPE step, 320–325

Reverse Breakdown, 232

rich text format (RTF), 235

RISE_EDGE, 169

RLCG lumped line segment models, 244, 245*f*

root mean square (RMS), 197, 200–202, 314

row format Property Editor, 80, 81*f*

RTOL tolerance property, 362

S

safe operating limits (SOL), 395

   power dissipation, 403, 403*f*

   resistor with, 404, 404*f*

Sallen and Key filters, 183–184, 184*f*

Save Bias Point, 44–45

Save Check Points, 119. *See also* checkpoints

Save data from, Monte Carlo simulations, 154

saved designs, 24

Save Symbol, 228, 327

Saving a project, 23–25

SCHEDULE, 118–119, 124

SCHEDULE, 118–119

Schematic Editor, 224, 225*f*, 238

Schematic to Schematic utility (SVS), 331

Schematic View, 287, 293

Search, 64–66*f*, 333–334, 334*f*

search commands, 179

Search For, 93–94*f*, 94

Search for Parts, 12–15, 13*f*, 60*f*, 64–66*f*, 93, 93*f*

Search Online, 15

secondary coils, 320–321

Secondary Sweep, 65, 214, 215*f*

Select derating type, 397

Select Matching, 327

self-heating effect, 398

semiconductor PSpice models, 394

semiconductors, 165

   convergence problems, 133–134

   MOSFETs, 198–199*t*, 226, 238

   noise analysis, 198

   temperature analysis, 209

Sensitivity Analysis, 351–365

   components, 352, 352*f*, 354–357

   log file, 360, 361*f*

   parameter tolerances, 354–357

   Worst Case analysis, 351, 365

sensitivity analysis, 165–167, 174–175, 348–349. *See also* worst-case analysis

Sensitivity Component Filter, 361

Session Log, 136

SFFM Attributes, 106*f*

SFFM (single-frequency FM), Stimulus Editor, 105–106

Shape, 299

short circuit transmission lines

   RL replaced with a short circuit, 250–252

   SWR, 254–257

shot noise, 198–199

   noise; semiconductors

sigma (σ), 380

signal power loss, 243–258

simple library, 6

Simulation Output Variables, 76, 212

Simulation Profile Settings, 44, 54, 134–138

Simulation Status window, 132

SIN Attributes, 105*f*

sinewaves, 105, 108*f*

SIN (sinusoidal), 105, 108*f*

single switch forward converter topology, 20f, 311–330

SIN (sinusoidal), Stimulus Editor, 105, 108, 108f

Skip the initial transient bias point calculation, 118

small signal response of a circuit, 70–78

Smoke analysis, 349, 375, 385–423, 386f
  Assign Tolerance window, 386, 387f
  bipolar transistors, 393–395
  capacitor, 391–392
  derating files, 395–398
  inductor, 389–390
  log file, 408, 409f, 410, 411f
  maximum operating conditions (MOCs), 385–386
  resistor, 387–389, 399, 399f

Smoke_Inductor, 406

software install path, 7–8, 235, 327

SPB, 7

Spice voltage-controlled sources, 186.
    See also analog behavioral models (ABMs)

Split Part Section Input spreadsheet, 297

spreadsheets
  results-view MPE step, 320–325
  Split Part Section Input spreadsheet, 297

Standard Derating, 403

standing wave ratio (SWR), 254–258
  open circuit loads, 258
  short circuit loads, 254–257

Start Menu, 394

Start saving data after, 118

Start value, 97, 97f

stationary waves, 254–258

statistical analysis, Monte Carlo simulations, 151–164

Statistical Information, 383

step-down/up transformers, 144–148.
    See also transformers

Stimulus Editor, 100–115

stimulus files, 100–115

stl files, 107, 111

Stop time, 117

subcircuits, 218, 219t, 221–222, 324, 325f, 327

Subparam, 289, 290f, 302

Sweep type, 65, 97, 97f

Sweep Variable, 54, 60, 65, 97, 97f

SWG wire standard, 320

switched-mode power supplies, templates, 19

symbols, 326–328, 326–328f
  Hierarchical Symbols, 283–310

Symbol Viewer, 15

T

T_ABS, 221

TBFiles, 331

temperature coefficients (TCs), 39, 209–210, 212f, 214, 214f

temperatures, 79

Test Bench, 306

test benches, 306
  comparing/updating master design differences, 335–336, 343–346
  creating, 336
  Diff and Merge, 343–346
  Digital Counter hierarchical design, 306
  errors, 333, 342
  modified component values, test benches, 331, 336
  selection of parts, 332–333
  unconnected floating nets, 333–335

TestBenches folder, 337–338

Test_Clock/Test Bench, 332–346, 332–333f

Test_Counter, 332, 332f, 341–342, 341–342f

Test Osc125Hz, 300–301, 302f, 305

Text Editor, 235

text files, time–voltage text files, 120–122

Text to Search Box, 333, 334f, 339, 339f

thermal noise, 198–199. *See also* noise

thermal resistances (RTH), 419–420

Time Domain (Transient), 248, 248f

Time Domain (Transient), 110, 118, 118f, 123, 123f

time steps, 117, 122–124, 124f, 247. *See also* transient analysis

time-voltage data
  Stimulus Editor, 107
  transient analysis, 106, 120–122

timing hazards, digital simulations, 262

Timing Mode, 262

timing violations, digital simulations, 259, 262

Tip for New Users, 3

TKNEE, 399, 410–411, 410f, 415

Tline distributed models, 243–258

TLOSSY PSpice device, 246, 246f

TLUMP lumped line segment models, 243

T_MEASURED, 221

TNOM, 221

TOLERANCE property, 359
Tolerances, 165–175
tolerances, 348–349
    SPICE model, 379
    subcircuit parameter, 379
top-down designs, 283–310. *See also* Hierarchical
    Blocks
topologies, 19, 311–330
Toroid, 315–318
total noise contributions, noise analysis,
    198–199
Total Points, 256, 256*f*
Total Points, 70–72
Trace, 74–76
Trace Expressions, 264, 376
traces, 54–67, 74, 76, 86
transfer functions
    ABMs, 185–191
    noise analysis, 208
transformers, 141–149, 311–330
transient, 329
transient analysis, 36, 45, 79, 118–124, 126–129,
    340, 342, 359
transient sweeps, 79
transistor-transistor logic (TTL), 259–273
transmission line delay (TD), 243–258
transmission lines, 243–258
    different load terminations, 247–258
    ideal transmission lines, 243–244
    lossy transmission lines, 243–247
    matched load for RL, 247–250
    RLCG lumped line segment models, 244,
        245*f*
    RL replaced with an open circuit, 252–253
    RL replaced with a short circuit, 250–252
    SWR, 254–258
    SWR for open circuit loads, 258
    SWR for short circuit loads, 254–257
    types, 243–247
T_REL_GLOBAL, 221
T_REL_LOCAL, 221
TRN, 220*t*
twin T notch filters, AC analysis, 70, 70*f*, 72–78, 73*f*,
    92, 92*f*
twisted wire pair models, 244
txt files
    saved bias point data, 44–45, 45*f*
    time–voltage text files, 120–122
Type, 299, 308

U

UADC, 220*t*
UDAC, 220*t*
UDLY, 220*t*
UEFF, 220*t*
UGATE, 220*t*
UIO, 220*t*
unconnected floating nets, test benches, 333–335
Undo Warning, 37, 37*f*
Update Cache, 305
Update Schematic, 112
Use Device Characteristic Curves, 223, 240
Use distribution, 153–154
User Defined, 66
User Properties, 299–300, 299*f*
Use Template, 223, 240–241
UTGATE, 220*t*
UU, 315–318

V

V (voltage source) prefix, PSpice implementation
    definitions, 219, 219*t*
Variables part, 355–356, 355*f*, 416
Variables symbol, 362
variance, 151–152
VCC_CIRCLE symbol, 8–9
VCC_CIRCLE symbol, 96
Vendor Name, 315–318
Vendor Part, 318, 318*f*
V(INOISE), 199
V(ONOISE), 199
VNTOL, 122
voltage-controlled current sources (VCCSs),
    185–191. *See also* analog behavioral models; G
    device
voltage-controlled voltage sources (VCVSs),
    185–191. *See also* analog behavioral models; E
    device
voltage pairs, transient analysis, 120–122
voltage regulators, 221
voltage source or inductor loop errors,
    132, 138
Vpulse, 112
VPWL, 104. *See also* PWL (piecewise linear),
    Stimulus Editor
VPWL, 120–122, 121–122*f*, 127, 128*f*, 129
VSIN, 145
VSS, 260

VSTIM, 100–115
VSWITCH, 220*t*

W

warning messages, 46, 46*f*, 56*f*, 139, 262.
    See also errors; mistakes
Waveform, 54–55, 57, 126–128
websites, downloading models from vendors,
    228–229
Width, 288
wildcards, 93–94*f*, 94
Wire Drag, 31
world icon, 233
Worst Case All Devices, 174–175
worst-case analysis, 165–175
    Sensitivity Analysis, 351, 365
Worst Case Direction, 171
Worst Case Summary, 175, 175*f*

X

X Axis, 216
XFRM_LINEAR linear transformer, 144, 144*f*
XFRM_NONLINEAR, 327, 327*f*
XGrid, 67
X (subcircuit) prefix, PSpice implementation,
    238

Y

YAxis, 66, 115
YGrid, 67
YMAX, 154–155, 168

Z

zener diodes, 231–234, 233–234*f*
Zetex library, 423–424
Z (IGBT) prefix, PSpice implementation
    definitions, 219, 219*t*

Printed in the United States
By Bookmasters